TECHNOLOGY, EDUCATION, and PRODUCTIVITY

Early Papers with Notes to Subsequent Literature

Zvi Griliches

Basil Blackwell

Copyright ©Zvi Griliches 1988

First published 1988

Basil Blackwell Inc.
432 Park Avenue South, Suite 1503
New York, NY 10016, USA

Basil Blackwell Ltd
108 Cowley Road, Oxford, OX4 1JF, UK

Library of Congress Cataloging in Publication Data

Griliches, Zvi, 1930–
Technology, education, and productivity : early
papers with notes to subsequent literature/Zvi Griliches.
p. cm.
Bibliography: p.
ISBN 0-631-15614-3
1. Technological innovations – Economic aspects. 2. Education –
Economic aspects. 3. Industrial productivity. 4. Labor
productivity. I. Title.
HC79.T4G695 1988
338′.06 – dc19   88-2985
CIP

British Library Cataloguing in Publication Data

Griliches, Zvi 1930–
Technology, education, and productivity :
early papers with notes to subsequent literature.
1. Econometrics
I. Title
330′.028

ISBN 0-631-15614-3

Typeset in 10 on 11½ Times
by Dobbie Typesetting Limited, Plymouth, Devon
Printed in Great Britain by
Bookcraft Ltd., Bath, Avon

Contents

Acknowledgements

I am indebted to my teachers and fellow students at the University of Chicago for the support and inspiration that led me to and kept me working on this range of topics. I would like to mention especially Theodore Schultz, Arnold Harberger, H. Gregg Lewis, Yehuda Grunfeld, Marc Nerlove, and Lester Telser. In subsequent years I benefitted greatly from interactions with my colleagues, friends, students, and collaborators at Chicago, Harvard, and elsewhere. An incomplete list would acknowledge the help and influence of Arne Amundsen, Gary Becker, Yoram Ben Porath, Gary Chamberlain, Robert Evenson, Trygve Haavelmo, Bronwyn Hall, Dale Jorgenson, Yoav Kislev, Simon Kuznets, G. S. Maddala, Jacques Mairesse, Ariel Pakes, Jacob Schmookler, Hans Theil, Manuel Trajtenberg, and many others. Extensive comments on the draft of the Introduction and the various Postscripts were provided by T. Bresnahan, E. Berndt, E. Denison, B. H. Hall, D. W. Jorgenson, M. King, J. Mairesse, A. Pakes, M. Trajtenberg, and J. Triplett. I am grateful for all the comments I received even though I was not able to take all of them into account in the revision. I would like also to thank Elizabeth Johnson for the encouragement that led to the development of this volume and Jeanette DeHaan who typed, retyped, and kept track of the various pieces of this manuscript. The National Science Foundation has supported most of my work in this area, beginning in 1957 and continuing through the following thirty years. Last, but most importantly, I want to thank my wife, Diane Asséo Griliches, for providing the warmth, the stable home base, and the emotional support which made all this work possible.

1
Introduction

This volume is a collection of my more important and interesting early papers. They all reflect my continued interest in technological change as the main source of long-run economic growth and my attempt to understand it, to comprehend what happened and why, through the collection and analysis of relevant economic data. The questions I asked about the determinants of the diffusion of new technology, the measurement of physical capital, the role of education in economic growth, and the contribution of research and development were all to be components in an ultimately richer understanding of the sources and processes of economic growth. The style of this work, however, is less general. It tends to focus on the particular and to take measurement and measurement issues seriously. Detail matters. It reflects my belief that one cannot get much insight from aggregate facts unless one appreciates their components, how they are constructed, and how they interact.

A unifying thread that runs through many of the papers is the view that technological change is itself an economic phenomenon and hence also an appropriate topic for economic analysis. Too much of economic work tends to take it as exogenous, as a manna from heaven which descends on us at a constant (or sometimes changing) rate. Early on I tried to argue that this is not necessarily so, that measurement frameworks can be expanded to bring more aspects of technological change into the domain of "standard" economics, removing thereby some of the mystery from this range of topics. This kind of quantitative work, however, takes much effort, is heavily data-dependent, and is rarely definitive. At its best it opens up new subjects rather than providing closure. It shows, by example, what can be done and what might be interesting to do more of; and often the question is as interesting as the possible answers.

The volume starts with the diffusion of hybrid corn article, based on my PhD thesis, and continues with a series of papers on the measurement of productivity and its various components, which constitute, essentially, an outline of my research program for the years to come. While trying to understand and improve the measurement of productivity I ran into a number of issues, each with its own significant research agenda: the treatment of product quality change in the construction of price indexes, for which I revived the "hedonic regression" solution; the problem of measuring capital and its accumulation and depreciation correctly, and the problem of measuring the contribution of education, which led me perilously close to the nature–nurture debate. I also investigated the impact of public and private investments in research on the rate of technological advance, a topic which has been at the center of my research activities during the past decade.

This collection is incomplete in two important aspects. First, it excludes my somewhat more technical econometric methodology papers; papers dealing with specification bias, aggregation, distributed lags, errors-in-variables, panel data, and related topics. These will be collected in a separate volume at a later date. Second, it does not contain any of my more recent work. The latest papers included in this volume were written in 1971. This is due, in part, to space limitations, to the yet incomplete nature of some of my current work, and to my desire to use some of this material in a more focused specific volume on patents and R&D which I plan to write in the not too distant future. Nevertheless, I do believe that the essays included here provide a good introduction to the range of my interests and the style of my work. In this Introduction I hope to compensate for the lack of immediacy by discussing where I think the subject has moved to since, pointing out the relevant newer literature and describing some of my later work on these topics. Besides such a general overview, more specific notes to the subsequent literature are appended as "postscripts" to some of these papers.

I

The goal of the first paper in this collection, "Hybrid Corn: An Exploration in the Economics of Technological Change," was to show that the process of diffusion of an important technological innovation was amenable to economic analysis. It was rather common, in formal economic analysis of the time, to put such events outside the scope of normal "equilibrium" theory. Technological change was, and often still is, treated as an exogenous event, something to be taken as given from outside the economic system which needs to be explained by it. The paper indicates that such a dichotomy was not necessary.

The paper interprets the innovation process, that is, the supply of new technology in the form of specific hybrids adaptable to particular areas, and also the diffusion of technology, the speed with which it was being adopted, as being both under the influence of economic variables. It shows how observed differences in the timing of such processes across states and regions can be rationalized by measurable differences in economic incentives. Using the logistic growth curve to summarize the spread of hybrid corn in the various regions of the US, it focused on the estimates of its three main parameters (origin, slope, and ceiling) in these regions as different aspects of the diffusion process to be explained by other economic variables.

A number of issues raised or implied in this paper have reverberated through the subsequent literature: Is the logistic the "right" functional form for the study of diffusion processes? Do most new technologies diffuse in a similar pattern? What is the appropriate model to use in describing and observing diffusion processes? How important are considerations of information diffusion and uncertainty about the qualities of the technology as compared to considerations of size, access to funds, and personal characteristics of the actors in these events?

When I chose the logistic form to analyze diffusion behavior I did it both because the data in front of me looked as if the logistic would fit quite well and because one could give a reasonable theoretic interpretation to it, either as an information-spread phenomenon based on mathematical epidemic models, or as a learning under uncertainty process based on sequential sampling or Bayesian considerations. But I did not claim then, and in fact I do not believe it to be the case now, that the logistic function represents some underlying invariant "law" of diffusion behavior. That is why I have been somewhat nonplussed by the various efforts to derive "the" model of diffusion or to argue at length about particular modifications of the functional form, adding more parameters or changing to another growth curve family, such as the Gompertz. If I were to return to this topic today I would take a more "dynamic" point of view and respecify the model so that the ceiling is itself a function of economic variables that change over time. I tried something like that in Appendix B of my thesis, but the state of econometric technology at that time prevented me from pursuing it very far.

Diffusion research emphasizes the role of time (and information) in the transition from one technology of production or consumption to another. If all variables describing individuals and affecting them were observable, one might do without the notion of diffusion and discuss everything within an equilibrium framework. Since much of the interesting data are unobservable, time is brought in to proxy for a number of distinct forces: (1) the decline over time in the real cost of the new technology due to decreasing costs as the result of learning by doing and to cumulative improvements in the technology itself; (2) the fall in price charged for the new technology due to rising competitive pressures faced by its original developers and the growth in the overall market for it; (3) the dying-off of old durable equipment, slowly making room for the new; and, (4) the spread of information about the actual operating characteristics of the technology and the growth in the available evidence as to its workability and profitability. In the work on hybrid corn I focused on the fourth "disequilibrium" interpretation, and emphasized the importance of differences in profitability both as a stimulus toward closing the disequilibrium gap and as the determinant of the time it takes to become aware of its existence. ("Disequilibrium" here means that additional change, diffusion, will happen even if prices and incomes do not change further, driven by changes in the information available to individual decision-makers.) Alternatively (see, e.g. David, 1969),* one can focus on reasons (1) or (3), in which case the existing size distribution of firms or the existing age distribution of the equipment to be replaced becomes one of the major determinants of the rate of "diffusion" and explanation of how and why "ceilings" shift over time. The relative importance of these forces varies from technology to technology, and the optimal mode of analysis is likely to be quite sensitive to that and to the kinds of data available to the analyst. In any case, all such approaches lay stress on the economic determinants of diffusion although they differ in the emphasis that they put on them.

In my original paper I emphasized differences in "profitability" as the major determinant of the rate of diffusion, and claimed in a final footnote that all other possible determinants such as various personal variables suggested by sociologists could be given an economic interpretation. This led to some controversy in the pages of *Rural Sociology* (see Brandner and Strauss, 1959, and Havens and Rogers, 1961). One of my responses to such comments, "Congruence versus Profitability: A False Dichotomy," is reprinted next to give the flavor of this type of debate. If I were to rewrite my original paper today, I would still take the same position but add "and vice-versa" at the end of that footnote. This view was reflected in the conclusions to my second rejoinder in this series with the following comment:

> In general I see little point in pitting one factor against another as *the* explanation of the rate of adoption. The world is just too complicated for such an approach to be fruitful. Thus I regret some of my previous "all or none" remarks. . . . If one broadens my "profitability" approach to allow for differences in the amount of information available to different individuals, differences in risk preferences, and similar variables, one can bring it as close to the "sociological" approach as one would want to. The argument here is really not about different explanations of the same phenomenon, but about the usefulness of different languages in interpreting [it]. . . . While this is not a trivial issue, the same explanations can usually be stated in either language. Problems of terminology are ultimately of secondary importance. Terminology is a means not an end (Griliches, 1962, p. 330).

When I turned from hybrid corn to the analysis of other major changes that were happening in US agriculture, mechanization and the growth in fertilizer use, it was clear that I needed a different model, one with shifting "ceilings," where these ceilings were themselves functions of economic ("profitability of use") variables. This led me to become, not entirely accidentally, one of the early "adopters" of the partial-adjustment distributed lag model. Marc Nerlove was a student at Chicago at about the same time,

* References to the subsequent literature mentioned in this Introduction and in the various Postscripts are to be found at the end of this volume in the "Additional References" section. The references in the reprinted papers appear in their original format.

Phil Cagan had just finished his dissertation there, and Hans Theil, who was one of my teachers, had brought Koyck's (1954) model to Chicago. In this model the "desired" level of use, which I identified with the "ceiling" of my earlier model, is a function of the underlying, more permanent, economic variables, while actual use approaches this level only gradually, both because of uncertainty about its exact location and because of costs of adjustment and other inertia factors. I used this model rather successfully to analyze the rise in fertilizer use and the growth in the demand for tractors, showing that their spread could be explained as a response to falling real prices plus a reasonable lag in adjustment to such changes (see Griliches, 1958, 1959, and 1960). The switch to a "demand" model was not thought by me as an abandonment of the diffusion model. It was rather an alternative approach to the same type of problem:

> Let this be clear. The author's argument is not that there was no learning involved in effecting these large increases in fertilizer consumption, only that the process should not be treated as exogenous. This learning process can be viewed largely as a the result of changing relative prices and should be treated within the framework of economics and not outside of it. Any substantial change in relative prices will always involve some "learning" on the part of the entrepreneurs. While things are stable there is no point in knowing anything more than the physical and economic facts in the neighborhood of the current equilibrium. Only when there are substantial changes in the relevant variables is there an incentive to get busy and learn "new" things (Griliches 1958, pp. 605–6; see also Atkinson and Stiglitz, 1969, for a much later working out of similar ideas).

Such models are subject, however, to a number of serious econometric difficulties, which accounts for my interest in the methodology of distributed lag estimation (see Griliches, 1959 and 1967a). It is also interesting to note that what is essentially a "hedonic" price regression makes its first appearance in a footnote to the 1958 paper (p. 599), a topic to which I shall return below.

II

Most of the other papers in this volume grew out of my interest in productivity measurement and attempts to do it "right," or at least better than was being done at the time. The productivity literature of the time computed rather large unexplained "residuals" in output growth and attributed them largely to "technological change." I was dissatisfied with this state of affairs. As an econometrician I found the spectacle of our economic models yielding large residuals uncomfortable, even when we fudged the issue by renaming the residual as technical change and then claimed credit for "measuring" it. Also the link between such a residual and my earlier work on actual changes in techniques, e.g. hybrid corn, was tenuous and unclear. My own approach to this problem could be best described as a version of specification analysis, a topic which became the major concern in my work on econometric methodology. I wanted to examine both the specification of the model used to compute such "residuals" and the ingredients, the data, used in their implementation. My first published paper in this area, "Measuring Inputs in Agriculture," reproduced next in this volume, already sounds this note, though it is focused primarily on issues associated with the measurement of some of the most important inputs in agriculture: fertilizer, farm machinery, and labor. All of the issues raised in this paper, especially the correct measurement of capital, the right price indexes for deflating output and input, and the role of education in affecting the quality of the labor force, continued to preoccupy me in the years to come.

It may be useful, at this point, to sketch out a more explicit statement of the productivity measurement problem. It provides a framework and a motivation for much of the research that was to come. A conventional measure of residual technical change (TFP) in an industry can be written as

$$\hat{t} = y - sk - (1-s)n$$

where y, k, and n are percentage rates of growth in output, capital, and labor respectively; s is the share of capital in total factor payments, and the relevant notion of capital corresponds to an aggregate of actual machine hours weighted by their respective base period (equilibrium) rentals. This procedure assumes that all the variables are measured correctly, that all the relevant variables are included, and that factor prices represent adequately the marginal productivity of the respective inputs. The last assumption is equivalent to the assumption of competitive equilibrium and constant returns to scale. To analyze t, the "unexplained" part of output growth, it is useful first to think in terms of a more general underlying production function:

$$y^* - f = \alpha(\,k^* - f) + \beta(n^* - f) + \gamma z + t$$
$$y = y^* + u$$

where the "true" production function is defined in terms of correctly measured outputs and inputs (the starred magnitudes) and at the technologically more relevant plant or firm level. That is, f is the rate of growth of the number of plants (firms) in the industry and implicitly, the production function is defined at the average plant level; α and β are the true elasticities of output with respect to capital and labor, while γ is the elasticity of output with respect to the z's, the inputs (or, rather, their rate of growth) which affect output but are not included in the standard accounting system. These could be services from the accumulated stock of past private research and development expenditures or services from the cumulated value of public (external) investments in research and extension in agriculture and other industries, or measurable disturbances such as weather or earthquakes. The measurement error in output is u. It differs from *t* in being more random and transitory while the forces behind *t* are thought to be more permanent and cumulative. The α, β, and γ coefficients need not be constants. If they are we have the Cobb–Douglas case. The whole framework can be complicated and generalized by adding square terms in rates of growth as approximations to a CES or translog type production function.

Defining two more shorthand terms: $s^* = \alpha/(\alpha + \beta) = \alpha/(1 \simeq h)$ where s^* is the true relative share of capital, and $h = \alpha + \beta - 1$ is a measure of economies of scale with respect to the conventional inputs k and n; the production function can be rewritten in terms of the "true" residual measure of technical change *t*, as

$$t = (y - u - f) - (1 + h)[s^*k^* + (1 - s^*)n^* - f].$$

Subtracting this from the conventional measure of residual technical change $\hat{t}$, we get an expression for the total "error" in our usual measures of total factor productivity growth:

$$\hat{t} - t = s(k^* - k) + (1 - s)(n^* - n) + (s^* - s)(k^* - n^*) \\ + h[s^*k^* + (1 - s^*)n^* - f] + \gamma z + u$$

The various terms in this formula can be interpreted as follows. The first term is the effect of the rate of growth in the measurement error of conventional capital measures on the estimated "residual." The second term reflects errors in the measurement and definition of labor input. The third term reflects errors in assessing the relative contribution of labor and capital to output growth. It would be zero if factor shares were in fact proportional to their respective production function elasticities or if all inputs were growing at the same rate (then the relative weights do not matter). The fourth term is the economies of scale term. It would be zero if there were no underlying economies of scale in production ($h = 0$) or if the rate growth in the number of new firms (plants) just equalled the growth in total (weighted) input. The fifth term (γz) reflects the contribution of left-out inputs (private or public), while the sixth term (u) represents the various remaining errors in the measurement of output.

This list of issues dominated my subsequent research activities, and provides a unifying framework for what at times appear to be rather disparate strands of work. My interest in the measurement of "correct" prices started with difficulties in the measurement of fertilizer and machinery capital in agriculture and led me to resurrect the "hedonic regression" approach to the measurement of quality change in the Stigler Committee Staff Report reprinted next in this volume. This paper was quite influential, and a whole literature developed in its wake, influencing the measurement of real-estate prices, wage equations, environmental amenities, and other aspects of "qualitative differences." I myself wrote a number of additional papers (Griliches, 1964, and Ohta and Griliches, 1976 and 1986), supervised several dissertations in this area and edited a book of essays on this range of topics (Griliches, 1971). The introduction to the 1971 book, reproduced here, provides a review of this literature as of the end of the subsequent decade. In a new postscript I comment briefly on the highlights of this strand of work as it continued to expand in the next two decades. I did not stay in the center of this field for long. But it was gratifying to observe that the methods outlined in this paper have recently received, with a lag of 28 years or so, their official "approval." The newly revised US national income accounts incorporate a new index of computer prices based on such "hedonic" methods (see the January 1986 issue of the *Survey of Current Business*).

This work connected also to my more general interest in the measurement of capital in the context of the measurement of productivity change. The next paper, reprinted from the Yehuda Grunfeld memorial volume, represents an early and rather complete statement of my position on this matter, which was later to be refined in the joint work with Jorgenson. The difficulty with the available capital measures then, and to a great extent still now, was, in my view, the fact that they were being over-deflated and over-depreciated, that items with different expected lives were being added together in a wrong way, and that no allowance was being made for changes in the utilization of such capital. The over-deflation issue was already alluded to in the discussion of my work on price deflators; it was fed by the strong suspicion that the various available machinery and durable equipment price indexes did not take quality change into account adequately, if at all. This issue connects also to the "embodied" technical change idea (Solow, 1960) and the literature that flowed from it. My view on overdepreciation remains controversial (see Miller, 1983). I myself turned early to the evidence of used-machinery markets to point out that the official depreciation numbers were too high, that they were leading to an underestimate of actual capital accumulation, but I also argued that the observed depreciation rates in second-hand markets contain a large obsolescence component, induced by the rising quality of new machines. This depreciation is a valid subtraction from the present value of a machine in current prices, but it is not the right concept to be used in the construction of a constant quality notion of the flow of services from the existing capital stock in "constant prices." The fact that new machines are better does not imply that the "real" flow of services available from the old machines has declined, either potentially or actually. In several places I tried also to explore some of these issues econometrically (see especially Pakes and Griliches, 1984) but the available data on types of machinery in place and their actual age structure have been rather sparse, and there has been less progress in this direction than I might have hoped at the time I was writing the reprinted paper.

III

The next three papers reflect my interest in education, its contribution to economic growth, its impact on economic inequality, and the issues involved in trying to assess the quantitative magnitude of such effects. It grew out of my work on adjusting productivity measures for the changing quality of the labor force, especially its level of schooling, and my association at Chicago with Theodore Schultz, Jacob Mincer, and Gary Becker who were developing the Human Capital idea at that time. The labor quality adjustments described in the first two papers use observed earnings differences by level of schooling to weight different workers. Such computations assume that (a) differences in earnings correspond to differences in contributions (marginal products) to national or sectoral output; (b) that they are in fact due to schooling and not to other factors such as native ability or family background which happen to be correlated with schooling; and (c) that the production function can be, in fact, written in such a way that the various types of labor can be aggregated into one total quality-adjusted labor input index. These assumptions were obviously controversial and were challenged from many directions. I was uneasy myself with "adjustments" for which there was no direct evidence. I embarked, therefore, on a series of econometric studies whose basic purpose was to investigate the validity of such estimates of the contribution of education to productivity and economic growth.

The first paper reproduced here, "Notes on the Role of Education. . .," discusses almost all of the issues raised by such computations. The first issue to be tackled is whether differences in schooling do indeed have productivity consequences. The only way I saw of testing it, rather than just assuming that wage differences are proportional to differences in marginal products, was to include an education-based measure of labor quality separately when estimating a production function, and check on its statistical significance and economic importance. If one defines a multiplicative quality index and uses a Cobb–Douglas production function framework, one has the additional implication that the coefficient of such a "quality" variable should be approximately equal to the coefficient of the labor quantity variable. This is the rationale behind a series of empirical studies of productivity differences in agriculture and manufacturing summarized in the first section of this paper. Besides reviewing the work on regional differences in agricultural productivity, some of which is also included in this volume, it also summarizes my later work on state–industry productivity differences in manufacturing (Griliches, 1967b and 1968). In general it supports the use of schooling-per-worker indexes as a productivity-relevant quality dimension of labor input. The inclusion of such variables in production functions results in coefficients that are statistically "significant" and have the right order of magnitude. This work also anticipates and responds, in a way, to a new strand of criticism which was to arise later in the literature under the label of "signalling" (Spence, 1974). It provides direct evidence on the "productivity" of education without using the *a priori* theoretical assertion that wage differences reflect marginal product differences. (A survey of similar later work can be found in Jamison and Lau, 1982.)

The most direct challenge to the original estimates of the contribution of schooling to economic growth was the issue of "ability." To what extent did observed income differentials exaggerate the contribution of schooling because of a positive correlation between native ability and the levels of schooling achieved by different parts of the population? There were rather conflicting views on this at that time (early to mid-1960s). Denison (1964), for example, claimed at first (on the basis of very little data) that as much as 40 percent of the observed income differentials could be due to the correlation of schooling with ability (though his later estimates do not use this type of adjustment). This seemed rather high to me, and I embarked on a search for data that would throw some light on this topic. I have still to emerge from this search.

The first and most obvious thing to do is to find data on both schooling and ability, and to hold ability constant in calculating the appropriate returns to schooling. The few scattered bits of evidence that were available on this topic at that time led me to the conclusion that "ability bias" could not be as high as asserted by Denison. I kept looking, however, for more representative and convincing data. The first large data set analyzed by me and W. Mason, and reprinted next here, was based on the NORC-CPS 1964 match of US Army veterans with data on earnings, schooling before and after service in the Armed Forces, family and other demographic background data, and AFQT (Armed Forces Qualification Test) scores. Using AFQT as the "ability" measure resulted in the conclusion that the bias in the estimated returns to schooling from this source was smaller than had been expected: on the order of 10–15 percent or about one percentage point of the estimated rate of return of about 6 percent. This conclusion was later supported by Gary Chamberlain's (1977) reanalysis of these same data using more advanced statistical techniques, and my own work using IQ scores and data on siblings from the National Longitudinal Survey of Young Men (see Griliches, 1976b, 1977, 1978, 1979a, 1980a and b; and Bound, Griliches and Hall, 1986, and the discussion in the postscript on Education and Economic Growth).

During the mid-1960s, when much of this work was initiated, I was writing against the background of a rapidly expanding educational system. Very few large industries were growing at anywhere near the same rate. I began to worry, therefore, whether this increase in the number of highly educated workers would not drive down their market price and the associated rate of return to higher education. The superficial signs were all positive, however. The observed relative constancy of skill and educational wage differentials during the 1950s and 1960s could be interpreted as implying that the demand for education workers is highly elastic, allowing one to hope that since past expansions in the educated labor force had not reduced them, neither would the current and future expansions.

But I had not noticed that the big demographic swing and the associated accelerator-like expansion in the demand for teachers had greatly distorted the observed data and made the persistence of such trends into the future quite unlikely. Neither did I realize that the big government-financed space–defence–R&D boom was something that may not last forever, or appreciate the fact that much of the impact of the growth in higher education on the labor force was delayed because of the large expansion in graduate education. It was not until 1968 that "net" production of BAs (those entering the labor market rather than continuing on to graduate school) began rising above its 1952 levels. By the early 1970s an annual wave of about an additional million highly educated workers began arriving at the doors of the full-time labor force and the rate of return to schooling did decline significantly (Freeman, 1976).

One of the possible explanations offered for the then-perceived constancy of educational wage differentials and a source of optimism for the future was the "capital–skill complementarity" hypothesis. This is the idea that educated labor is complementary to, rather than a substitute for, various advanced forms of machinery and that therefore a rapid rate of capital accumulation (and innovation) will increase the demand for such labor and prevent the fall in its relative price. This argument is already outlined in my "Notes on the Role of Education . . ." paper, and a more serious attempt at testing it is attempted in my 1969 "Capital–Skill Complementarity" note reprinted next. It uses data for US manufacturing industries in 1954 and 1963, and shows that skilled labor is more complementary with physical capital than unskilled labor. Further work along these lines has been done by Dougherty (1972), Weiss (1974), Welch (1970), Fallon and Layard (1975), Welch, Murphy and Plant (1985) and Morrison and Berndt (1981).

A major recent strand of criticism of human capital work is that schooling does little more than sort individuals according to ability, and that the resulting private returns to schooling overestimate significantly the rather meager social returns from such an activity (Spence, 1974 and Arrow, 1973). There are two versions of this argument:

1 Schools do little but sorting. No additional "real" human capital is embodied (augmented) during the schooling process itself.
2 Sorting itself is not particularly socially productive, it only exacerbates income inequality, and leads people to over-invest in activities such as schooling which are supposed to certify (signal) their potential abilities to ignorant employers.

There is a major important, though not all that novel, theoretical point contained in these criticisms. In a world of uncertainty in which information and the appearance of information has value, the private returns from the production and dissemination of information can easily exceed the social returns from the same activity. But the empirical import of such criticisms appears to me to be quite exaggerated.

Screening and signalling, like ability and motivation, are generalized concepts with few observable empirical counterparts that one could get one's teeth (or computers) into. If the returns to schooling were largely due to the informational content of the certificate and not to the process of schooling itself, one would expect that (1) cheaper ways of testing and certifying would be developed by employers and employees; (2) the returns to schooling would be lower among the self-employed versus wage and salary workers since, presumably, the self-employed would not pay themselves (or be able to collect) for a false signal (cf. Wolpin, 1977); and (3) the returns to schooling should decline with age as more experience is accumulated by employers about the "true" worth of their employees and as the initial signal provided by schooling fades away into insignificance. None of these effects is observed in the data, leading one to question the empirical import of the "schooling as a signal" hypothesis.

The issue of who gets schooling was not adequately considered in the earlier literature, and the radical critics have rightfully, I think, focused their attention on the exaggerated hopes for egalitarianism that were implicit in some of the more extreme "schooling is good for everything" positions. First, it is relatively easy to see that a general expansion in schooling need not result in any reduction in inequality. Second, the large public investments in schooling, like much of other governmental investment, were not as progressive as advertised. Much of the benefit of such subsidies redounded to the children of the middle classes, and was dissipated in the induced rise of teacher salaries. Nevertheless, the critics seem to underestimate the amount of social mobility that did occur in the past and the opportunity provided by the schooling system for class turnover (see Blau and Duncan, 1967, Hauser and Featherman, 1977, and Jencks et al., 1979, on schooling as a source of social mobility in the United States), and the use of this route for social mobility by specifically disadvantaged ethnic groups such as Jews in the earlier part of this century and Blacks more recently (cf. Freeman, 1973, and Smith and Welch, 1986). Nor do they give enough credit to schooling for the rise in the *average* standard of living in the economy as a whole, and the concomitant driving-up of the price of human time, the one resource distributed relatively equally throughout our economic system.

Finally, there is the strand of criticism that takes the position that none of this can matter much, since little of the total observed inequality of incomes (say the variance in the logarithm of income or wages) can be accounted for by differences in schooling (Jencks, 1972). At the individual level, detailed regression equations (income-generating functions) explain only between 30 and 50 percent of the observed inequalities, and the partial role of schooling differences in such an accounting is much smaller (on the order of 10–15 percent). There are a number of responses to be made to such criticisms. First, some of the more prominent studies claiming to have shown the importance of "luck" and the negligible impact of schooling are marred by not taking adequate account of the rather large transitory variations in income at the individual level and by ignoring age, life-cycle effects and on-the-job training differences as a source of *ex-post* but not *ex-ante* inequality. Second, in my Lewis volume paper (Griliches, 1976b), I show that, while schooling by itself does not account for a great deal of the observed variance in wages or income, additional schooling could be and has been used in the United States in the 1960s to overcome social class handicaps and to compensate for and eliminate some of the *systematic* sources of observed income inequality. The estimated model implies that two youngsters who are one standard deviation apart on each of a list of family background variables *and* IQ would find themselves, other things equal, about 0.75 and 0.4 of a standard deviation apart on schooling and wages, respectively, implying a rather strong regression toward the mean. But if the youngster with the lower family background and IQ managed somehow to acquire an extra 4 years of schooling (e.g. went on to and completed college), which would be equal to an additional 1.5 standard deviation units of schooling, this would essentially wipe out his original handicap. If, in fact, he had an equal IQ to start out with, he would need only about 2 more years of schooling to compensate him for his lower social class start.

While these various interpretations and intellectual positions were being debated inside the profession, the world outside was changing, and with it the economic fortunes of the educational system. The slowdown in population growth and in economic growth may bring to an end a 200-year-long era of rising importance of, and rewards in, education. The era was initiated by the "enlightenment" period, fed by the rapid rise in population which started in the West in the eighteenth century and by the great burst in economic growth and technical change that ran through the nineteenth and twentieth centuries. With both population and economic growth slowing down, and with the educational industry seriously over-expanded, it may be in for hard times, at least for a while.

There are other changes that may be also impinging on the scarcity value of knowledge. The technology revolution in communication and copying has reduced significantly the transmission costs of knowledge, and made it even harder to appropriate. It has also increased significantly the flow of information (both relevant and irrelevant) that inundates employers, consumers, and public officials. Whether or not this will lower or raise the price of knowledge "handlers," if not of knowledge "possessors," is still an open question. It is also quite likely that part of the economic value of education arises out of its interaction with technical change. That is, better-educated entrepreneurs and managers use new technology quicker and current technology better, in the sense of allocating the available resources more optimally. This line of thought (due to Nelson and Phelps, 1966 and Welch, 1970) would imply that much of the returns to schooling would evaporate if there were no new information to process. Up to now these ideas have been tested only on agricultural data (Huffman, 1974 and 1977). If they are correct, and if we are entering a no- or slow-growth era including a slowdown in the rate of economic and technical change, this also may contribute to a decline in the scarcity value of education. I am less pessimistic, however. In spite of some scattered signs to the contrary I see no deceleration in the rate of technical change in the economy in the immediate future, and hence I remain optimistic about the demand for, and the value of, education in our economy.

IV

The next two papers represent my continued interest in the economics of research, development, and technology. If one is to treat technological change as endogenous, as something that is being "produced" by the economic system and the actors in it and not like some manna from heaven, one needs to look for its sources, for the activities that cause it, directly or indirectly. Organized R&D activity is clearly one such source, perhaps the major one, though clearly not the sole one. One way of showing this is to compute the social returns from such investments and show that they are positive and sizeable. Another way is to try to extend the national growth accounts to incorporate R&D as another investment activity and reinterpret them in this light. The reprinted articles explore both of these approaches.

The 1958 *JPE* paper presents the first detailed calculation of social returns from a public research program. It was a by-product of my thesis research on hybrid corn diffusion. As part of that research I had also collected information on hybrid corn research expenditures in agricultural experiment stations and private seed companies. I used these data, together with an estimate of the value of the additional corn yielded by hybrids, to construct an estimate of social returns from this activity and compare them to the social cost incurred in generating them.

The ratio of returns to costs turned out, perhaps unsurprisingly so, to be very high, and the resulting number, "743 percent," was used rather widely in subsequent research funding debates. This number is a benefit–cost ratio. The comparable internal rate of return was estimated at 40 percent per annum, still a rather high number. The paper also presents similar computations for agricultural research as a whole and a projection of social returns from hybrid sorghum research, which was then only in the beginning of its diffusion phase. The methodology of this kind of calculation was taken over, extended, and improved by others. A number of similar calculations were made for other agricultural research programs (a summary of some of these can be found in Evenson et al., 1979). A related approach to the computation of private and social returns for a number of manufacturing innovations was developed by Mansfield and his co-workers (Mansfield et al., 1977). In general, most efforts to trace the results of individual public and private research endeavors have found rather high social returns to them.

Such individual invention or innovation cost and returns calculations are very data- and time-expensive, and are always subject to attack for not being representative, since they tend to concentrate on the prominent and the successful, and for attributing all the results to the particular research program examined without ever being able to trace back all of the other possible contributors to its success. This is why in later work I turned away from individual event studies to direct econometric estimation of the contribution of R&D to productivity using one or another version of the production function approach to this problem. This approach abandons much of the interesting specific detail about individual innovations, concentrating instead on estimating the relationship between total output or productivity and various measures of current and past R&D expenditures (and other variables). All productivity growth (to the extent that it is measured correctly) is related to all expenditures on R&D, and an attempt is made to estimate statistically the part of productivity growth that might be attributed to R&D (and/or to its various components). The difficulties associated with this approach are surveyed in my *Bell Journal* paper (Griliches, 1979b). Examples of this type of analysis can be found in the two agricultural production function papers reprinted next, and will be discussed in somewhat more detail below.

The issues that arise in trying to extend this type of analysis out of the agricultural sector and onto the national scene are discussed in the next paper on "Research Expenditures and Growth Accounting," which was written for an IEA conference in St Anton in 1971. It provides a brief review of the literature as it had evolved up to that time, estimates the social rate of return to R&D in manufacturing at about 30 percent per annum, using Census of Manufactures data on 85 industries during 1958–63, and discusses how such results could be incorporated into an amended national accounting framework. Besides the usual econometric problems involved in estimating production functions, it raised a number of issues which have continued to worry me over the years to come: (1) Difficulties in the measurement of R&D output in the public sector which is its largest "consumer" (defense, space, and health research). (2) Difficulties in measuring the output of, and quality improvements in, technologically complex commodities such as computers and communication equipment. (3) Accounting problems arising from the fact that R&D labor and capital expenditures are already included once in the conventional labor and capital figures (see Schankerman, 1981). (4) The problem of externalities and "spillovers," the contribution of research in other laboratories, firms, industries, and countries to the success of a particular research project.

In spite of these difficulties I have kept working on this range of problems to this day because I believe in the importance of science and organized research activity to economic growth in this country and the world at large, and in the necessity to comprehend it. In the late 1960s I initiated a large collaborative project with the Bureau of the Census to match the R&D data they had collected for the NSF at the firm level with other economic data on these same firms. The results of this work on R&D returns at the firm level were published in Griliches (1980d). A subsequent round of replication and extension of this work to a later period occupied me for a good part of the 1970s and is summarized in Griliches (1986). That paper also addresses the issue whether the role of R&D has declined in recent years (concluding in the negative) and emphasizes the contribution of basic research to productivity growth in the 1970s. Recently I have been engaged in a large-scale study trying to use patent data to measure some aspects of R&D "output," and to understand the invention–innovation process better. For an interim summary of results from this range of studies see the NBER *R&D, Patents, and Productivity* book (Griliches, 1984) and Griliches, Pakes, and Hall (1987).

The role of R&D, if any, in the pervasive productivity growth slowdown in the 1970s and 1980s was the topic of concern in a number of my other papers. I have come to a negative conclusion on this topic (see Griliches and

Lichtenberg, 1984 and Griliches, 1986) but not because of the unimportance of R&D. It is my belief that one cannot see the impact of R&D on the production possibilities frontier of the economy in periods when the economy is not on this frontier or close to it. When the economy is not operating at full capacity it is hard to tell what is happening to its potential for production and growth. It may indeed be the case that the cessation of growth in real R&D expenditures in the 1960s did contribute to our problems in the 1970s and 1980s. Or so I thought in 1971 when I wrote: "This [contribution of R&D to growth in 1970] is about two-thirds of the comparable number for 1966 and reflects a rather significant slowdown in the rate of growth of R&D expenditures. Given the lags involved, this reduction may not show up, however, until the mid seventies" (this volume, p. 260). But one cannot decide that on the basis of the currently available data. The energy price shock induced worldwide recessions of 1974–5 and 1979–80 make it difficult to find traces of the R&D growth slowdown in the aggregate data; especially since much of the direct effect of such expenditures is not captured in these kind of data in the first place.

V

The last section of this volume contains papers which try to bring all these strands of work together in some kind of a more complete accounting of the sources of economic growth. Probably the best known of these is the 1967 joint paper with Jorgenson on "The Explanation of Productivity Change." Given its twentieth anniversary in 1987 it may be worthwhile to review some of the issues raised there. In a sense, though, the first two papers in this section, on the explanation of productivity growth in agriculture, come closer to representing my own point of view on this topic, then and now.

The 1967 Jorgenson and Griliches paper argued that a "correct" index number framework and the "right" measurement of inputs would reduce greatly the role of the "residual" ("advances in knowledge," total factor productivity, disembodied technical change, and/or other such terms) in accounting for the observed growth in output. It brought together Jorgenson's work on Divisia indexes, on the correct measurement of cost of capital, and on the right aggregation procedures for it, with my own earlier work on the measurement of capital prices and quality change and the contribution of education to productivity growth. It produced the startling conclusion, already foreshadowed in my agricultural papers, that an adjustment of conventional inputs for measurement and aggregation error may eliminate much of the mystery that was associated with the original findings of large unexplained components in the growth of national and sectoral outputs. It did this with a "Look Ma! No hands!" attitude, neither using additional outside variables such as R&D, nor allowing for economies of scale or other disequilibria (e.g. differences in rates of return to different private and public investments). This did indeed attract attention and also criticism. The most penetrating criticism came from Denison (1969), and led to an exchange between us in the May 1972 issue of the *Survey of Current Business*. An excerpt from our "Reply" is also reprinted here.

Denison found a number of minor errors and one major one in our computations. By trying to adjust for changing utilization rates we used data on energy consumption of electric motors in manufacturing, a direct measure of capital equipment utilization in manufacturing, but extrapolated it also to non-equipment components of capital in manufacturing and to all capital outside of manufacturing, including residential structures. There was also the uneasy issue of integrating a utilization adjustment within what was otherwise a pure equilibrium story. Once we conceded most of the utilization adjustment, our "explanation" of productivity growth shrank from 94 to 43 percent, and with it also our claim to "do it all" (without mirrors).

I still believe, however, that we were right in our basic idea that productivity growth should be "explained" rather than just measured, and that errors of measurement and concept have a major role in this. But we did not go far enough in that direction. We offered improved index number formulae, a better reweighting of capital input components, a major adjustment of the employment data for improvements in the quality of labor, revisions in investment price indexes, and estimates of changes in capital utilization. The potential orders of magnitude of the adjustments based on the first two contributions, index number formulae and the reweighting of capital components, are not large enough to account for a major part of the observed "residual." The labor quality adjustment was not really controversial, but the capital price indexes and utilization adjustments deserve a bit more discussion. We argued for the idea that technical change could be thought of, in a sense, as being "embodied" in factor inputs, in new machines and human capital, and that a better measurement of these inputs via the non-tautological route of hedonic index numbers for both capital and labor, could account for most of what was being interpreted as a "residual." It became clear, however, that without extending our framework further to allow for increasing returns to scale, R&D, sectoral disequilibria, and other externalities we were unlikely to approach a full "explanation" of productivity change (see the last paragraphs of our "Reply to Denison"). Such a wider, less equilibrium-based approach, was already pursued in my earlier agricultural productivity papers. Before I turn, however, to a discussion of these papers, there is still some more to be said about the price indexes and utilization adjustments.

It may appear that adjusting a particular input for mismeasured quality change would not have much of an effect on productivity growth measurement, since one would need also to adjust the output figures for the corresponding industry. But as long as the share of this industry in final output is less than the elasticity of output with respect to this input, the two adjustments will not cancel themselves out. Since the share of investment in output is significantly lower than reasonable estimates of the share of capital in total factors costs, adjusting capital for mismeasurement of its prices does lead to a net reduction in the computed residual. Empirically it is clear that, even without considering any of the potential externalities associated with new capital, there are enough questions about the official price indexes in these areas to make further work on this topic a high priority (see the evidence presented in the "Measuring Inputs in Agriculture" and the "Hedonics" papers, and in the subsequent literature discussed in the relevant postscripts).

The utilization adjustments fit uneasily within the rather strict competitive equilibrium framework of the Jorgenson–Griliches paper. The analogy was made to labor hours, calling for the parallel concept of machine hours as the relevant notion of capital services. We also had in mind the model of a continuous process plant where output is more or less proportional to hours of operation. Since we were interesting primarily in "productivity" change as a measure of "technical" change, a change that is due to changes in techniques of production, fluctuations in "utilization," whether a plant worked one shift or two, 10 months or 12, were not really relevant for this purpose. But while labor unemployment was happening off-stage as far as business productivity accounts were concerned, capital "underemployment" was difficult to reconcile with the maximizing behavior with perfect foresight implicit in our framework.

There are two somewhat separate "utilization" issues. Productivity as measured is strongly pro-cyclical. Measured inputs, especially capital and labor services, fluctuate less than reported output. The resulting fluctuations in "productivity" do not make sense if we want to interpret them as a measure of the growth in the level of technology or the state of economically valuable knowledge of an economy. The US economy did not "forget" 4 percent of its technology between 1974 and 1975. Nor was there a similar deterioration in the skill of its labor force. (National welfare did go down as the result of OPEC-induced world-wide rise in energy prices, but that is a separate story.)

What is wrong with the productivity numbers in this case is that we do not measure accurately the actual amount of labor or machine hours used rather than just paid for. Since both capital and labor are bought or hired in anticipation of a certain level of activity, and on long-term contracts, actual factor payments do not reflect their respective marginal products except in the case of perfect foresight and only in the long run. Underutilization of factors of production is the result of unanticipated shifts in demand and various rigidities built into the economic system due to longer-term explicit and implicit contracts (and other market imperfections) between worker and employer and seller and buyer. If our interest is primarily in the "technological" interpretation of productivity measures, we must either ignore such shorter-run fluctuations or somehow adjust for them. This was the rationale behind our original use of energy consumed by electric motors (per installed horsepower) as a utilization adjustment.

We used energy consumption as a proxy for the unobserved variation in machine hours, and not on its own behalf as an important intermediate input.

Used in the latter fashion it is a produced input which would cancel out at the aggregate level (as was pointed out by Denison in his comment on our paper). Alternatively, one could adjust the weight (share) of capital services in one's total input index, to reflect the fact that underutilization of this existing stock of resources should reduce significantly the shadow price of using them (this is the approach suggested in Berndt and Fuss, 1986). Unfortunately it is difficult to use the observed factor returns for these purposes, both because prices do not fall rapidly enough in the face of unanticipated demand shocks and because of a variety of longer-run contractual factor payments arrangements which break the link between factor rewards and their current productivity.

In a sense this reflects the failure of the assumption of perfect competition on which much of the standard productivity account is based. The actual world we live in is full of short-run rigidities, transaction costs, immalleable capital, and immobile resources, resulting in the pervasive presence of quasi-rents and short-term capital gains and losses. While I do not believe that such discrepancies from "perfect" competition actually imply the presence of significant market power in most industries (as argued, for example, by Hall, 1986), they do make productivity accounting even more difficult.

The other aspect of utilization is the longer-run trend in shift-work, length of the work-week, and changes in hours of operation per day by plants, stores, and service establishments. Consider, for example, a decline in overtime or night-shift premia due, say, to a decline in union power. This would reduce the price of certain types of capital services and expand their use. If capital is not measured in machine hours we would show a rise in productivity even though there has been no "technological" change in methods of production. I would prefer not to include such changes in the productivity definition, since I interpret them as movements along (or toward) a stable production possibilities frontier. But there did occur an organizational change which allowed us to get more "flow," more hours per day or year, from a given stock of equipment or other resources. One can think of this as a mixture of two types of activities: output production which rents machine and labor hours and the supply of capital (and also effective labor hours) from the existing stock of resources. A decline in overtime premia would be similar to a decline in the tariff on a certain kind of imported input. It would lead to an improvement in "productivity" but not necessarily to a "technical" change.

It is still my belief that we need to adjust our data for such capacity utilization fluctuations for a better understanding of "technical" change, the issue that brought us to this in the first place. A consistent framework for such an adjustment will require, however, the introduction of adjustment costs and ex-post errors in the productivity measurement framework. (See Morrison, 1985, and the literature cited there for recent developments in this area.) It is not clear, however, whether one can separate longer-run developments in the utilization of capital from changes in technology and the organization of society. Much of capital is employed outside continuous process manufacturing and there the connection between productivity and its utilization is much looser. The rising cost of human time and the desire for variety and flexibility have led to much investment in what might be called "standby" capacity with rather low utilization rates. The hi-fi system in my home is operating only at a fraction of its potential capacity. Much inventory is held in many businesses to economize on other aspects of labor activity. Nor is it clear that an extension of store hours with a resulting decline in productivity per square-foot-hour of store space is necessarily a bad thing. Thus it is difficult to see how one could separate long-run trends in utilization from changes in production and consumption technologies. It is, however, a topic worth studying and a potentially important contributor to "explanations" of apparent swings in measured productivity statistics.

Whether we include or exclude such changes from our "productivity" concept will affect our ability to "account" for them. But that is not the important issue. We do want to measure them, because we do want to understand what happened, to "explain". The rest is semantics.

Many of the problems discussed here arise because we do not aggregate adequately and do not describe the production process in adequate detail. A model which would distinguish between the use of capital and labor at different times of the day and year, and would not assume that their shadow prices are constant between different "hours" or over time, would be capable of handling these kind of shifts. We do not have the data to implement such a program, but it underscores the message of our original: much of what passes for productivity change in conventional data is the result of aggregation errors, the wrong measurement of input quantities, and the use of wrong weights to combine them in "total factor input" indexes.

Something more should be said about the rather vague notions of "explanation" and "accounting." National Income accounts and associated index numbers are economic constructs, based on an implicit model of the economy and a variety of more or less persuasive logical and empirical arguments. It is not well adapted to "hypothesis testing" or debates about causality. In proposing a better measure of, say, labor, we rely on the evidence of market wage differentials in offering up our improved measure. By bringing in more evidence on this topic we are not just reducing the "residual" tautologically. But the fact that it goes down as the result of such an adjustment does not make it right either. A different kind of evidence is required to provide a more persuasive justification for such adjustments. That is why I turned early on to the use of production functions for econometric testing. Without moving in such a direction one tends to run into various paradoxes. For example, capital growth accelerated in the 1970s in many industries without a comparable increase in the growth of output. In the index number sense of growth accounting, capital "explained" a larger fraction of the growth of output and we did, indeed, have a smaller residual. But in spite of this "accounting," the mystery only deepened.

The "econometric" approach to growth accounting evolves one in the estimation of production functions. This allows one to test or validate a particular way of measuring an input or adjusting it for quality change; to estimate and test the role of left-out public good inputs such as R&D and other externality-generating activities; to estimate economies of scale; and to check on the possibility of disequilibria and estimate the deviation of "true" output elasticities from their respective factor shares. Production function estimation raises many problems of its own, including issues of aggregation, errors of measurement and simultaneity, but it is one of the few ways available to us for checking the validity of the attribution of productivity growth to its various suggested "sources."

The two agricultural production function papers represent my most successful attempts to accomplish this. The first paper (*JPE*, 1963) was based on cross-sectional data for 68 regions of the US in 1949. It used the production function framework to show that education of the farm labor force was an important contributor to productivity, validating this particular adjustment to the measurement of labor input; that the estimated role (elasticity) of farm equipment and machinery was higher and that of farm labor was lower than was implied by their respective factor shares; and that there was evidence of significant economies of scale in agricultural production. These estimates were then applied to the aggregate output and input series for US agriculture in 1940 and 1960, adjusting also the official farm capital series for errors in their deflators, with the result that they "explained" all, and even somewhat more, of the rise in agricultural output between these years. The next paper replicated this approach on state-level data for 1949, 1954, and 1959, and added a measure of public investments in agricultural research and extension to the estimation equation. The results were rather similar: "education" was significant, and so also was the estimated, though somewhat smaller, economies of scale parameter. The major new finding was the rather large and significant contribution of public R&D and extension expenditures with a rather high implied social rate of return to them. When these estimates were used to analyze aggregate agricultural productivity growth between 1949 and 1959 they could essentially account for all of it, with about one-third to one-half of the total "explanation" coming from the growth in the scale of the average agricultural enterprise, about one-third coming from public investments in research and extension, and the rest being divided about equally between adjustments in the measurement of conventional inputs and adjustments in their relative weights.

This work left me with the conviction that education, investment in research, and economies of scale (both at the level of the firm and at the level of the market) were the important sources of productivity growth in the long run. Since in the paper with Jorgenson we had not allowed for the two latter sources of growth, I was not too surprised or disheartened when it turned out that we could not really explain all of aggregate productivity change by formula and labor quality adjustments alone. It was clear, however, that one would need more and better data to make such additional adjustments more reliable and convincing. I turned, therefore, to trying to amass more data and more evidence on these topics. The task proved harder than I had anticipated, the data sparser and more brittle than one might have wished, and hence my sojourn in this purgatory much longer than I had expected. It is not clear whether we have yet the data to do an adequate and convincing accounting of productivity change at the aggregate level (see, for example, the continued and still unsettled debate about the causes of the recent slowdown in productivity growth). Progress has been made, however, in several directions, and we have now a much better understanding of the measurement issues in the various areas, and also a deeper appreciation of the difficulties involved in saying something definitive about them.

I have already discussed my subsequent work on the productivity of education and the measurement of returns to R&D. The latter topic, which I am still pursuing, suffers especially from the difficulty of tackling the externalities question econometrically. It is difficult, if not impossible, to get a measure of the relative contribution of university science to different industries or of knowledge "leakages" from one industry or firm to another. But that is where most of what passes for exogenous disembodied technical change may be coming from.

I did pursue the issue of economies of scale quite a bit further. In Griliches (1967b and 1968) I analyzed US manufacturing data by state and industry for 1954, 1958, and 1963, and found persistent and significant but relatively small traces of economies of scale (on the order of 1.05). This work was based on per-establishment averages for different states and industries. Since I thought of economies of scale as primarily a micro-phenomenon, occurring at the plant or firm level, I kept looking around for relevant micro-data to pursue this topic further. In the late 1960s I gained access to the micro-data from the Norwegian Census of Manufactures, and together with Vidar Ringstad produced a detailed study of it (Griliches and Ringstad, 1971). There too significant but not very large economies of scale (about 1.05) appeared to be present. In spite of the *a priori* conviction in their importance, it was much harder to find significant traces of economies of scale in manufacturing than in agriculture. The main difficulty arises from the fact that different size plants and firms, even in well-defined industries, rarely produce the same type of product or sell it at the same price. And there is, usually, no adequate price or product detail available at the plant level in census-type data. Thus, despite much work, little convincing evidence has been produced on this topic either in the US or elsewhere. The main evidence on the potential importance of economies of scale has come from data on regulated utilities, where the product is much more homogeneous and the data are more plentiful (see Nerlove, 1963, and McFadden, 1978a for examples of such work, and the various papers in Fuss and McFadden, 1978 for a more extensive discussion of some of these issues).

Even though we now have more data, more advanced econometric technology, and better computer resources, the overall state of this field has not advanced greatly in the past 20 years. We are really not much closer to an "explanation" of the observed changes in the various productivity indexes. A tremendous effort was launched by Jorgenson and his co-workers

(Christensen, Fraumeni, Gollop, Nishimizu, and others) to improve and systematize the relevant data sources, to produce and analyze a consistent set of industry-level total factor productivity accounts, to extend and generalize our original labor quality adjustments, and to extend all of this also to international comparisons of productivity. In the process, however, rather than pursuing the possibly hopeless quest for a complete "explanation" of productivity growth, they chose to focus instead on developing more precise and detailed productivity measures at various levels of aggregation and devising statistical models for their analysis. Denison (1974 and 1979), in parallel, was pursuing his quest for a more complete accounting of the sources of growth, putting together as many reasonable scraps of information as were available, but not embedding them in a clear theoretical framework or an econometrically testable setting. The incompleteness of both approaches, and the unsatisfactory state of this field as a whole, was revealed by the sharp and prolonged slowdown in the growth of measured productivity which began in the mid-1970s. Despite the best attempts of these and other researchers it has not been possible to account for this slowdown within the standard growth accounting framework without concluding that the "residual" had changed, that the growth rate of total factor productivity growth rate fell some time in the late 1960s or early 1970s (see Denison, 1984; Griliches, 1980e; Kendrick, 1983, and many others).

I do not believe, however, that this slowdown can be interpreted as implying that the underlying rate of technical change has slowed down, that we have exhausted our technological frontiers. In my opinion it was caused by misguided macro-policies induced by the oil price shocks and the subsequent inflation and the fears thereof. Without allowing for errors in capital accumulation (which continued initially at a rather high rate, in spite of the sharp declines in aggregate demand) and widespread underutilization of capacity, it is not possible to interpret the conventional productivity statistics. Surely "knowledge" did not retreat. Moreover, I do not believe that one can use statistics from such periods to infer anything about longer-term technological trends. If we are not close to our production possibilities frontier we cannot tell what is happening to it, and whether the underlying growth rate of an economy's "potential" has slowed down or not. We need a better-articulated theoretical framework, one that would allow for long-term factor substitution and short-term rigidities and errors, for dynamics and for adjustment costs, before we are able to understand what has happened to us recently. We also need better data, especially on output and input prices and various aspects of labor and capital utilization.

In the long run productivity grows when we either acquire more resources or figure out better ways of using them. Better ways can mean moving available resources into more productive uses and eliminating various obstacles to their full utilization. It also means finding entirely new ways of satisfying human needs and desires through new products, processes, and new organizational arrangements. All such activities are affected by economic forces and will repay economic study. In the work collected in this volume I have tried to approach this elephant from various different directions. While I have learned much from this work, and also from the work of many others on this same range of topics, there is still much that we do not know, especially since the world keeps changing while we try to observe it. That is why I have continued working on this set of issues. Hope springs eternal.

Zvi Griliches

Part I

Technology and the Measurement
of Input and Output

2

Hybrid Corn: An Exploration in The Economics of Technological Change*

This is a study of factors responsible for the wide cross-sectional differences in the past and current rates of use of hybrid seed corn in the United States.

Logistic growth functions are fitted to the data by states and crop reporting districts, reducing differences among areas to differences in estimates of the three parameters of the logistic: *origins, slopes*, and *ceilings*.

The lag in the development of adaptable hybrids for particular areas and the lag in the entry of seed producers into these areas (differences in *origins*) are explained on the basis of varying profitability of entry, "profitability" being a function of market density, and innovation and marketing cost.

Differences in the long-run equilibrium use of hybrid corn (*ceilings*) and in the rates of approach to that equilibrium (*slopes*) are explained, at least in part, by differences in the profitability of the shift from open pollinated to hybrid varieties in different parts of the country.

The results are summarized and the conclusion is drawn that the process of innovation, the process of adapting and distributing a particular invention to different markets and the rate at which it is accepted by entrepreneurs are amenable to economic analysis.

Introduction

The work presented in this paper is an attempt to understand a body of data: the percentage of all corn acreage planted with hybrid seed, by states and by years. By concentrating on a single, major, well-defined, and reasonably well-recorded development – hybrid corn – we may hope to learn something about the ways in which technological change is generated and propagated in US agriculture.

*Reprinted from *Econometrica*, vol. 25, no. 4, October 1957. ©Copyright, 1957, The Econometric Society. All rights reserved.

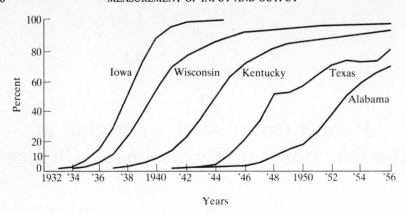

Figure 2.1 Percentage of total corn acreage planted with hybrid seed. (Source: USDA, *Agricultural Statistics*, various years.)

The idea of hybrid corn dates back to the beginning of this century and its first application on a substantial commercial scale to the early 1930s. Since then it has spread rapidly throughout the Corn Belt and the rest of the nation.[1] There have been, however, marked geographic differences in this development (see figure 2.1). Hybrid corn was the invention of a method of inventing, a method of breeding superior corn for specific localities.[2] It was not a single invention immediately adaptable everywhere. The actual breeding of adaptable hybrids had to be done separately for each area. Hence, besides the differences in the rate of adoption of hybrids by farmers – the "acceptance" problem – we have also to explain the lag in the development of adaptable hybrids for specific areas – the "availability" problem.

In the following sections I shall first outline a method used to summarize the data. Essentially it will consist of fitting trend functions (the logistic) to the data, reducing thereby the differences among areas to differences in the values of a few parameters. Then I will present a model rationalizing these differences and illustrate it with computational results. Finally, I shall draw some conclusions on the basis of these results and other accumulated information.

The Method and the Model

A graphical survey of the data by states and crop reporting districts along the lines of figure 2.1 led to the conclusion that nothing would be gained by trying to explain each observation separately, as if it had no antecedent.[3] It became obvious that the observations are not points of equilibrium which may or may not change over time, but points on an adjustment path, moving more or less consistently towards a new equilibrium position. Hence we should phrase our questions in terms of the beginning of the movement, its rate, and its destination. This led to the decision to fit some simple trend functions to the data and concentrate on the explanation of the cross-sectional differences in the estimates of their parameters.

The choice of a particular algebraic form for the trend function is somewhat arbitrary. As the data are markedly S-shaped, several simple S-shaped functions were considered. The cumulative normal and the logistic are used most widely for such purposes. As there is almost no difference between the two over the usual range of data,[4] the logistic was chosen because it is simpler to fit and in our context easier to interpret. While there are some good reasons why an adjustment process should follow a path which is akin to the logistic, I do not want to argue the relative merits of the various S-curves.[5] In this work the growth curves serve as a summary device, perhaps somewhat more sophisticated than a simple average, but which should be treated in the same spirit.

The logistic growth curve is defined by $P = K/1 + e^{-(a+bt)}$, where P is the percentage planted with hybrid seed, K the ceiling or equilibrium value, t the time variable, b the rate of growth coefficient, and a the constant integration which positions the curve on the time scale. Several features of this curve are of interest: it is asymptotic to 0 and K, symmetric around the infection point, and the first derivative with respect to time is given by $dP/dt = -b/(P/K) (K-P)$.[6] The rate of growth is proportional to the growth already achieved and to the distance from the ceiling. It is this last property that makes the logistic useful in so many diverse fields.[7]

There are several methods of estimating the parameters of the logistic.[8] The method chosen involves the transformation of the logistic into an equation linear in a and b. The dividing both sides of the logistic by $K - P$ and taking the logarithm, we get its linear transform, $\log_e [P/K - P)] = a + bt$, allowing us to estimate the parameters directly by least squares.[9] The value of K, the ceiling, was estimated crudely by plotting the percentage planted to hybrid seed on logistic graph paper and varying K until the resulting graph approximated a straight line. After adjusting for differences in K, the logistic was fitted to the data covering approximately the transition from 5 to 95 percent of the ceiling. The observations below 5 and above 95 percent of the ceiling value were discarded because they are liable to very large percentage errors and would have had very little weight anyway in any reasonable weighting scheme. The period included in the analysis, however, accounts for the bulk of the changes in the data.

The procedure outlined above was used to calculate the parameters of the logistic for 31 states and for 132 crop reporting districts within these states.[10] The states used account for almost all of the corn grown in the US (all states except the West and New England). Out of a total of 249 crop reporting districts only those were used for which other data by crop reporting districts were readily available. Districts with negligible amounts of corn and unreliable estimates of hybrid corn acreage were also left out.[11]

The results of these calculations are presented in tables 2.1 and 2.2. Table 2.1 summarizes the state results, table 2.2 the results by crop reporting districts. Time is measured from 1940, and $(-2.2 - a)/b$ indicates the date

Table 2.1 Hybrid corn logistic trend functions by states

States	Origin $\dfrac{-2.2-a}{b}$	Rate of acceptance b	Ceiling K	r^2
NY	− 0.89	0.36	0.95	0.99
NJ	− 1.48	0.54	0.98	0.90
Pa.	− 1.29	0.48	0.95	0.98
Ohio	− 3.35	0.69	1.00	0.97
Ind.	− 3.13	0.91	1.00	0.99
Ill.	− 4.46	0.79	1.00	0.99
Mich.	− 1.44	0.68	0.90	0.98
Wisc.	− 3.52	0.69	0.91	0.99
Minn.	− 3.06	0.79	0.94	0.99
Iowa	− 4.34	1.02	1.00	0.99
Mo.	− 3.32	0.57	0.98	0.98
ND	− 0.65	0.43	0.65	0.96
SD	− 0.40	0.42	0.85	0.95
Neb.	− 0.60	0.62	0.97	0.99
Kan.	0.42	0.45	0.94	0.97
Del.	0.21	0.47	0.99	0.98
Md.	− 0.73	0.55	0.98	0.97
Va.	1.60	0.50	0.92	0.97
WVa.	− 0.23	0.39	0.85	0.98
NC	5.14	0.35	0.80	0.89
SC	5.72	0.43	0.60	0.96
Ga.	7.92	0.50	0.80	0.99
Fla.	2.89	0.38	0.90	0.93
Ky.	0.08	0.59	0.90	0.99
Tenn.	2.65	0.34	0.80	0.97
Ala.	7.84	0.51	0.80	0.99
Miss.	4.75	0.36	0.60	0.98
Ark.	1.46	0.41	0.78	0.99
La.	4.89	0.45	0.53	0.99
Okla.	3.57	0.56	0.80	0.98
Tex.	3.64	0.55	0.78	0.98

Notes: $P = \dfrac{K}{1 + e^{-(a+bt)}}$; $\log_e \left(\dfrac{P}{K-P} \right) = a + bt$; $t_{1940} = 0$; $N = -6$ to 16; Max $S_b = 0.06$;

$Origin$ = Date of 10 percent = $\dfrac{-2.2-a}{b}$ measured from 1940, e.g. $-4.0 = 1936, +7.0 = 1947$;
$Rate\ of\ acceptance = Slope = b$; and $Ceiling = K$

at which the function passed through the 10 percent value. This date will
be identified below with the date of *origin* of the development. Several things
are noteworthy about these figures: the high r^2's indicate the excellent fits
obtained.[12] The b's, representing the slope of the transform or the rate of
adjustment, are rather uniform, becoming lower as we move towards the
fringes of the Corn Belt. The values of $(-2.2-a)/b$, the dates of *origin*,

indicate that the development started in the heart of the Corn Belt and spread, rather regularly, towards its fringes.[13] The ceiling – K – also declines as we move away from the Corn Belt.

In this section we have succeeded in reducing a large mass of data to three sets of variables – *origins, slopes*, and *ceilings*. "Thus on the basis of three numbers we are prepared, in principle, to answer all the questions the original data sheet can answer provided that the questions do not get down to the level of a single cell. . . . This is saying a great deal."[14]

The economic interpretation of the differences in the estimated coefficients will be developed in the following sections. The values of the different parameters are not necessarily independent of each other, but for simplicity will be considered separately. Variations in the date of origin will be identified with supply factors, variations in slopes with factors affecting the rate of acceptance by farmers, and variations in ceilings with demand factors affecting the long-run equilibrium position. In each case we shall consider briefly the implicit identification problem.

The Supply of a New Technique

There is no unique way of defining the data of *origin* or of "availability." Hybrid corn was not a single development. Various experimental hybrids were tried until superior hybrids emerged. After a while these were again superseded by newer hybrids. Nor is there a unique way of defining *origin* with respect to growth curve. The logistic is asymptotic to zero; it does not have a "beginning." Nevertheless, it is most important to distinguish between the lag in "availability" and the lag in "acceptance." It does not make sense to blame the Southern farmers for being slow in acceptance, unless one has taken into account the fact that no satisfactory hybrids were available to them before the middle 1940s.

I shall use the date at which an area began to plant 10 percent of its ceiling acreage with hybrid seed as the date of *origin*.[15] The 10 percent date was chosen as an indicator that the development had passed the experimental stage and that superior hybrids were available to farmers in commercial quantities. The reasonableness of this definition has been borne out by a survey of yield tests in various states and it has been supported by conversations with various people associated with developments in hybrid corn in the experiment stations and private seed companies.[16]

"Availability" is the result of the behavior of agricultural experiment stations and private seed companies. If we include the growers of station hybrids in the general term "commercial seed producers," then availability is the direct result of the actions of seed producers with the experiment stations affecting it through the provision of free research results and foundation stocks. The activities of the experiment stations serve to reduce the cost of innovation facing the seed producers but the entry decisions are still their own. The date at which adaptable hybrids became available in an area is

Table 2.2 Hybrid corn logistic trend functions by crop reporting districts*

State and CR district		Origin	Rate of acceptance	Ceiling	r^2
Pa.	1	0.15	0.41	0.85	0.99
	2	1.16	0.49	0.90	0.99
	3	0.76	0.46	0.91	0.98
	4	0.44	0.47	0.92	0.99
	5	0.11	0.62	0.95	0.98
	6	-1.02	0.55	0.95	0.99
	7	-0.63	0.40	0.90	0.98
	8	-1.04	0.54	0.96	0.99
	9	-2.35	0.60	0.98	0.97
Ohio	1	-3.22	1.25	1.00	0.99
	2	-2.73	0.99	1.00	0.98
	3	-1.77	0.75	0.95	0.98
	4	-3.00	0.90	1.00	0.98
	5	-3.19	0.77	1.00	0.98
	6	-3.14	0.69	0.95	0.94
	7	-2.69	0.88	1.00	0.98
	8	-1.78	0.60	0.95	0.99
	9	-1.80	0.73	0.95	0.97

State and CR district		Origin	Rate of acceptance	Ceiling	r^2
Wisc.	5	-2.54	0.61	0.90	0.98
	6	-3.03	0.87	0.78	0.99
	7	-4.16	0.89	0.98	0.99
	8	-3.55	0.88	0.95	0.99
	9	-3.21	0.72	0.95	0.98
Minn.	7	-3.08	1.36	1.00	0.99
	8	-3.66	1.14	1.00	0.99
	9	-3.04	1.01	1.00	0.99
Iowa	1	-4.39	1.01	1.00	0.99
	2	-4.78	1.05	1.00	0.99
	3	-4.46	1.00	1.00	0.99
	4	-3.71	1.12	1.00	0.99
	5	-4.70	1.13	1.00	0.99
	6	-5.15	1.09	1.00	0.99
	7	-2.74	1.25	1.00	0.99
	8	-3.61	1.07	1.00	0.99
	9	-4.15	1.10	1.00	0.99

Ind.	1	-3.82	1.15	1.00	0.99	Mo.	1	-1.37	1.19	1.00	0.97
	2	-3.60	1.10	1.00	0.99		2	-1.33	1.15	1.00	0.95
	3	-3.12	1.15	1.00	0.99		3	-1.27	1.15	1.00	0.96
	4	-3.24	0.95	1.00	0.99		4	-1.51	0.66	0.95	0.98
	5	-2.85	1.07	1.00	0.99		5	-0.64	0.78	0.93	0.99
	6	-2.63	1.12	1.00	0.99		6	-1.11	0.72	0.97	0.97
	7	-1.67	0.87	1.00	0.96		7	0.16	0.46	0.90	0.99
	8	-1.57	0.82	1.00	0.98		8	0.63	0.63	0.87	0.99
	9	-1.88	0.76	1.00	0.98		9	0.94	0.64	0.97	0.99
Ill.	1	-4.81	1.13	1.00	0.99	ND	9	0.40	0.74	0.85	0.96
	3	-4.59	0.98	1.00	0.99	SD	3	-0.53	0.57	0.90	0.99
	4	-4.16	1.08	1.00	0.99		6	-0.71	0.85	0.93	0.99
	4a	-2.65	1.09	1.00	0.99		9	-1.72	0.75	0.95	0.99
	5	-4.68	1.17	1.00	0.99	Neb.	3	-2.48	0.90	1.00	0.99
	6	-4.25	1.18	1.00	0.99		5	0.36	0.82	0.93	0.99
	6a	-2.46	0.91	1.00	0.99		6	-2.18	0.85	1.00	0.99
	7	-0.81	0.64	1.00	0.97		7	2.33	0.90	0.95	0.99
	9	-0.58	0.78	1.00	0.97		8	1.60	0.94	0.95	0.99
Mich.	7	-1.12	0.77	0.92	0.97		9	0.77	0.91	1.00	0.97
	8	-1.04	0.89	0.92	0.98	Kan.	1	2.68	0.41	0.95	0.95
	9	-1.70	0.78	0.92	0.98		2	1.52	0.66	1.00	0.98
Wisc.	1	-2.17	0.81	0.85	0.99		3	-0.88	0.72	1.00	0.99
	2	-2.22	0.97	0.70	0.99		6	-0.88	0.68	0.92	0.99
	3	-2.42	0.93	0.60	0.99		9	0.73	0.41	0.95	0.99
	4	-3.24	0.67	0.95	0.96						

(continued)

Table 2.2 *(continued)*

State and CR district		Origin	Rate of acceptance	Ceiling	r^2
Md.	1	2.92	0.37	0.97	0.94
	2	−1.12	0.64	1.00	0.97
	8	0.88	0.48	0.98	0.98
	9	0.40	0.60	1.00	0.93
Va.	2	0.87	0.68	1.00	0.99
	4	1.51	0.61	0.98	0.93
	5	2.37	0.68	0.95	0.99
	6	2.06	0.63	0.97	0.96
	7	1.21	0.29	0.80	0.85
	8	2.18	0.40	0.85	0.90
	9	1.04	0.50	0.95	0.96
Ky.	1	0.67	0.89	0.95	0.97
	2	−0.42	0.72	0.98	0.99
	3	0.49	0.61	0.90	0.97
	4	−0.36	0.83	0.92	0.99
	5	−0.77	0.78	0.90	0.99
	6	1.94	0.62	0.60	0.98
Tenn.	1	0.76	0.29	0.85	0.97
	2	1.88	0.33	0.55	0.99
	3	2.64	0.39	0.70	0.97
	4	2.53	0.43	0.75	0.96
	5	3.43	0.35	0.80	0.91
	6	2.94	0.33	0.70	0.95

State and CR district		Origin	Rate of acceptance	Ceiling	r^2
Ala.	1	7.73	0.56	0.60	0.98
	2	6.33	0.57	0.99	0.99
	2a	8.80	0.45	0.90	0.97
	3	7.68	0.54	0.95	0.98
	4	7.45	0.42	0.50	0.95
	5	8.08	0.49	0.70	0.95
	6	8.15	0.39	0.60	0.95
	7	7.84	0.58	0.85	0.97
	8	8.24	0.45	0.70	0.97
	9	8.53	0.55	0.90	0.99
Ark.	1	0.41	0.37	0.75	0.97
	2	1.98	0.40	0.82	0.98
	3	0.68	0.50	0.85	0.99
	4	2.24	0.42	0.77	0.94
	5	1.89	0.35	0.85	0.95
	6	1.54	0.35	0.80	0.99
	7	1.66	0.32	0.55	0.93
	8	2.41	0.37	0.70	0.92
	9	1.88	0.33	0.85	0.99
Okla.	3	2.61	0.49	0.80	0.97
	5	3.62	0.55	0.90	0.97
	6	3.17	0.52	0.88	0.93
	7	4.05	0.39	0.80	0.97
	8	4.85	0.67	0.90	0.98
	9	4.08	0.52	0.75	0.95

* I am indebted to the Field Crop Statistics of the Agricultural Market Service for the unpublished data by crop reporting districts.

viewed as the result of seed producers ranking different areas according to the expected profitability of entry and deciding their actions on this basis.[17] The relative profitability of entry into an area will depend on the size of the eventual market in that area, marketing cost, the cost of innovating for that area, and (given a positive rate of interest) the expected rate of acceptance.[18]

It is extremely difficult to define "market size" operationally. The definition is not independent of marketing cost or of the particular characteristics of the innovation (the area of adaptability of a particular hybrid) and is complicated by the arbitrariness of the political subdivisions used as the geographic units of analysis. The problem of the "right" geographic unit of analysis, however, will be postponed to the end of this section. As an approximation to the size of the market I used the average corn acreage in the area at about the time of the date of entry, adjusted for differences in ceilings. That is, the average corn acreage was multiplied by 0.9 if that was the estimate of the fraction of the corn acreage which would be ultimately planted with hybrid seed. Because the political subdivisions are of various and sundry sizes, to make them more comparable the adjusted corn acreage was divided by total land in farms. The resulting variable $-X_1 = $ (average corn acreage) $\times K \div$ total land in farms – is a measure of "market density" rather than of "market size."[19] If the areas are not too different in size and in the range of adaptability of their hybrids, market density will closely approximate a relevant measure of market size. Also, in its own right, it is important as a measure of marketing cost, the relative cost of selling a given supply of seed in different areas. Under the name of "market potential," such a variable was, in fact, used by at least one major seed company in its decision making process. The importance of a variable of this sort was strongly emphasized, in private conversations, by executives of the major seed companies.

The importance of marketing cost is underscored by the striking differences in marketing methods of hybrid seed in different parts of the country. While almost 90 percent of all the seed in the Corn Belt is sold by individual salesmen who call on each farmer, almost all of the seed in the South is sold through stores where the farmer must come and get it. The small size of the corn acreage per farm, the relative isolation of the small farm, and the large proportion of corn on noncommercial farms make the type of marketing used in the Corn Belt prohibitively expensive in the South. The cost of selling a given amount of seed is quite different in various parts of the country, as many more farmers have to be reached in one area than in another. As a measure of "average size of sale," I used average corn acres per farm reporting corn, X_3.

The estimated slope coefficient, b, was used as a measure of the expected rate of acceptance in different areas. This assumes that producers were able to predict reasonably well the actual rate of acceptance.

There is no good way of estimating the relative costs of innovation. It is probably true that there are no substantial differences in the cost of developing a hybrid from scratch for any corn growing area of the country and, if there

Table 2.3   Correlation coefficients on the state level – $n = 31$

	X_1	X_3	b	X_4	X_{10}
Y	-0.44	-0.35	-0.62	-0.89	0.82
X_1		0.52	0.77	0.55	-0.39
X_3			0.46	0.28	-0.36
b				0.68	-0.51
X_4					-0.79

Notes: Y = Date or *origin*. The date an area reached 10 percent, computed. See tables 2.1 and 2.2.

X_1 = Market density. For states: average corn acreage 1937–46 times K, divided by land in farms in 1945. Similar for crop reporting districts but averaged over different periods, depending on the availability of data. Source: *Agricultural Statistics, Census of Agriculture*, and published and unpublished materials from state agricultural statisticians.

X_3 = For states, average corn acres per farm, 1939. Source: *Census of Agriculture*. By crop reporting districts: the same average corn acreage as in X_1, divided by the 1939 or 1945 census number of farms reporting corn, depending on availability of data.

b = The slope of the logistic transform, computed.

X_4 = "Corn Beltliness." The proportion of all inbred lines accounted by "Corn Belt" lines in the pedigrees of recommended hybrids by areas. Source: C. B. Henderson, "Inbred lines of corn released to private growers from state and federal agencies and double crosses recommended by states," 2nd revision, Illinois Seed Producers Association, Champaign, April 15, 1956; and unpublished data from the Funk Bros. Seed Co., Bloomington, Ill.

X_{10} = Earliest date of origin in the immediate neighborhood.

were some, they would be swamped by the large differences in returns. A difficulty arises, however, from the fact that a hybrid may be adaptable in more than one area, allowing the cost of innovation to be spread over several areas, and because the experiment stations have borne a substantial part of the innovation cost by developing and releasing inbred lines and whole hybrids. That is, the actual cost of innovating for an area will depend on whether or not hybrids which have already been developed for other areas prove adaptable in this area, and on whether or not the experiment stations have produced and released inbred lines or hybrids adaptable to this area.

Since most of the early research was done for the area known as the "Corn Belt," other areas benefitted from the availability of these research results to a varying degree, depending on the adaptability of Corn Belt inbred lines to those areas. A measure of the degree to which other areas are different from the Corn Belt with respect to the adaptability of Corn Belt lines can be approximated by taking the published pedigrees of the recommended hybrids in different areas in 1956 and computing the percentage of all inbred lines represented by "Corn Belt" lines. An index of "Corn Beltliness," X_4 was defined as the number of Corn Belt inbred lines in the pedigrees of the recommended hybrids for that area, divided by the total number of lines.[20]

To take other aspects of the "complementarity" problem into account, another variable, X_{10}, was defined as the earliest date of entry (*origin*) in the immediate (contiguous) neighborhood of the area under consideration.[21] X_{10} was introduced on the assumption that it may be cheaper, both from the point of view of the additional research needed and from the point of view of setting up a marketing organization, to enter an area contiguous to an area already entered even though the "market potential" there may be lower than in some other area farther away.

Using either the number of released inbred lines or hybrids or the reported research expenditures, several unsuccessful attempts were made to measure the relative contribution of the various experiment stations. To some extent, however, the impact of this variable is already accounted for by our measures of the "market." The contribution of the various experiment stations is strongly related to the importance of corn in the area. In the "good" corn areas the stations did a lot of work on hybrids and in the marginal areas, less.[22]

The simple correlation coefficients between these variables, on the state level and on the crop reporting district level, are presented in tables 2.3 and 2.4 respectively. All of the correlation coefficients with Y have the expected sign and most of them are also significantly different from zero. The intercorrelation among the independent variables, however, prevents us from successfully estimating their separate contributions from these data. Almost all sets and subsets of independent variables in these tables were tried without yielding more than one significant coefficient in each multiple regression.[23] These results are disappointing, particularly because the highest correlations are with the rather artificial variables X_4 ("Corn Beltliness") and X_{10} (the "spatial trend").[24] Hence another approach to the problem was sought.

The trouble with the above approach is that it does nothing about the problem of the "right" geographic unit of analysis. Considering only the "market density" variable, it is obvious that it does not always measure what we want. Markets are continuous. While some areas are poor by themselves they may be a part of a larger market. Also an area may be entered because it is a springboard to other areas rather than on its own grounds. One

Table 2.4 Correlation coefficients on the crop reporting district level – $n = 132$

	X_1	X_3	b	X_4	X_{10}
Y	-0.56	-0.35	-0.70	-0.73	0.95
X_1		0.69	0.73	0.57	-0.57
X_3			0.54	0.40	0.36
b				0.67	-0.73
X_4					-0.76

See *Notes* to table 2.3.

way of taking these considerations into account is to define the "market potential" of an area as a weighted average of the "market densities" in *all* areas, densities in other areas weighted inversely to the distance from the area under consideration.[25] Given more than a few areas, however, the calculation of such a variable becomes impracticable.[26]

The trouble with our geographic units arises because states are too large while crop reporting districts are too small, and neither corresponds either to technical regions of adaptation of particular hybrids or to the decision units of the private seed companies. It is possible, however, to ask a more modest question: What were the characteristics of the areas entered in a particular year as compared with the characteristics of areas entered in another year? It is possible to aggregate areas according to the year of entry and test the "market potential" hypothesis on these aggregates. I shall define areas according to the year of entry, i.e. all districts with the *origin* in 1939 will make up one such area, and aggregate the data by crop reporting districts into such areas. Given our "10 percent" definition of *origin*, we have 16 such areas, 1935 to 1950. Alternatively, we would like to define areas according to the adaptability of particular hybrids. However, most hybrids overlap geographically and there are almost no data on the geographical distribution of particular hybrids, but there are breakdowns of the country into "maturity regions." A major seed company breaks down the US and its line of hybrids into 11 "maturity groups," locating the areas of adaptation of these groups on a map. It is possible to aggregate the crop reporting districts into these "technical" regions and ask whether high market areas were entered earlier than others.

The results of these calculations are presented in table 2.5a. In the aggregation by year of origin, to simplify the calculations, the actual "10 percent or more" year rather than the calculated date from the logistic was used. For the technical regions the computed origins by districts were used, weighted by the average corn acreage in the district and adjusted for

Table 2.5a Correlation coefficients between the aggregates of Y, X_1, X_4, and X_{10}

Aggregation by	X_1	X_4	X_{10}
"Date of origin": all areas ($n = 16$)			
Y	−0.82	−0.98	0.95
X_1		0.82	−0.64
X_4			−0.93
"Date of origin": all areas except the Southeast ($n = 13$)			
Y	−0.94	−0.97	0.96
X_1		0.90	−0.86
X_4			−0.97
"Technical regions": ($n = 12$)			
Y	−0.69	−0.82	0.95
X_1		0.90	−0.59
X_4			−0.78

differences in ceilings. That is, aggregate $Y = \Sigma YAK/\Sigma AK$, where A stands for average corn acres. Aggregate X_1 was defined as $\Sigma AK/\Sigma L$, where L stands for total land in farms. Because of the simplicity of the computations involved in this particular approach, 90 more crop reporting districts were added at this point to the analysis, raising the number of included districts to 222. Where separate logistic curves were not computed, Y was estimated by linear interpolation. As the technical regions overlap, each of the aggregates includes a few districts also included in the neighboring aggregates.[27]

To make the results comparable with those presented in table 2.4, similar calculations were also performed on X_4 and X_{10}. For the aggregation procedure by "date of origin," X_{10} was defined as the earliest date of origin in the immediate neighborhood of the area defined by the procedure, and X_4 as a simple unweighted average for the districts included in the aggregate. For the aggregate by "technical regions," X_{10} was defined as the lowest weighted average date of origin among the neighboring "technical regions." No aggregation had to be performed on X_4, as it had been originally defined and computed for these regions.

The results presented in table 2.5a indicate a strong association between the date of *origin* and average market density in the area, and suggesting that the market density variable is much more important than is indicated by the results in table 2.3 and 2.4. The association is higher if we exclude the Southeast from the aggregation procedure. This is explained by the relative lateness of the research contributions of the southeastern experiment stations and by the various obstacles put in the way of private seed companies there. Also, after we come down to a certain low level, it does not really pay to discriminate between areas on the basis of X_1 because the differences are too small, and other factors predominate. This is brought out when we ask the same question about the association of Y and X_1 within each technical region separately. When regressions of Y on log X_1 were computed for each of the technical regions separately, nine had the expected sign and were significantly different from zero, while the other three were not significantly different from zero. This result is significant on a sign test alone. But more interestingly, the r^2's were 0.66 rank correlated with the mean value of X_1, indicating that the explanatory power of this variable declined for areas with low average X_1 values.

The aggregation procedure, besides indicating that X_1 is a better variable than is implied by tables 2.3 and 2.4, also helps us with the collinearity problem. Before aggregation, the partial correlation coefficient of Y with X_1, holding X_{10} constant, was -0.24 on the state level and only -0.08 on the crop reporting district level. Now it becomes -0.90 for the aggregates by date of origin, -0.84 when we leave out the Southeast, and -0.64 for the data by technical regions. The regressions of Y on X_1 and X_{10} are presented in table 2.5b. The coefficient of X_1 has the expected sign and is significantly different from zero for the aggregates by "date of origin" and is almost twice the size of its standard error for the aggregates by technical

Table 2.5b　Regressions of Y on X_1 and X_{10}

	Coefficients of		
Aggregation by	X_1	X_{10}	R^2
"Date of origin": all areas	−17.8	1.02	0.982
	(2.5)	(0.07)	
"Date of origin": all areas except the			
Southeast	−16.5	1.03	0.077
	(3.4)	(0.07)	
"Technical region":	−10.5	0.88	0.925
	(5.6)	(0.12)	

Note: Figures in parenthesis are the calculated standard errors.

region. This indicates that it is possible to separate the contributions of X_1 and X_{10} if we define our area units correctly.

While these results may not be too conclusive, together with information gathered in conversations with executives in the industry and a graphical survey of the data, they leave little doubt in my mind that the development of hybrid corn was largely guided by expected pay-off, "better" areas being entered first, even though it may be difficult to measure very well the variables entering into these calculations.

The Rate of Acceptance

Differences in the *slope* or adjustment coefficient b will be interpreted as differences in the rate of adjustment of demand to the new equilibrium, and will be explained by variables operating on the supply side.[28] Actually, the path traced out is an intersection of short-run supply and demand curves. It is assumed, however, that while shifts on the supply side determine the origin of the development, the rate of development is largely a demand, or "acceptance," variable.[29] The usefulness of this assumption is due to a very elastic long-run supply of seed and is supported by the fact that only local and transitory seed shortages were observed. On the whole, the supply of seed was not the limiting factor.[30]

Differences in the rate of acceptance of hybrid corn, the differences in b, are due at least in part to differences in the profitability of the changeover from open pollinated to hybrid seed. This hypothesis is based on the general idea that the larger the stimulus the faster is the rate of adjustment to it.[31] Also, in a world of imperfect knowledge, it takes time to realize that things have in fact changed. The larger the shift the faster will entrepreneurs become aware of it, "find it out," and hence they will react more quickly to larger shifts.[32]

My hypothesis is that the rate of acceptance is a function of the profitability of the shift, both per acre and total. Per acre profitability may be defined as the increase in yield due to the use of hybrid seed, times the price of corn, and minus the difference in the cost of seed. As there is very little relevant cross-sectional variation in the price of corn, the seeding rate, or the price of seed, these will be disregarded and only differences in the superiority of hybrids over open-pollinated varieties taken into account.[33]

I shall use two measures of the superiority of hybrids over open pollinated varieties: (1) the average increase in yield in bushels per acre, based on unpublished mail questionnaire data collected by the Agricultural Marketing Service, X_7, and (2) the long-run average pre-hybrid yield of corn, X_8. The latter measure was used on the basis of the widespread belief that the superiority of hybrids can be adequately summarized as a percentage increase.[34] A variation in pre-hybrid yields, given a percentage increase, will also imply a variation in the absolute superiority of hybrids over open pollinated varieties. Twenty percent is the figure quoted most often for this superiority.[35] Average corn acres per farm, X_3, were used to add the impact of total profits per farm.

As the value of b depends strongly on the ceiling K, to make them comparable between areas, the b's had to be adjusted for differences in K. Instead of b, $b' = bK$ was used as the dependent variable, translating the b's back into actual percentage units from percentage of ceiling units. Alternatively, one should have adjusted the independent variables to correspond only to that fraction of the acres which will eventually shift to hybrids. But there are no data for making such an adjustment; hence b was adjusted to imply the same actual percentage changes in different areas.[36]

Linear and log regressions were calculated for the data from 31 states and 132 crop reporting districts. The results are presented in table 2.6.[37] The figures speak largely for themselves, indicating the surprisingly good and uniform results obtained. The long form and X_8 rather than X_7 did somewhat better but not significantly so. The similarity of the coefficients in comparable regression is striking. For example, compare the coefficients of X_8 and X_7 in the log regressions and all the coefficients in the similar regressions on the state and crop reporting district levels. These results were also similar to those obtained in preliminary analyses using b rather than b' as the dependent variable[38] (see table 2.7).

An attempt was made to incorporate several additional variables into the analysis. Rural sociologists have suggested that socioeconomic status or level-of-living is an important determinant of the rate of acceptance of a new technique.[39] The United States Department of Agriculture level-of-living index for 1939, when added to the regressions by states, had a negative coefficient in the linear form and a positive coefficient in the logarithmic form. In neither case was the coefficient significantly different from zero.

A measure of the "importance" of corn – the value of corn as a percentage of the value of all crops – was added in the belief that the rate of acceptance may be affected by the relative importance of corn within the farmer's enterprise. However, its coefficient was not significantly different from zero. Nor was the coefficient of total capital per farm significantly different from

Table 2.6 Regressions of *slopes* on "profitability" variables

	Coefficients of			
Regression	X_3	X_7	X_8	R^2
By states ($N=31$)				
$b' = c_0 + c_3 X_3 + x_8 X_8$	0.006		0.017	0.66
	(0.002)		(0.005)	
$\log b' = c_0 + c_3 \log X_3$	0.30		0.66	0.67
$\quad + c_8 \log X_8$	(0.08)		(0.11)	
By crop reporting districts ($N=132$)				
$b' = c_0 + c_3 X_3 + c_7 X_7$	0.0073	0.079		0.57
	(0.008)	(0.009)		
$b' = c_0 + c_3 X_3 + c_8 X_8$	0.0076		0.016	0.61
	(0.0007)		(0.002)	
$\log b' = c_0 + c_3 \log X_3$	0.44	0.70		0.61
$\quad + c_7 \log X_7$	(0.04)	(0.09)		
$\log b' = c_0 + c_3 \log X_3$	(0.44)		0.57	0.69
$\quad + c_8 \log X_8$	(0.03)		(0.05)	

Notes: Figures in parentheses are the calculated standard errors.

X_3 = Average corn acres per farm reporting corn.

X_7 = The average difference between hybrid and open pollinated yields by districts tabulated only from reports showing both and averaged over 4 to 10 years, depending on the overlap of the available data with the adjustment period (10 to 90 percent). Based on unpublished AMS "Identicals" data.

X_8 = Pre-hybrid average yield. Usually an average for the 10 years before an area reached 10 percent in hybrids. Sometimes fewer years were used, depending on the available data. Source: for states, *Agricultural Statistics*; for crop reporting districts, various published and unpublished data from the AMS and from state agricultural statisticians.

Table 2.7 Regressions of unadjusted *slopes* on "profitability" variables

	Coefficients of			
Regression	X_3	X_7	X_8	R^2
$b = c_0 + c_3 X_3 + c_7 X_7$ ($N=65$)	0.005	0.06		0.40
	(0.001)	(0.01)		
$b = c_0 + c_3 X_3 + c_8 X_8$ ($N=32$)	0.005		0.022	0.75
	(0.001)		(0.002)	

zero. The latter variable was introduced in an attempt to measure the impact of "capital rationing."[40]

The rate of acceptance may be also affected by the "advertising" activities of the extension agencies and private seed companies. There are no data, however, which would enable us to take this into account. There is also some evidence that the estimated rate of acceptance will be affected by the degree of aggregation and the heterogeneity of the aggregate. Heterogeneous areas imply different component growth curves and hence a lower aggregate slope coefficient. This is exhibited by the lower state values for b as compared to the values for the individual crop reporting districts within these states. No way has been found, however, to introduce this factor into the analysis.

Nevertheless, our results do suggest that a substantial proportion of the variation in the rate of acceptance of hybrid corn is explainable by differences in the profitability of the shift to hybrids in difference parts of the country.

The Equilibrium Level of Use

I am interpreting the *ceilings* as the long-run equilibrium percentages of the corn acreage which will be planted to hybrid seed. Differences in the percentage at which the use of hybrid seed will stabilize are the result of long-run demand factors. It is assumed that in the long run the supply conditions of seed are the same to all areas, the same percentage increase in yield over open pollinated varieties at the same relative price. However, this same technical superiority may mean different things in different parts of the country.

The ceiling is a function of some of the same variables which determine b, the rate of acceptance. It is a function of average profitability and of the distribution of this profitability. With the average above a certain value no farmer will be faced with zero or negative profitability of the shift to hybrids. With the average profitability below this level some farmers will be facing negative returns and hence will not switch to hybrids. In marginal corn areas, however, "average profitability" may become a very poor measure. Its components lose their connection with the concepts they purport to represent. Yield variability may overshadow the average increase from hybrids. The relevance of the published price of corn diminishes. In many marginal corn areas there is almost no market for corn off the farm. The only outlet for increased production is as an input in another production or consumption process on the farm. But on farms on which corn is a marginal enterprise, with little or no commercial livestock production, the use of corn is limited to human consumption, feed or draft animals, a cow and a few chickens. The farmers is interested in producing a certain amount of corn to fill his needs, having no use for additional corn. It will pay him to switch to hybrid corn only if he has alternative uses for the released land and other resources which would return him more than the extra cost of seed. But in many of these areas corn is already on the poorest land and uses resources left over from other operations on the farm. Also, there may already be substantial amounts of idle land in the area. All these factors may tend to make hybrids unprofitable although they are "technically" superior. Similarly, in areas where capital rationing is important the recorded market rate of interest will be a poor measure of the opportunity costs of capital. While the returns to hybrid corn may be substantial, if corn is not a major corp, the returns to additional investments in other branches of the enterprise may be even higher.

Ceilings are not necessarily constant over time. Even without any apparent change in the profitability of the shift from open pollinated to hybrid corn, a change in the relative profitability of corn growing, an improvement in the functioning of the market for corn, or an increase in storage facilities may change them. Also, in areas where there are large year-to-year changes in the corn acreage, the percentage planted to hybrid seed may fluctuate as a result of the differential exit and entry in and out of corn of farmers using hybrid or open pollinated seed. These changes may occur without any "real" changes in the relative profitability of hybrids or in farmers' attitudes towards them. It is very difficult to deal statistically with a development composed of a series of adjustments to shifting equilibrium values.[41] As a first approximation I shall ignore this problem. Only in the marginal corn areas is this a problem of some importance. For most of the Corn Belt the assumption of an immediate ceiling of 100 percent is tenable. In the fringe areas ceiling values somewhat lower than 100 percent fit very well. There are some indications that in the South ceilings may have shifted over time, but I doubt that this is important enough to bias seriously our results.

In spite of all these reservations and the crudeness with which the ceilings were estimated in the first place, it is possible to explain a respectable proportion of their variation with the same "profitability" variables that were used in the analysis of *slopes*. Because there is a ceiling of 1.00 to the possible variation in K, the logistic function was used again, giving us logit $K = \log_e [K/(1-K)]$ as our dependent variable. As there were a substantial number of areas with $K = 1.0$, a value not defined for the transform, two approximations were used. On the state level all values of $K = 1.0$ were set equal to 0.99, while on the crop reporting level, where there was no problem of degrees of freedom, these values were left out of the analysis. X_3, average corn acres per farm, and X_8, pre-hybrid yield, were used as "profitability" measures, and X_{11}, capital per farm, was added to take "capital rationing" into account. The results of these calculations are presented in table 2.8. They indicate that differences in measures of average profitability, differences in average corn acres and pre-hybrid yields, can explain a substantial proportion of the variation in *ceilings*, the long-run equilibrium level of hybrid seed use. The proportion of the variation explained on the state level is substantially higher, indicating that additional variables which may be at work at the crop reporting district level may cancel out at the state level. For example, the coefficient of capital investment per farm, a measure of "capital rationing," is significant at the crop reporting district level but not at the state level. Undoubtedly this analysis could be improved by the addition of other variables but I would not expect it to change the major conclusion appreciably.

Limitations, Summary and Conclusions

The above analysis does not purport to present a complete model of the process of technological change. Rather the approach has been to break down

Table 2.8 Regressions of logit K on "profitability" variables

Regression	Coefficients of			
	X_3	X_8	X_{11}	R^2
By states $(N=31)$				
$c_0 + c_3 X_3 + c_8 X_8$	0.03	0.11		0.71
	(0.01)	(0.02)		
$c_0 + c_3 \log X_3 + c_8 \log X_8$	1.94	5.88		0.71
	(0.56)	(0.80)		
$+ c_{11} \log X_{11}$	1.55	5.25	0.71	0.72
	(0.84)	(1.30)	(1.14)	
By crop reporting districts $(N=86)$				
$c_0 + c_3 \log X_3 + c_8 \log X_8$	1.09	2.22	1.35	0.39
$+ c_{11} \log X_{11}$	(0.48)	(0.61)	(0.64)	

Notes: Figures in parentheses are the calculated standard errors.
X_3 = Average corn acres per farm.
X_8 = Pre-hybrid yield.
X_{11} = On the state level, value of land and buildings per farm, 1940. Source: *Statistical Abstract of the United States, 1948*, p. 600. On the crop reporting district level, total capital investment per farm, 1949. Computed from Table 11, E. G. Strand and E. O. Heady, "Productivity of resources used on commercial farms," USDA, Technical Bulletin no. 1128, Washington, November 1955, p. 45.

the problem into manageable units and to analyze them more or less separately. I have concentrated on the longer-run aspects of technological change, interpreting differences in the pattern of development of hybrid corn on the basis of the long-run characteristics of various areas, and ignoring the impact of short-run fluctuations in prices and incomes. This limitation is not very important in the cases of hybrid corn because the returns from the changeover were large enough to swamp any short-run fluctuations in prices and other variables.[42] It might, however, become serious were we to consider other technical changes requiring substantial investments, and not as superior to their predecessors as was hybrid corn. Nor can we transfer the particular numerical results to the consideration of other developments. Nevertheless, a cursory survey of trends in the number of cornpickers and tractors on farms, and of trends in the use of fertilizer, does indicate that it might also be possible to apply a version of our approach to their analysis.

I hope that this work does indicate that at least the process of innovation, the process of adapting and distributing a particular invention to different markets and its acceptance by entrepreneurs, is amenable to economic analysis. It is possible to account for a large share of the spatial and chronological differences in the use of hybrid corn with the help of "economic" variables. The lag in the development of adaptable hybrids for particular areas and the lag in the entry of seed producers into these areas can be explained on the basis of varying profitability of entry. Also, differences in both the long-run equilibrium use of hybrids and in the rate of approach to that equilibrium level are explainable, at least in part, by differences in the profitability of the shift from open pollinated to hybrid varieties.

Looking at the hybrid seed industry as a part of the specialized sector which provides us with technological change, it can be said that both private and public funds were allocated efficiently within that sector.[43] Given a limited set of resources, the hybrid seed industry expanded according to a pattern which made sense, allocating its resources first to the areas of highest returns.

The American farmer appears also to have adjusted rationally to these new developments. Where the profits from the innovation were large and clear cut, the changeover was very rapid. It took Iowa farmers only four years to go from 10 to 90 percent of their corn acreage in hybrid corn. In areas where the profitability was lower, the adjustment was also slower. On the whole, taking account of uncertainty and the fact that the spread of knowledge is not instantaneous, farmers have behaved in a fashion consistent with the idea of profit maximization. Where the evidence appears to indicate the contrary, I would predict that a closer examination of the relevant economic variables will show that the change was not as profitable as it appeared to be.[44]

Acknowledgements

This research was begun during my tenure as a Research Training Fellow of the Social Science Research Council. It has been supported by the Office of Agricultural Economics Research at the University of Chicago and is being supported by a generous grant from the National Science Foundation. I am indebted to Professor T. W. Schultz for arousing my interest in this problem and for encouraging me in my work, to Professors H. G. Lewis and A. C. Harberger for their valuable advice and guidance, and to the members of the Public Finance Workshop and other members of the Department of Economics at the University of Chicago, both faculty and students, for their suggestions and criticisms. I owe to the generosity of the Field Crop Statistics Branch of the Agricultural Marketing Service a large part of the unpublished data used in this paper. I also want to acknowledge and thank the people directly connected with hybrid corn, both in the Agricultural Experiment Stations and in the private seed companies, for their complete cooperation. This article is based on my unpublished PhD dissertation, "Hybrid Corn: An Exploration in Economics of Technological Change," on file at the University of Chicago Library.

Notes

1 A popular history of hybrid corn can be found in A. R. Crabb, *The Hybrid Corn Makers: Prophets of Plenty*, Rutgers University Press, 1948. See also, F. D. Richey, "The lay of the corn huckster," *Journal of Heredity*, 39(1), 1946, 10–17; P. C. Mangelsdorf, "The history of hybrid corn," *loc. cit.*, 39, 1948, 177–80; G. F. Sprague, "The experimental basis for hybrid maize,"

Biological Reviews, 21, 1946, 101–20; M. T. Jenkins, "Corn improvement," *US Department of Agriculture Yearbook*, 1936, 455–522; and H. A. Wallace and W. L. Brown, *Corn and Its Early Fathers*, Michigan State University Press, 1956.

2 "Hybrid corn is the product of a controlled, systematic crossing of specially selected parental strains called 'inbred lines.' These inbred lines are developed by inbreeding, or self-pollinating, for a period of four or more years. Accompanying inbreeding is a rigid selection for the elimination of those inbreds carrying poor heredity, and which, for one reason or another, fail to meet the established standards." " [The inbred lines] are of little value in themselves for they are inferior to open-pollinated varieties in vigor and yield. When two unrelated inbred lines are crossed, however, the vigor is restored. *Some* of these hybrids prove to be markedly superior to the original varieties. The development of hybrid corn, therefore, is a complicated process of continued self-pollination accompanied by selection of the most vigorous and otherwise desirable plants. These superior lines are then used in making hybrids." First quote is from N. P. Neal and A. M. Strommen, "Wisconsin corn hybrids," Wisconsin Agricultural Experiment Station, Bulletin 476, February 1948, p. 4; and the second quote is from R. W. Jugenheimer, "Hybrid corn by Kansas," Kansas Agricultural Experiment Station, Circular 196, February 1939, pp. 3–4.

3 This conclusion was also supported by the results of an attempt to fit a model in which the year-to-year changes in the percentages planted to hybrid seed were to be explained by year-to-year changes in the price of corn, price of hybrid seed, the superiority of hybrids in the previous year or two, etc. The trend in the data was so strong that, within the framework of this particular model, it left nothing of significance for the "economic" variables to explain.

4 For a comparison, see C. P. Winsor, "A comparison of certain symmetrical growth curves," *Journal of the Washington Academy of Sciences*, 22, 1932, 73–84, and J. Aitchison and J. A. C. Brown, *The Lognormal Distribution*, Cambridge University Press, 1957, pp. 72–5.

5 It may be worthwhile to indicate why it is reasonable that the development should have followed an S-shaped growth curve. The dependent variable can vary only between 0 and 100 percent. If we consider the development to be an adjustment process the simplest reasonable time-path between 0 and 100 percent is an ogive. While the supply of seed can increase exponentially, the market for seed is limited by the total amount of corn planted, and that will act as a damping factor. Also, if we interpret the behavior of farmers in the face of this new, uncertain development as if they were engaged in sequential decision making, the ASN curve will be bell-shaped, and the cumulative will again be S-shaped. See also H. Hotelling, "Edgeworth's taxation paradox and the nature of demand and supply curves," *Journal of Political Economy*, 40, October 1932. The argument for the logistic is given by R. Pearl, *Studies in Human Biology*, Baltimore, 1924, 11, 558–83, and S. Kuznets, *Secular Movements in Production and Prices*, Houghton Mifflin, Boston, 1930, pp. 59–69.

6 For a more detailed description of the logistic and its properties, see Pearl, op. cit.

7 Perhaps the simplest interpretation of the logistic is given by A. Lotka, *Elements of Physical Biology*, Williams and Wilkins, Baltimore, 1925, p. 65. We are interested in the general adjustment function, $dP/dt = F(P)$. Using a Taylor series approximation and disregarding all the higher terms beyond the quadratic we get a function whose integral is the logistic. The logistic is the integral of the quadratic approximation to the adjustment function.

8 See Pearl, op cit.; H. T. Davis, *The Theory of Econometrics*, Principia Press, 1941, chapter II; and G. Tintner, *Econometrics*, John Wiley & Sons, 1952, pp. 208–11, and the literature cited there.

9 This is a simplification of a method proposed by Joseph Berkson. Berkson's method is equivalent to weighted least squares regression of the same transform with $P(K - P)$ as weights. J. Berkson, "A statistically precise and relatively simple method of estimating the bioassay with quantal response, based on the logistic function," *Journal of the American Statistical Association*, 48 (1953), 565–99, and "Maximum likelihood and minimum chi-square estimates of the logistic function," loc. cit., 50 (1955), 130–62. Berkson proposed this procedure in the context of bio-assay. It is not clear, however, whether the bio-assay model is applicable in our context, nor is it obvious, even in bio-assay, what system of weights is optional. See also J. Berkson, "Estimation by least squares and by maximum likelihood," *Proceedings of the Third Berkeley Symposium on Mathematical Statistics*, vol. I, University of California Press, pp. 1–11. Hence no weights were used. In view of the excellent fits obtained, it is doubtful whether alternative weighting systems would have made much difference.

10 Each state is usually divided into nine crop reporting districts numbered in the following fashion:

$$
\begin{array}{ccc}
 & N & \\
1 & 2 & 3 \\
W \quad 4 & 5 & 6 \quad E \\
7 & 8 & 9 \\
 & S & \\
\end{array}
$$

11 It should be noted that the sum of logistics is not usually a logistic. However, the logistic is also valid for an aggregate, as long as the components are similar in their development. See L. J. Reed and R. Pearl, "On the summation of logistic curves," *Journal of the Royal Statistical Society*, 90 (new series), 1927, 729–46. How good the approximation is in fact is indicated by the results below.

12 These r^2's should be taken with a grain of salt. They are the r^2's of the transform rather than of the original function and give less weight to the deviations in the center. Also, they do not take into account the excluded extreme values. Nevertheless, an examination of the original data indicates that they are not a figment of the fitting procedure.

13 *Origin* is measured from 1940. Hence, the *origin* in Iowa is placed approximately in 1936, and in Georgia in 1948.

14 R. R. Bush and F. Mosteller, *Stochastic Models for Learning*, John Wiley and Sons, New York, 1955, p. 335.

15 The date at which the fitted logistic passes through 10 percent is given by $Y = (-2.2 - a)/b$. As the variation of b is small relative to that of a, small changes in the definition of Y will be in the nature of an additive constant and will rarely change the ranking of the data of *origin* in different areas.

16 This is essentially a definition of "commercial" availability. An attempt was made to measure the date of "technical" availability by going through yield tests and other official publications and noting the first year in which hybrids clearly outyielded the open pollinated varieties. The rank correlation between this technical definition and the "10 percent" definition was 0.93. The average lag between the technical and the commercial availability was approximately 2 years. Also, preliminary explorations with 1 and 5 percent definitions, and with the rank of an area rather than the absolute date, indicated that the results are not very sensitive to changes in definition.

17 Implicitly, we have assumed here that the lag between the entry decision and actual availability is approximately constant or at least independent of other variables under analysis.

18 Throughout the paper it is assumed that the price of hybrid seed is given and approximately uniform in different areas. This is a very close approximation to reality and a result of a very elastic long-run supply curve of seed.

19 Differences in seeding rates have been disregarded here. There is, however, some evidence that the results would have been somewhat better if X_1 were adjusted for these differences.

20 On the state level, a published list of recommended hybrids and their pedigrees was used, with Iowa, Illinois, Indiana, Ohio, and Wisconsin lines defined as "Corn Belt" lines. See C. B. Henderson, "Inbred lines of corn released to private growers from State and federal agencies and double crosses recommended by states," 2nd revision, Illinois Seed Producers Association, Champaign, April 15, 1956. On the crop reporting district level I used unpublished data from the Funk Bros Seed Co. listing their hybrids by "maturity groups" and giving coded pedigrees.

21 This is analogous to the introduction of a lagged value of the dependent variable into the regression in time series analysis. Except that the "lag" here is spatial rather than a time lag.

22 There are a few exceptions to this statement. In the North, Connecticut, Wisconsin, and Minnesota contributed more than their "share," and so did Texas and Louisiana in the South.

23 Similar results were obtained when the logarithms rather than the actual values of the independent variables were used.

24 The good performance of X_{10} is not surprising. The smaller the geographic unit of analysis, the better will be the relationship between Y and X_{10}. This can be seen by comparing the correlation coefficients on the state and crop reporting district levels. There is, however, another way of rationalizing the performance of X_{10}. See note 27.

25 See W. Warnz, "Measuring spatial association with special consideration of the case of market orientation of production," *Journal of the American Statistical Association*, 51, December 1956, 597–604.

26 It does suggest, though, a reason for the good performance of X_{10}. Consider a simple model in which the date of origin is a function of the "true" market

measure, the "true" measure being a weighted average of the densities in all areas, weights declining with distance. This "true" measure can be approximated by the actual density in the area and the "true" measure in the immediate neighborhood. But the date of origin in the immediate neighborhood is a function of the "true" density there and can serve as its measure. This implies that X_{10} is another measure of the "market!" For a similar approach in a different context, see M. Nerlove, "Estimates of the elasticities of supply of selected agricultural commodities," *Journal of Farm Economics*, 38, May 1956, 500–3.

27 Because one of the "maturity" areas is much larger than the others, it was divided into two on a north-south basis. Hence we have 12 technical regions in our analysis.

28 The dimension of b, the adjustment coefficient, may be of some interest, b indicates by how much the value of the logistic transform will change per time unit. A value of $b = 1.0$ implies that the development will go from, e.g. 12 to 27 to 50 to 73 to 88 percent from year to year; i.e. the distance from 12 to 88 percent will be covered in 4 years. A value of $b = 0.5$ would imply a path: 12, 18, 27, 38, 50, 62, 73, 82, 88, etc., i.e. it would take twice the time 8 years, to transverse the same distance. If one thinks in terms of the cumulative normal distribution positioned on a time scale, which is very similar to the logistic, then b is approximately proportional to $1/\sigma$. A low standard deviation implies that it will take a short time to go from, e.g. 10 to 90 percent, while a higher standard deviation implies a longer period of adjustment.

29 Implicitly, we have the following model: the potential adjustment path of supply is an exponential function, which after a few years rises quickly above the potential adjustment function of demand. The demand adjustment function has the form of the logistic. The actual path followed is the lower of the two, which, after the first few years, is the demand path.

30 "Clearly it would have been physically impossible for a large percentage of operators to have planted hybrids in the early 1930s. There simply was not enough seed. It seems likely, however, that this operated more as a potential than an actual limitation upon the will of the operator, and that rapidity of adoption approximated the rate at which farmers decided favorably upon the new technique." B. Ryan, "A study in technological diffusion," *Rural Sociology*, 13, 1948, p. 273. Similar views were expressed to the author by various people closely associated with the developments in hybrid corn.

31 E.g. "The greater the efficiency of the new technology in producing returns, . . . the greater its rate of acceptance." – "How farm people accept new ideas," Special Report no. 15, Agricultural Extension Service, Iowa State College, Ames, November 1955, p. 6.

32 This is analogous to the situation in sequential analysis. The ASN (average sample number) is an inverse function of, among other things, the difference between the population means. That is, the larger the difference between the two things which we are testing, the sooner we will accumulate enough evidence to convince us that there is a difference. See A. Wald, *Sequential Analysis*, John Wiley & Sons, New York, 1947.

33 The apparent cross-sectional variation in the average price of hybrid seed is largely due to differences in the mix of "public" versus "private" hybrids

bought by farmers. The "public" hybrids sell for about $2.00 less per bushel. The rank correlation between the price of hybrid seed and the estimated share of "private" hybrids in 1956 was 0.73.

34 The data from experiment station yield tests indicate that this is not too bad an assumption. See Sprague, op. cit., and the literature cited there. It is unfortunate that these data are not comparable between states and, hence, cannot be used directly in this study.

35 "If an average percentage increase in yield to be expected by planting hybrids as compared to open pollinated varieties were to be computed at the present it would probably be near 20 per cent. . . ." – J. T. Scwartz, "A study of hybrid corn yields as compared to open pollinated varieties," Insurance Section, FCIC, Washington, April and May 1942, unpublished manuscript.

"Experience in other corn-growing regions of the United States shows that increases of approximately 20 per cent over the open pollinated varieties may be expected from the use of adapted hybrids. Results so far in Texas are in general agreement with this figure," J. S. Rogers and J. W. Collier, "Corn production in Texas," Texas Agricultural Experiment Station, Bulletin 746, February 1952, p. 7.

"Plant breeders conservatively estimate increase in yields of 15 to 20 per cent from using hybrid seed under field conditions. They expect about the same relative increases in both low- and high-yielding areas," USDA, *Technology of the Farm*, Washington, 1940, p. 22.

36 This adjustment affects our results very little. See table 2.7.

37 X_7 was not used on the state level because it was felt that the aggregation error would be too large. We want an average of differences while I could only get a difference between averages. For some states this difference exceeded the individual differences in all the crop reporting districts within the state.

38 These were calculated for subsamples of 65 and 32 crop reporting districts.

39 See "How farm people accept new ideas," op. cit., and E. A. Wilkening, "The acceptance of certain agricultural programs and practices in a Piedmont community of North Carolina," unpublished PhD thesis, University of Chicago, 1949, and "Acceptance of improved farm practices in three coastal plain counties," Tech. Bull. no. 98, North Carolina Agricultural Experiment Station, May 1952.

40 The failure of the last two variables is due largely to their strong intercorrelation with the included variables. "Importance" is highly correlated with average yield and capital with corn acres per farm. When used separately, these two variables did as well on the state level as yield and corn acres per farm.

41 I am aware of only one attempt in the literature to deal with this kind of problem. See C. F. Roos and V. von Szelisky, "Factors governing changes in domestic automobile demand," particularly the section on "The concept of a variable maximum ownership level," pp. 36–8, in General Motors Corporation, *Dynamics of Automobile Demand*, New York, 1939.

42 Estimates made for Kansas data indicate returns from 300 to 1000 percent on the extra cost of seed.

43 Some minor quibbles could be raised about the allocation of public funds, but the returns to these funds have been so high that the impact of the existing inefficiencies is almost imperceptible.

44 In this context one may say a few words about the impact of "sociological" variables. It is my belief that in the long run, and cross-sectionally, these variables tend to cancel themselves out, leaving the economic variables as the major determinants of the pattern of technological change. This does not imply that the "sociological" variables are not important if one wants to know which *individual* will be first or last to adopt a particular technique, only that these factors do not vary widely cross-sectionally. Partly this is a question of semantics. With a little ingenuity, I am sure that I can redefine 90 percent of the "sociological" variables as economic variables. Also, some of the variables I used, e.g. yield of corn and corn acres per farm, will be very highly related cross-sectionally to education, socio-economic status, level-of-living, income, and other "sociological" variables. That is, it is very difficult to discriminate between the assertion that hybrids were accepted slowly because it was a "poor corn area" and the assertion that the slow acceptance was due to "poor people." Poor people and poor corn are very closely correlated in the US. Nevertheless, one may find a few areas where this is not so. Obviously, the slow acceptance of hybrids on the western fringes of the Corn Belt, in western Kansas, Nebraska, South Dakota, and North Dakota was not due to poor people, but the result of "economic" factors, poor corn area.

3
Congruence versus Profitability:
A False Dichotomy*

In a note in this journal Brandner and Strauss have argued that the recent history of hybrid sorghum in Kansas proves that "congruence to an existing cultural pattern" rather than "profitability" is the basic element in the diffusion process.[1] Reference was made to my hybrid corn study,[2] with the implication that the facts about hybrid sorghum in Kansas are somehow inconsistent with the hypotheses advanced in that study. I would like to point out, first, that the diffusion pattern of hybrid sorghum in Kansas is very much what my study would have predicted and, second, that "congruence" and "profitability" are not and should not be alternative and mutually exclusive explanations of the same behavior.

The hybrid corn study concluded that one of the major factors accounting for the difference in the rate of acceptance of hybrid corn in different areas was the difference in the *absolute* profitability of the shift over from open pollinated to hybrid varieties. Thus even though the *relative* superiority of hybrids may have been the same in different areas (on the order of 20 percent), it meant much more to farmers in higher corn yield areas. We know little about the properties of hybrid sorghum varieties as yet, but it is probably safe to assume that, in this case as in the case of hybrid corn, their superiority over open pollinated varieties can be described well as a percentage increase in yield of about the same magnitude at different yield levels.[3] If this is true, then a difference in the pre-hybrid level of sorghum yields would indicate differences in the absolute profitability of the shiftover to hybrids. Thus we would expect hybrid sorghum to spread faster in high-sorghum-yield areas than in the lower-yield areas, and that is what is happening. Hybrid sorghums are spreading much faster in east and northeast Kansas than in west and southwest Kansas, and this pattern is clearly consistent with the difference in long-term yield levels of sorghum in different parts of Kansas. In fact, the Spearman rank correlation coefficient for Kansas crop reporting districts between the percentage of all grain sorghum acres planted with hybrid varieties

*Reprinted from *Rural Sociology*, 25(3), September 1960.

in 1957 and the average yield of sorghum harvested for grain in 1957 was 0.97 ($N=9$).[4] A similar high correlation is also found if longer period sorghum yield averages are used instead.[5]

What then is the evidence that is in conflict with the "profitability" hypothesis? Brandner and Strauss do not present any explicit profitability calculations. In fact, their main argument is based on the very different concept of "economic need." They argue that sorghum is more important in southwestern Kansas than in the northeast, that sorghum is better than corn in the southwest but that the opposite is true in the northeast, and that hybrid sorghum "would seem to fill an economic need" in southwest Kansas that is in some sense (never clearly defined) more urgent than it is in northeast Kansas. It may be true that sorghum is better than anything else in the southeast, and relatively worse than corn in the northwest, but it is also clear and crucial that sorghum in the northeast is better than sorghum in the southwest.[6] It is the absolute level of sorghum yield that matters, not its relative yield to something else. The relative comparison may tell us how much of the available acreage will be devoted to sorghum as opposed to the other crops, but it does not tell us how fast new sorghum production techniques will be accepted in different sorghum areas.

The trouble with "need," even "economic need," is that as a concept it is not operational. Profitability, on the other hand, can be given an operational content. To make clear the distinction between the two, consider an example from a very different area. Most people would probably agree that in some sense the "economic need" for an education is higher among Negroes than among whites in this country. On the other hand, it seems to be true that the rate of return on a dollar invested in higher education by Negroes appears to be smaller on the average than the rate of return on the same investment by whites. An economist given this fact, while saddened by it, would nevertheless predict that Negroes are likely to invest relatively less in higher education than whites even at comparable income and asset levels.

While the available evidence is thus quite consistent with the "profitability" hypothesis, this does not imply that "congruence" is not important. What Brandner and Strauss are saying is that hybrid sorghum may have been accepted faster in northwest Kansas because farmers there have had a more extensive and more favorable experience with hybrid corn than the farmers in southeast Kansas. This may be a very important factor, but it is not in conflict with the profitability hypothesis. More generally one can consider the problem of choosing between hybrid and open-pollinated varieties as a problem of decision making under uncertainty. The amount of experimentation and evidence necessary to convince one that there is an important difference between hybrid and open-pollinated varieties will depend on the actual size of this difference, its variance, and the relative weight attached to Type I and Type II errors in accepting or rejecting the hypothesis.[7] The first two factors are "objective" variables; the last is "subjective" and may well reflect the previous experience gained in testing similar hypotheses. Given previous favorable experience with "hybrids" in a different crop, one may be willing to accept the hypothesis of hybrid sorghum superiority on the basis of somewhat flimsier evidence (lower significance levels) than if one had no previous experience with "hybrids." Thus "congruence" could be an important variable, within the more general rational decision making under uncertainty model, supplementing rather than supplanting the other "profitability" variables.

Notes

1 L. Brandner and M. A. Strauss, "Congruence versus Profitability in the diffusion of hybrid sorghum," *Rural Sociology* 24(1959), 381–3.
2 Zvi Griliches, "Hybrid corn: an exploration in the economics of technological change," *Econometrica*, 25(1957), 501–22.
3 Experimental data indicate that good, practical grain (sorghum) hybrids – those having proper height, maturity, and other characters – outyielded standard varieties by 25 to 30 per cent." *1956 Kansas Agriculture*, 39th report of the Kansas State Board of Agriculture (Topeka, 1956), p. 58.
4 The data are from Kansas State Board of Agriculture, *Farm Facts 1957–58* (Topeka, State Printing Plant, 1958).
5 The 1950–8 average yield of sorghum harvested for grain by crop reporting districts has a rank correlation coefficient of 0.83 with the percentage of grain sorghum acres planted to hybrid varieties in 1957, 0.68 with the percentage planted with hybrids in 1958, and 0.58 with the percentage planted with hybrids in 1959. The correlation gets weaker with time, since the higher the level of hybrid use (66 percent in 1959 in Kansas as a whole) the poorer is the level of hybrid use as a measure of the *rate* of acceptance of hybrids by farmers. The figures used in these calculations are from various issues of *Kansas Farm Facts* and an October 19, 1959 release from the Kansas Crop and Livestock Reporting Service.
6 Actually the evidence solely with respect to the two districts considered by Brandner and Strauss is even more favorable to the "profitability" hypothesis than is indicated by the rank correlations above. For the previous nine years (1950–8) the yield of grain sorghum in northeast Kansas has exceeded sorghum yields in southwest Kansas by an average of 47 percent.
7 The analogy to sequential sampling is intended. If one used a Bayes decision model instead, "congruence" considerations would enter in via their effect on the *a prior* probability attached to the hypothesis that hybrids are better.

4

Postscript on Diffusion

The economic characteristics of a new technology and the state of the economy define a maximum feasible penetration level ("ceiling") for it. It is the amount of the new technology that would be purchased (used) at that particular point of time if there were no uncertainty about its true characteristics and profitability. It is defined as holding current prices and the current distribution of the stocks of earlier technologies and sizes of firms constant. The approach (diffusion toward such an equilibrium level) may be approximated adequately by a logistic curve, representing the pattern of information spread and learning over time.

One can interpret what I did in the 1957 *Econometrica* paper as analyzing the rate of diffusion toward such "ceiling" as determined by the first main wave of hybrid corn varieties (the early hybrids defined the availability date). Since new hybrid varieties were ultimately developed and improved to fit various remaining nooks and crannies, and since the supplies of the old technology (open-pollinated seed) eventually dried up, ceilings shifted and the actual numbers did not follow a single logistic curve all the way, but rather a logistic type curve with a shifting ceiling. (This was the brunt of Dixon's 1980 comment on this paper.) The later slow upper tail is thus not really the property of the "acceptance" (diffusion) process, but rather a reflection of the long lags in the availability of well-adapted hybrids for specific small regions of various states. Because of this I would now use a model with an endogenous and shifting ceiling parameter, something that I believe can be implemented given the current state of econometric art.

The economic approach to the study of the diffusion of new techniques was developed further and applied to a number of industrial innovations by Mansfield in several important papers (see Mansfield, 1961 and the papers collected in Mansfield, 1968). David (1969) emphasized the importance of size of the firm and the shape of existing distribution of the previous technology. Pakes (1976) outlines a model with shifting "ceilings." More recent contributions in this vein include Davies (1979) and Stoneman (1976). A major comparative study of the diffusion of a number of industrial innovations in different countries is reported in Nabseth and Ray (1974). Trajtenberg (1985) is an example of a recent study of the diffusion of CAT-scanners.

The sociological literature is represented at its best by the Coleman, Katz, and Menzel (1966) study of the diffusion of a new drug. Reviews of the literature can be found in Rogers (1983), Mahajan and Peterson (1985), Feder, Just, and Zilberman (1985), and Thirtle and Ruttan (1986). For recent contributions to the theory of diffusion see Jensen (1982) and Bhattacharaya, Chatterjee, and Samuelson (1986).

5
Measuring Inputs in Agriculture: A Critical Survey*[1]

1 Introduction

1.1 In this preliminary report the measurement of inputs in agriculture is discussed solely from the point of view of aggregate productivity analysis. Statistical series that are criticized here may be perfectly adequate for their original purposes and no reflection is intended on their "producers." The comments are necessarily brief and selective. Some of the conclusions may appear extreme, because they are stated without the detailed explanation and qualification that will be possible in the full report of the study.

1.2 The discussion is limited mainly to conceptual problems, and ignores the important statistical problems of sources of data, randomness, size of sample, and so on. Most of the attention is paid to two problems: the changing quality of agricultural inputs and the measurement of the services and value of capital equipment on farms.

2 The Changing Quality of Agricultural Inputs

2.1 There is little doubt that the quality of most agricultural inputs has changed substantially during the last 20 years. Tractors have increased in size and versatility; the plant nutrient content of fertilizers has gone up by about 50 percent; and there has been an increase in the educational level of the farm labor force. It is clear that we want our input measures to take some of these changes into account, e.g. changes in the average capacity of farm trucks. Whether or not we want the input measures to cover all possible quality changes is a semantic rather than substantive issue. Hybrid seed corn can be viewed either as an improvement in the quality of seed or as "technical change." Since we are interested in explaining the growth of agricultural

* Reprinted from *Journal of Farm Economics* XL22(5), December 1960.

output, it does not matter much whether we put it into the "input change" category or the "productivity change" category as long as we put it somewhere and know where it is. But it matters very much that we should try to measure it the best we can since it is such an important source of output change.

2.2 Since many of our input series are based on deflated expenditure series, if the deflators fail to take quality change into account, so will also the resulting "constant price" estimates. All of the deflators used for these purposes have been based on USDA collected price statistics and are components or a recombination of components of the Prices Paid Index. While most price indexes do poorly as far as quality change is concerned, the Prices Paid by Farmers Index has been especially affected by the official USDA position on these matters. The USDA, instead of recognizing that it is a difficult problem with no perfect solution, but one worth fighting for, has taken the position that nothing *should* be done about it, that it is not *desirable* to hold quality constant when pricing items bought by farmers, and thus has made the quality change problem much more serious for the USDA Prices Paid Index than is the case for similar indexes in other sectors of the economy.[2] As the result of this position, most commodity definitions in the Prices Paid Index are quite vague, e.g. the quality or brand priced is the one "most commonly bought by farmers." Also, the insistence on pricing items with all "the customarily bought attachments" could result in a substantial bias in the USDA estimates of prices paid for more complicated pieces of machinery. Since the USDA is fortunately not completely consistent on this matter, a shift from Chevrolets to Cadillacs will not show up in the index, but if farmers were to shift to automatic transmissions, power steering, or built-in air conditioning, those shifts would show up as an increase in the price paid by farmers for automobiles.

That this is a serious problem is indicated by a comparison of the trend of USDA and BLS (Wholesale Price Index) prices for comparable items presented in table 5.1. For most items, the USDA prices have gone up more than comparable prices in the WPI. The difference is small or negative for relatively simple items such as moldboard plows, grain drills, hammer mills, and wagons, but is quite large for more complicated pieces of equipment such as tractors, cornpickers, and milking machines.[3] It is true that the USDA prices are "retail" whereas the WPI prices are "wholesale," but for this to explain the consistently higher USDA prices dealer markups should have risen substantially. Considering that 1947–9 were still relatively "tight" postwar years, and that by 1958 discounting had become the rule rather than the exception, dealer margins probably narrowed rather than widened during this period. If so, the figures in table 5.1 may actually underestimate the upward drift in USDA prices due to incomplete commodity specifications. Thus, about a quarter of the rise in specified USDA prices could be due to a shift to higher quality and "more attachment" machines rather than to any real price change.

Table 5.1 A comparison of BLS (WPI) and USDA prices for comparable farm machinery items in 1958

| Machine | Price index in 1958 (1947–9 = 100) | |
	WPI[a]	USDA[b]
Tractor, wheel, 30–9 belt h.p.	127	148
Plow, moldboard, two-bottom	159	142
Plow, moldboard, three-bottom	177	153
Disc harrows, tandem	140	161
Cultivator, two-row, mounted	147	162
Manure spreader, two-wheel rubber tires	145	165
Corn planter, two rows	153	186
Grain drill, 20-tube	148	147
Mower	145	169
Hay rake, side delivery	156	207
Combine, self-propelled, 12 ft	140	147
Combine, 5–6 ft PTO	154	161
Cornpicker, two-row	129	174
Hammer mill	158	156
Milking machines	119	135
Wagons	126	129
Power sprayer	144	132
Weighted average[c]	139	149

[a] *Wholesale Prices and Price Indexes*, 1958, Bu. Labor Stat. Bul. no. 1257, July 1959.
[b] USDA, *Agricultural Statistics*, 1958 and 1959, and various issues of USDA, *Agricultural Prices*.
[c] Weighted by USDA Index of Prices Paid 1955 weights. From B. R. Stauber *et al., "The January 1959 revision of the price indexes," Agr. Econ. Res.*, April–July 1959.

Index makers and economists have by and large accepted the assumption that if an item has a separate price (at least for awhile), and if people are buying it, it must be worth what they are paying for it, at least on the average. Thus the appearance of automatic transmissions on the market at $200 extra will not raise the price of automobiles in the consumer price index or wholesale price index even though eventually almost all cars are sold with it and their base price incorporates it as "standard equipment."

The USDA Prices Paid Index could be substantially improved if it were to accept the same standards and methods of commodity specification and chaining used in the construction of the CPI and WPI. These indexes at least try to adjust for those quality changes to which a price can be attached.

2.3 Unfortunately, only few of the observed quality changes come in discrete lumps with an attached price tag. Most of the changes are gradual, are not priced separately, and are likely to be missed even by relatively well "specified" indexes such as the CPI. Nevertheless, not all is hopeless. Many dimensions of quality change can be quantified (e.g. horsepower, weight, top speed, durability, fuel consumption per unit of output for tractors);

a variety of models with different specifications can be observed being sold at different prices; using multiple regression techniques on these data one can derive implicit prices per unit of the chosen additional dimension of the commodity; and armed with these "prices" one can proceed to adjust the observed price per "average item" for the changes that have occurred in its specification. There are many technical problems to be solved, e.g. the use of beginning versus end period prices, but the main idea is quite simple: Derive implicit specification prices from cross-sectional data on the price of various "models" of the particular item and use these in pricing the time series specification change in the chosen (average or representative) item.

An example of the possible magnitude of such an adjustment is provided by an analysis of US passenger car prices and specifications in the periods 1937–50 and 1950–9.[4] It appears that "quality change" accounted for between one third and two thirds of the actual change in list prices of Chevrolet, Ford, and Plymouth four-door passenger sedans during these periods. The relative importance of quality change appears to have been substantially greater in the more recent 1950–9 period than during the 1937–50 period. As compared to the USDA price index for automobiles, the index of list prices adjusted for quality change rose by about 30 percent less from 1937 to 1950 and by about 40 or more percent less from 1950 to 1959. Since discounting has become prominent during the last few years and is presumably caught by the USDA, the "true" difference between these indexes may be even greater. Thus, the use of price data collected with very little control over what is being priced, and the lack of almost any adjustment for quality change in the official price indexes, has probably biased seriously some of the agricultural inputs series.

2.4 While for "constant dollar" expenditure series the quality problem enters via the price indexes used to deflate them, it is more direct and explicit for inputs that are measured in quantity or "physical" units. Labor, for example, is measured in number of workers or man-hours worked. In most measures used it does not matter whether the labor unit is composed of men, women, or children, or illiterates or college graduates, of native whites, Negroes or imported foreign workers. It is true that in one sense, a shift from or towards the use of child labor would reflect itself in labor "productivity." But it is still very interesting to distinguish between productivity increases due to a change in the labor force "mix" and those due to an increase in the productivity of a laborer of a given, narrowly specified, type or quality.

It is impossible here to enter into all the problems raised by "the changing quality of the human agent." One aspect of this problem is illustrated by the changing level of formal education in the agricultural labor force. My approach here is basically the same as before; find some indication of the relative market value of these qualities at a point or points in time and apply this information (or "weights") to the changing quality mix over time. This is more difficult for education than for some other qualities, since we have very little data that hold "other things constant" adequately. In such a situation,

Table 5.2 Quality of labor input in agriculture as measured by changes in education: the rural-farm population

School years completed	Males, 25 years and older					Females, 25 years and older				
	Education, rural-farm population (percentage) in				Median income, 1950, total population	Education, rural-farm population (percentage) in				Median income, 1950, total population
	1940[a]	1950[b]	1957[c]	1959[d]		1940[a]	1950[b]	1957[c]	1959[d]	
None	5.2	3.6	3.7	3.2	1,108	4.1	2.6	2.7	2.0	518
Grade school										
1–4	17.5	15.8	14.1	13.5	1,365	13.6	11.4	9.6	8.1	547
5–7	26.5[f]	23.9	20.7	19.3	2,035	25.8[f]	21.9	18.6	18.7	725
8	28.7[f]	26.9	26.0	25.2	2,533	27.4[f]	25.1	22.6	23.5	909
High school										
1–3	10.8	12.8	13.8	14.6	2,917	13.1	15.3	16.2	16.6	1,086
4	6.2	10.7	15.2	16.7	3,285	9.1	14.7	21.3	20.9	1,584
College										
1–3	2.5	3.1	3.7	3.5	3,522	4.4	5.6	5.6	5.9	1,660
4 or more	1.2	1.8	2.0	2.8	4,407	1.4	2.2	2.8	3.4	2,321
Not reported	1.4	1.8	0.8	1.2	2,329	1.1	1.3	0.6	0.9	1,084
Quality measure[g]										
Dollars	2,250	2,379	2,452	2,500		934	1,020	1,085	1,102	
Index 1940 = 100		106	109	111			109	116	118	

[a] *1940 Census of Population*, vol. II, part I.
[b] *1950 Census of Population*, vol. II, part I.
[c] US Bu. Census, *Census of Population Reports*, Ser. P-20, no. 77.
[d] US Bu. Census, *Census Population Reports*, Serv. P-20, no. 99, Feb. 4, 1960.
[e] 1950 Census of Population, vol. IV, Special Reports, *Education*, P-E, no. 5-B.
[f] 1940 breakdown was 5–6 and 7–8 years of grade school completed. 1950 data were available for 7 and 8 separately. The 1950 relationship of 7 and 8 in the 7–8 category was used to break down the 1940 7–8 category. Because of the upward trend in education, the adjustment overestimates the percentage completing 8 years of grade school in 1940, and hence it results in an underestimate of the index for 1950 and 1957.
[g] US median income in 1950, by school years completed, weighted by the respective distribution of the rural-farm population by school years completed.
Note: The 1940 adjustment, the use of medians rather than means, and the restriction to 25 years or older, all contribute to an underestimate of the indexes for 1950, 1957, and 1959.

we can get at best limits on the possible order of magnitude of this type of quality change using a variety of weights. The educational distribution of the rural-farm population, 1950 US income weights, and the resulting "quality" indexes are presented in table 5.2. When the males and females are put together, these results indicate that the "quality" of the farm population has improved since 1940 by about 7 percent.[5] If we use 1958 income weights instead, for the 14-years-and-over population, we get an estimated 9 percent increase in quality using total population income weights, and a 12 percent increase if we use the rural farm median incomes by education in 1958 as weights.[6] One gets similar or slightly higher quality increases if one uses the educational distribution of farmers and farm managers and farm laborers and foremen instead of the rural-farm population. The only difference is a substantially higher increase in the quality of female farm laborers.[7] No matter how we weight it, there appears to have been a noticeable, albeit not very large, improvement in the quality of the farm labor force as a result of rising educational levels.

2.5 For some inputs the USDA has tried to take quality change into account. But even these measures could be improved upon. For example, given the tremendous increases in the average concentration (analysis) of fertilizers sold in this country, it would have made little sense to use tons of fertilizer as measure of fertilizer inputs. The USDA has used instead tons of "plant nutrients" as its input measure, counting the weight of the active ingredients rather than the total weight of fertilizers, and giving *equal* weight to the three most important plant nutrients: nitrogen, phosphoric acid, and potash. But even this approach may prove unsatisfactory in the long run. The three plant nutrients have somewhat different roles and are sold at different prices per plant nutrient unit. Since nitrogen is the most expensive nutrient and since there has been a trend towards relatively higher nitrogen consumption, even the plant nutrient measure will underestimate the "true" increase in fertilizer inputs. What is needed here is a "weighted" plant nutrient concept with the weights reflecting the relative prices of the different nutrients in some base period. Using 1955 implicit price weights (1.62 for N, 0.93 for P, and 0.45 for K) a weighted plant nutrient measure of fertilizer increases by about 6 percent more between 1940 and 1959 than the official USDA unweighted plant nutrient series.[8] Since the trend to nitrogen is only just now beginning to accelerate, this bias in the official measures is likely to become more serious in the future.

3 Measuring the Services of Capital Equipment

The common practice in productivity measures has been to measure changes in the services of capital equipment by changes in the *value* of the *stock* of capital in "constant" prices. The assumption was made (a) that the *flow of services* is proportional to the *stock* of capital equipment, and (b) that the relevant measure of stock for these purposes is the net (depreciated) *value* of equipment in "constant" prices. There is a variety of reasons why such measures may not be good approximations to the relevant flow of services concept, but most of them can be summarized by pointing out that the value of the stock of capital at any point in time is the current valuation of current *and* all *future* services expected from this stock, whereas what we are interested in for estimating current productivity is only the value of *current* services from this stock. The two measures will diverge if the average age of equipment is changing, and that is happening all the time due to cycles in net investment, and if there are anticipations of future price changes or technological advances. In addition, the deflators used to convert current stock values into "constant" prices may not be right for performing the same task on the flow of service figures. In particular, they do not adjust for what is essentially a relative price change: the decline in value of older machinery due to technical obsolescence.

The relevant measure of capital services for productivity comparisons would be approximated, in a perfect market, by the rental price per machine hour times machine hours used. The rental price would be a "constant" price which would not only adjust for fluctuations in the general price level but also for price changes in the new and used machinery markets. It would change only if the physical flow of services were to deteriorate due to wear and tear (and age) or change its character of use due to the appearance of new and different machines.

The easiest and not unreasonable assumption to be made is that there is no deterioration with age, and that the service flow is constant over the life span of a machine. In this case, the conversion from stock to flow is relatively simple. Assume that all machinery has a fixed life of 15 years and that during its lifetime, the flow of services is the same each year. On its 15th birthday it just falls apart and has no scrap value. If such a machine is worth $100 when new, and if the going rate of interest is 6 percent, then the annual flow of services during each of these 15 years must equal $10.30 (a 15-year annuity of $10.30 per year will have a present value of $100 if the rate of interest is 6 percent). If the rate of interest is 10 percent, the annual flow of services from this $100, 15-year machine equals $13.15. Put another way, this is what the annual depreciation and interest charges will come to. Since under these assumptions the flow of services is constant, the sum of interest and depreciation is also constant, with the interest charges falling and the depreciation charges rising as the machine ages.

Given these assumptions, the service flow is a constant fraction of the *original* value of the machine throughout its life. Thus, a 15-year moving sum of past gross investment is then both the relevant "stock" or volume index for productivity purposes (though not for valuation) and the relevant figure that is to be converted at a constant rate into the flow of services concept.

That considerations of this type have a practical importance can be illustrated using USDA figures on gross expenditures on farm machinery, trucks, tractors, and 40 percent of automobiles and comparing our results with USDA service flow estimates (table 5.3).

The difference in the percentage change in the flows is not very great, but their absolute size is quite different. The absolute size of the flow of services under the "constant services" assumption (at 6 percent) is 14 percent greater than the estimated 1940 flow based on USDA methods, and is 19 percent higher in 1958. If we assume higher rates of interest, as I think we should, the difference between the two methods gets even bigger. At 10 percent, "my" flow estimates are 27 percent higher in 1940 and 30 percent higher in 1958. At 15 percent, the estimates differ by about 42 percent in 1940 and 43 percent in 1958.

Perhaps the most serious result of the use of the wrong capital measure is the appearance of spurious cyclical movements in productivity. That these various measures may move quite differently over the investment cycle is illustrated in figure 5.1, which plots the percentage changes in the official and the 15-year constant service estimates of capital on farms since 1950.

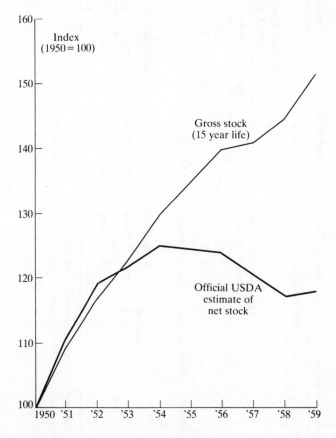

Figure 5.1 Recent developments in machinery and power on farms: different measures (tractors, trucks, other farm machinery, and 40 percent of farm automobiles).

Table 5.3 Comparison of constant service flow estimates and USDA service flow estimates for farm machinery and power[b]

(a) Constant service flow estimates (billion dollars, 1935–9 prices)

Year	15-Year sum of past gross investment	Annual flow of services at			Index (1940=100)
		6%	10%	15%	
1940	7.26	0.747	0.954	1.241	
1958	16.91	1.741	2.223	2.891	233

(b) USDA service flow estimates (billion dollars, 1935–9 prices)

Year	Value of of stock	Interest charge at			Depreciation	Depreciation plus interest at			Index (1940=100) Stock	Flow	
		6%	10%	15%		6%	10%	15%		6%	15%
1940	2.46	0.148	0.246	0.369	0.507	0.655	0.753	0.876			
1958	6.20	0.372	0.620	0.930	1.092	1.464	1.712	2.022	252	224	231

[a] Under "constant services" assumption – see text.
[b] Tractors, trucks, other farm machinery, and 40 percent of automobiles on farms.

While the official measure turns down in 1954, the service measure keeps rising through 1959.

The particular rate of interest chosen to "convert" stock into service flows will affect strongly the final results. We know very little, however, about the rates at which entrepreneurs discount the future services of new capital items. What is relevant for our computations is this "internal" discount rate rather than some market rate of interest and whatever evidence we do have points to a relatively high rate. Expected "pay-off" periods in industry are quite short, and the marginal cost of capital is reasonably high. In agriculture, about half of the new farm machinery is bought on credit, and the rate of interest paid on this type of debt is at least 10 percent. Since substantial price and technical uncertainty surrounds the future use of these machines and since at least some of the future obsolescence may be anticipated and "discounted," the average discount rate is probably substantially higher than 10 percent, perhaps around 15 percent.

Whatever little evidence we have on the "physical" productivity of capital in agriculture indicates also that it is (or was) quite high. Most production function studies in agriculture find that the marginal product of machinery is around 20 to 25 percent with 15 percent being close to the lower limit.[9] Similarly, studies of internal capital rationing indicate that the rate of return required to induce new investment is very high. One study found that of those interviewed farmers who would conceivably borrow more money for non-real-estate uses, 74 percent would not borrow more unless there was a high probability of a rate of return of 25 percent or higher.[10] The rate of return expected from internal funds perhaps would be lower, but it is unlikely to be very much so.

4 The Value of the Stock of Machinery on Farms

Having argued, in the previous section, that for productivity accounting we are interested in the value of the *services* of capital equipment rather than in the value of the *stock* of this equipment, we may still find it useful to examine the latter measure since it is a useful concept for many other purposes.

The value of the stock of equipment of a given kind on farms is estimated by the USDA by cumulating annual expenditures on the particular item, in constant prices, using a declining balance depreciation method. The estimated stock values depend then on three types of information: expenditure series in current prices, deflators, and the assumed method and rates of depreciation. Having discussed the deflators in a previous section, I would like to turn to the official depreciation method. Not much can be said about the other main input of data: the expenditure series in current prices. The methods used here seem to be the best possible, given budget and data limitations.[11]

The USDA describes as follows its depreciation assumptions and methods:

Depreciation is the estimated outlay in current prices which would be required if farmers were to replace exactly the plant and equipment *used up* during the year. The derivation of the estimates is based on the declining balance method of depreciation, in which a constant percentage, representing the annual rate of consumption of each type of capital, is applied to the value of a capital item at the beginning of each year to obtain depreciation during the current year. The rates of depreciation assumed result in a total charge-off of approximately 95 percent of total value over the period of years estimated as the useful life of the capital item, the remaining 5 percent representing scrap value. The annual depreciation rates employed for *recent* years are: . . . automobiles, 22.0 percent; motor trucks, 21.0 percent; tractors, 18.5 percent; and other farm machinery and equipment, 14.0 percent.[12]

. . . This method (declining balance) assigns larger depreciation changes in the earlier than in the later years of the life of a piece of equipment. It is the most nearly consistent with the curve of declining *market* values of used equipment in the case of automobiles, trucks, tractors, and other farm machinery.[13]

Ignoring the rather meaningless reference to the quantity of capital that is "used up" in production, the main question to be asked is, do the pattern and rates of depreciation used make sense? Since the USDA is interested in approximating the *market value* of the existing stock of capital,[14] using a constant percentage rate of depreciation is very much in accord with the few facts that we do have in this area. The evidence on the relative price of different age machines in the used machinery markets indicates clearly that the declining balance depreciation assumption approximates reality very well.

While the pattern of depreciation used by the USDA is the right one, the rates used seem to be too high for some of the items. The USDA has used the *pattern* of depreciation in the market place in deriving the pattern for its depreciation accounting but it has not followed this up to its logical conclusion by accepting also the market place evidence on *rates* to be used. Rather than using rates derived from used machinery prices, it has decided on rates by the arbitrary criterion that the resulting rate should depreciate the item to 5 percent of its original value by the time its "useful" (average) life is up.[15] While for some items such as automobiles and trucks the resulting rates of depreciation appear to be close enough to observed market rates, for other farm machinery and, in particular, tractors, the rates used by the USDA are much too high.

The used machinery market evidence on depreciation comes from annual tractor and other farm machinery trade-in guides ("blue books") and auction prices published in *Implement and Tractor*. Based on available "blue book" data,[16] except for the first year, the calculated depreciation rates are substantially below the USDA rates. The data point to a declining balance depreciation model, with a somewhat higher rate in the 1930s than in the 1950s. Over the whole period the depreciation rate computed from the relative price of different age tractors (leaving the first year out) was around

11 percent.[17] This seems to be also the average rate of depreciation in the market value of other farm machinery in recent years. Adding another percentage point for the "pure accident rate," the computed market rates of depreciation are still only around 12 percent and are substantially lower than the 18.5 (and higher) tractor rate and the 14 percent rate used on other farm machinery by the USDA.[18]

While the rates of depreciation used currently by the USDA may appear to be too high, the rates used in the past were even higher. In the twenties, the USDA depreciated farm automobiles and trucks at about 35 percent per annum and tractors at about 25 percent per annum. In the thirties the auto and truck depreciation rate was around 25 percent and the tractor rate around 24. The use of too high depreciation rates currently and even higher ones in the past leads to a substantial underestimate of the value of the stock of machinery on farms. Table 5.4 shows the magnitude of this underestimate for various items as of January 1957.

Another independent way of checking the USDA estimates of the value of capital on farms has become possible as the result of the 1956 National Farm Machinery Survey which provides detailed estimates of the number of different machines on farms and their age distribution. One can price each of these machines as of the beginning of 1957 at the average new price for each item, and then, applying estimated depreciation rates from used machinery market data to the survey age distributions, derive the average net (depreciated) value of each item on farms.[19] The resulting estimates are summarized with comparisons in table 5.4. They are about 36 percent higher than the comparable USDA estimates. The largest underestimate is for

Table 5.4 Different estimates of the value of the stock of tractors and other farm machinery on farms as of January 1, 1957 (million dollars)

Basis of estimate	Tractors	Other farm machinery
USDA[a]	3,319	7,636
Constant 12 percent declining balance depreciation rate[b]	4,609	8,721
1957 stock priced[c]	5,814	9,133
1957 stock priced[d]	6,307	
1957 stock priced[e]	4,448	

[a] From USDA, *The Balance Sheet of Agriculture, 1959*, table 11.
[b] Computed from unpublished USDA data, using the same data and methods as the USDA except for holding the depreciation rate constant at 12 percent. The USDA estimated 1920 values of stock were used to initiate the cumulation process.
[c] From NBER worksheets. Detailed tables to be published.
[d] The same as [c], but using the average price paid by farmers for new tractors in 1955 inflated to 1957 by the USDA price index of prices paid by farmers for tractors. The resulting average "new" tractor price is about 8.5 percent higher than the price used in [c].
[e] The number of tractors on farms multiplied by the average price paid in 1955 by farmers for a *used* tractor, inflated to 1957. The assumption made here is that tractors traded on the used market represent a random sample of all tractors on farms. The 1955 price for this and the previous footnote computation is taken from USDA, Stat. Bul. 243, table 72.

tractors, about 75 percent, but even "other farm machinery" is under-estimated by about 20 percent. Most of the difference, however, comes from the use of a different depreciation rate rather than from differences in coverage and pricing. The difference is very much smaller when the USDA data are recomputed using lower depreciation rates.

Looking at it in different ways, it seems quite clear to me that the existing USDA figures seriously underestimate the current value of the stock of machinery on farms. Since the used machinery price data, while poor, are better than no data, and are reasonably easily accessible, I see no reason why the USDA could not improve its estimates in this area by using this type of information in determining its depreciation rates.

5 Input Series Based on Implicit Productivity Assumptions

Besides the general paucity of information on a variety of items, one of the major difficulties with many input series in agriculture is the underlying lack of annual data and the unsatisfactory nature of the resulting compromises. For many inputs in agriculture the series are constructed as follows: Given a benchmark for the item in a particular year and its average use per unit of output or per unit of some other input for which more information is available, the benchmark values are extrapolated by the proxy series (output or other input) with which this input series is assumed to move in the same proportion. The most important example of such a series is the ARS estimate of manhours "used" in agriculture.[20] It is true that they are based on a very detailed set of productivity assumptions, e.g. on the estimated manhours required for harvesting a bale of cotton under different harvesting methods and information on the changing relative importance of the different harvesting methods; still they are essentially analytical constructs with many built-in productivity assumptions, and provide no independent information (since the "benchmarks" are also constructs) on actual manhours used in agriculture. Similarly, many expenditure series in agriculture such as building, automobile, and truck repair expenditures, and expenditures on veterinary medicine and on dairy supplies are estimated by extrapolating a benchmark by the number of automobiles, cows or milk production, and appropriate price indexes.

While these may be the best possible methods of estimating expenditure series and perfectly adequate for many purposes, they are less satisfactory if one is interested in productivity changes, and quite unsatisfactory if one is going to use these data for *annual* productivity comparisons. A series based on such "constructed" input series or deflated expenditure data consists of nothing more (between benchmark years) than the proxy series (usually output) and the implicit productivity assumption about the relation of the particular input to output or other proxy series. Thus, in the short run, such a series provides no independent information about movements in productivity. An output per unit of such an input series would be either constant between benchmark years or reproduce the productivity assumptions built into the input series by their estimators.

Besides producing irrelevant or "assumed" annual movements in the input series, such estimating procedures may also lead to biased estimates in the level of various inputs used unless the "benchmarks" are revised often. Unfortunately, for many items the benchmarks are far apart, and the possible cumulative bias over time is quite serious. The 1955 Farmers' Expenditure Survey disclosed a 10 percent underestimate in the official USDA production expenditure series (excluding depreciation). This underestimate was substantially larger for items for which little direct data are available and various indirect extrapolation techniques are used. Seed purchases, for example, were underestimated by 42 percent, and repair and operation of capital items by 26 percent.[21] Similarly, the ARS manhours series were revised downward recently by 17.5 percent.[22]

6 Conclusions

In a brief survey of this kind it is impossible to be comprehensive and point out the data gaps and special problems of all the agricultural inputs. For example, nothing has been said here, and very little is known, about the changing quality of land and the apparently large investments that have been and are being made in "farm structures other than buildings."[23] It is clear, however, from the evidence presented above, that some of our input series could be subject to definition and measurement errors of substantial magnitude. Since our "productivity" measures are residual measures, an error in some of the input series would result in a relatively much larger error in the productivity measures.

The possible empirical importance of some of the issues raised in this paper is illustrated in table 5.5. It presents the official input and productivity estimates for 1940–58, 1947–9 input weights, and the adjustments that are implied by *some* of our findings above. Four adjustments are made: (1) the machinery and power input index is given a 30 percent higher weight and the other weights are recomputed accordingly. This adjustment is based on our finding that the official capital series underestimate the absolute magnitude of capital services in agriculture (even if they are measured by value of stock) by about 30 percent, and the 26 percent underestimate of repairs and operation of capital equipment disclosed by the 1955 survey. (2) The machinery and power input index, since it is based on deflated expenditures, is assumed to be biased downward by 7 percent since 1947–9, and again by the same amount between 1940 and 1947–9. This is based on the estimated upward bias in the Prices Paid index 1947–9 (see table 5.1) and on the assumption that a similar amount of bias accumulated during the previous 1940–9 period. It does not allow for the type of quality changes that are not caught by indexes such as the WPI or CPI, discussed in section 2.3. (3) The index of labor input is increased by about 8 percent to reflect the

Table 5.5 US agriculture: changes in input and productivity, 1940–58

Category	Weight 1947–9[a]	Indexes (1947–9 = 100)				Adjusted weight[d]
		USDA[b]		Adjusted[c]		
		1940	1958	1940	1958	
Labor	0.45	122	66	117	69	0.43
Real estate	0.14	98	105	98	105	0.13
Power and machinery	0.16	58	137	54	147	0.20
Feed, seed, and livestock	0.09	63	141	63	141	0.09
Fertilizer and lime	0.03	48	166	47	171	0.03
Other	0.13	93	127	90	131	0.12
Total Input		97	100	92	106	
Output		82	125	82	125	
Productivity[e]		85	124	89	118	
1940 = 100			147		133	

[a] From R. A. Loomis, "Production inputs of US agriculture," *The Farm Cost Situation*, May 1960, p. 32.
[b] 1960 Economic Report of the President, tables D-67, 68, 69.
[c] 1940 labor index divided by 1.04, machinery and power index divided by 1.07, and fertilizer and lime and "other input" indexes divided by 1.03. For 1958 these indexes were *multiplied* by these same numbers. Total input index computed using "adjusted" weights.
[d] Computed by multiplying the weight of power and machinery by 1.3 and dividing all the weights by 1.048 to make them add up to 1.00.
[e] Computed from unrounded data used in this table. Will differ slightly from the official indexes due to rounding errors and the more detailed breakdown of inputs used in computing them.

increasing level of education of the farm labor force. (4) The fertilizer and lime index is increased by 6 percent to reflect improvement in quality not caught by the unweighted plant nutrient measure. The same 6 percent increase is applied to the "other inputs" category on the assumption that quality improvements in this category have been *at least* of the same order of magnitude as those not caught by an already quality oriented input measure of fertilizer. Since most of the series in this category are measured with very little regard for quality changes and are deflated by Prices Paid indexes that explicitly include quality changes, the 6 percent figure is very much an underestimate of the possible importance of quality change in this category. The last two adjustments are distributed equally between the pre- and the post-1947–9 periods. Taken together, these few adjustments "eliminate" (or account for) about 30 percent of the apparent increase in agricultural productivity since 1940.[24] This figure and the paper as a whole is not intended to measure the importance of total input quality change in agriculture. Rather, it should be taken as an indication that quality change *may* be important and that attempts to measure its importance are feasible and should be encouraged.

The divergent trends in the use of different inputs in agriculture result in a very great sensitivity of productivity measures to changes in the relative weight of different inputs.[25] It is impossible for me, given the time limitations, to discuss the problem of the "optimal" choice of weights here, but I would like to draw your attention, in closing, to a very important and difficult problem in that area. If one accepts the view that US agriculture has been in a chronic disequilibrium state most of the time, and that the marginal product of labor in agriculture has been below the going wage rate (on the average) and the marginal product of capital has been above its going price (due to capital rationing, among other things), it may be wrong to use "price" weights in aggregating our input indexes. What we need here are production function weights or coefficients, and price weights may provide a very poor approximation in a disequilibrium world.

Notes

1 A preliminary report on a study supported by a National Science Foundation grant. A full report on the study will be published at a later date. This preliminary report has not been approved for publication by the Board of Directors of the National Bureau of Economic Research.
2 See USDA Agr. Hb. 118, vol. 1, pp. 32–3.
3 For automobiles the USDA showed in 1958 an increase of about 56 percent in their price since 1947–9. In the same time period the consumer price index of new cars, which prices about the same models as the USDA, rose only 34 percent.
4 For details of this analysis see Zvi Griliches, "Hedonic price indexes for automobiles: an econometric analysis of quality change," Bu. of the Budget – NBER Price Statistics Review Committee, Staff Report no. 4, forthcoming.
5 The total increase is smaller than the separate quality increases for males and females, due to the rise of females from 11.4 to 18.6 percent of the farm labor force since 1940.
6 These income weights are taken from Current Population Reports, Ser. P-60, no. 33, Jan. 15, 1960.
7 The 1959 education data are taken from BLS, Special Labor Force Report no. 1, Feb. 1960.
8 These weights were derived from a 1955 cross-sectional analysis of prices different fertilizer mixtures and their specifications. For details see Z. Griliches, "Distributed lags, disaggregation, and regional demand functions for fertilizer," J. Farm Econ., Feb. 1959, pp. 91–2.
9 E.g. E. O. Heady and E. R. Swanson, Resource Productivity in Iowa Farming, Iowa Agr. Expt. Sta. Bul. no. 388, Ames, 1952, p. 756.
10 Heady and Swanson, op. cit., p. 771.
11 The 1955 Farmers' Expenditures Survey disclosed substantial errors in the official measures, particularly for items for which little direct information was available before this survey, such as trucks and automobiles. The survey results indicated that the AMS estimates of farmers' gross expenditures on the purchase of automobiles where too low by about 14 percent, that the AMS estimates of expenditures on trucks were too high by about 28 percent, that

the estimates for tractors were too low by about 3 percent, and that expenditures on "other machinery" were underestimated by about 18 percent. Since the survey, the AMS has revised the truck and tractor expenditure series accordingly, but not the automobile or other farm machinery series. It is not clear why different categories of expenditures were treated differently in the revision process. The survey figures are from *1954 Census of Agriculture*, vol. III, pt. 11. The unrevised AMS figures are taken from USDA, Agr. Hb. no. 118, vol. 3, table 8. The revised figures are from *The Farm Income Situation*, July 1959, table 19. Note that the 1955 Survey figures in subsequent publications, e.g. Stat. Bul. 224 and Stat. Bul. 243, are not well adapted for our purposes since they subtract the values of a traded-in truck from the purchase cost of an automobile, if the truck was traded for an automobile. But if a substantial amount of trucks should be traded for autos, the cumulation of such "net" expenditure series would result in an underestimate of the stock of automobiles on farms and an overestimate of the stock of trucks on farms.

12 USDA, Agr. Hb., no. 118, vol. 3, p. 17. Emphasis supplied.

13 USDA, *The Farm Income Situation*, Oct. 1955, p. 32. Emphasis supplied.

14 While this may not be too clear from the above quotations, the resulting stock estimates are used in *The Balance Sheet of Agriculture*, where the desired concept is clearly market value.

15 Besides being arbitrary, this approach suffers from the problem that as far as *social* accounting is concerned it is wrong to depreciate everything over the average life of the item. At the end of the average life span of an item, one half of the original population is still alive.

16 Auction-price data are difficult to summarize because of the very large variance in reported prices. On the whole, however, the implied depreciation rates in auction prices are very close to the "blue book" data with the possible exception of somewhat lower depreciation of old (ten years and older) tractors.
    Copies of tables summarizing some of the "blue book" evidence on depreciation for tractors and other farm machinery are available from the author on request.

17 The estimated first year depreciation rate is unreliable and may not be relevant. The denominator is the "list price," which very few people pay, and last year's model is usually more than one year old (probable median age around fifteen to sixteen months). Also, part of the first year's large drop in value is simply due to the "loss of virginity," the passage of the item from the "new" to the "used" category, which may be irrelevant for our calculations.

18 The used machinery market provides us with information on the decline in value of a *surviving* machine with age. But not all machines survive to the same age. The additional percentage point is intended to take care of the disappearance of some machines during the early part of their life. Actually, as far as declining balance accounting is concerned, this omission does not matter too much. By the time machines begin to die off in serious numbers, i.e. around age 15 or 20, even at the relatively low 12 percent rate of depreciation there is little left of the original value of the machine (about 15 percent).

19 While the depreciation rates used varied from item to item and from age to age, this variance was small, and one can describe the method used,

approximately, as applying an 11 percent rate of depreciation to the 1957 stock and age distribution of machinery valued at (mostly) average 1955 prices paid for new items inflated to 1957.

20 These series were called originally, and much more accurately, "required" manhours rather than manhours "used."

21 USDA Stat. Bul. no. 224, *Farmers' Expenditures in 1955 by Regions*, Washington, 1958, p. 3.

22 Without announcement or explanation.

23 In 1955, farmers' expenditures on "other farm improvements" amounted to $544 million. At the same time expenditures on farm building construction came only to $390 million. But the existing series include only the latter figure. These figures are from *1954 Census of Agriculture*, vol. III, pt. 11, table 4.

24 This percentage would have been very much higher if we had stopped our calculations as of 1957 rather than 1958. From 1957 to 1958 the output index went up by about 10 percent while the input index increased by only 1 percent. A little less than one-half of the difference between the two productivity indexes is due to giving a higher weight to the machinery and power index. The rest is the result of the other "quality" adjustments.

25 See also R. A. Loomis, "Effect of weight period selection on measurement of agricultural production inputs," *Agr. Econ. Res.*, Oct. 1957, 129–35.

6
Hedonic Price Indexes for Automobiles: An Econometric Analysis of Quality Change*

1 Introduction and Summary

"If a poll were taken of professional economists and statisticans, in all probability they would designate (and by a wide majority) the failure of the price indexes to take full account of quality changes as the most important defect in these indexes."[1] In spite of its potential importance, there is almost no published empirical work devoted explicitly to this problem. The only available book that deals with problems raised by changes in quality reaches essentially defeatist conclusions (Holfsten, 1952).

The main purpose of this paper is to investigate a relatively old, simple, and straightforward method of adjusting for quality change and find out whether (a) this method is feasible and operational, and (b) whether the results are promising and different enough to warrant the extra investment. It is standard practice in the price index industry to adjust for those quality changes to which a price can be attached. The appearance of automatic transmissions on the market at $200 extra will not raise the price of automobiles in the conventional indexes (except those of the US Department of Agriculture) even though eventually almost all cars are sold with it and the base price incorporates it as "standard equipment." However, only a few of the observed quality changes come in discrete lumps with an attached price tag. Most of the changes are gradual and are not priced separately. Nevertheless, many dimensions of quality change can be quantified (e.g. horsepower, weight, or length for automobiles); a variety of models with different specifications can be observed being sold at different prices at the same time; using multiple regression techniques on these data one can derive implicit prices per unit of the chosen additional dimension of the commodity; and armed with these "prices" one can proceed to adjust the observed

* From *Price Statistics of the Federal Government*, 1961. New York: National Bureau of Economic Research. Reprinted from the slightly corrected version in *Price Indexes and Quality Change*, Cambridge, MA: Harvard University Press, 1971.

price per "average item" for the changes that have occurred in its specification. There are many technical problems to be solved, but the main idea is quite simple: derive implicit specification prices from cross-sectional data on the price of various "models" of the particular item and use these in pricing the time series change in specifications of the chosen (average or representative) item.[2] Alternatively, one can interpret the procedure as answering the question of what the price of a new combination of specifications (or qualities) of a particular commodity would have been in some base period in which that particular combination was not available, by interpolating or extrapolating the apparent relationship of price to these specifications for models or varieties of the "commodity" that were available in that period. This latter interpretation avoids some of the more metaphysical problems involved in the notion of "quality" and "quality change."

In this paper I investigate the relationship of automobile prices in the US to the various dimensions of an "automobile" in 1937, 1950, and 1954 through 1960. A limited number of specifications or dimensions explain a very large fraction of the variance of car prices (as among different models) in any one of these years. Due to the high intercorrelation between some of these dimensions, there is some instability to the estimated "implicit prices" (the coefficients) of the dimensions. Also, there appears to have been a very substantial secular decline in the "price" of some of these dimensions (e.g. horsepower). Thus, estimates of the actual price change (after the quality change adjustments are made) differ markedly depending on whether they are based on beginning or end period weights. If we value the quality changes at their 1950 "implicit prices," we find that all the apparent increase in car prices between 1950 and 1960 can be explained by quality improvements, the hedonic price index actually falling during 1950–60. Valued at 1960 implicit quality prices, these same quality changes account for a little over half of the apparent price increase over this period. Over the whole period since 1937 the CPI may be overestimating the rise in automobile prices by at least a third. Since the CPI is a Laspeyres index the appropriate quality adjustment should also be based on "base" (beginning) period weights. If this is done, about three-fourths of the rise in automobile prices in the CPI since 1937 could be attributed to quality improvements.

Some limitations of this type of approach are explored in the last part of the paper and, in light of these, it is not yet recommended that such adjustment should be made routinely as part of the price index computations. Continuous studies of this sort, however, covering a wide range of commodities, would be of great value. They could provide us with estimates of the order of magnitude of the possible upward drift in the official price indexes due to their inability to cope adequately with the ever-present quality change problem. Moreover, they would spot for the price data collecting agencies what appear to be the more relevant dimensions or specifications of a commodity, providing them with a better basis for judging which specifications should be controlled in the pricing process.

2   Theoretical Considerations

It is impossible to deal here with all the index number problems raised by the changing quality of commodities.[3] Since we are interested in the effect of quality change on measured prices and price indexes, our first job is to find what relationship, if any, there is between the price of a particular commodity and its "quality."

Most commodities, particularly consumer and producer durables, are sold in many varieties or models. Thus at any one time we can observe a population of prices – p_{it} – where i is the index of varietal designation (e.g. no. 2 corn, or a Chevrolet Impala four-door hardtop with a V-8 engine) and t stands for the time period of observation. The reason why these different varieties or models sell at different prices must be due to some differences in their properties, dimensions, or other "qualities," real or imaginary. Thus we can write p_{it} as a function of a set of "qualities" X, and some additional small, and hopefully random, factors measured by the disturbance u.

$$p_{it}=f_t(x_{1it}, x_{2it}, \ldots, x_{kit}, u_{it}).\tag{1}$$

These qualities do not necessarily have to be numerical. Given a sufficient number of observations we can use variables which take the value one if the item possesses the particular quality and zero if it does not and derive the average contribution of this "quality" to the price of the item. Nor do they have to be desired for their own sake. It will suffice if they are well correlated with some more basic dimension which may be more difficult to measure. For example, for many commodities, and at least over some range, "size" and "capacity" are very important qualities. They are, however, quite elusive and difficult to measure. On the other hand, they can often be approximated quite well by variables such as volume, weight, or length, even though none of these "proxy" dimensions may be desirable *per se*.

The existence and usefulness of such a function is an empirical rather than theoretical question.[4] To estimate such a function we have to make additional assumptions about the number and kind of relevant qualities and the form in which they affect the price of the product. There is no a priori reason to expect price and quality to be related in any particular fixed fashion. This again is an empirical question. In this study, I have used the semilogarithmic form, relating the logarithm of the price to the absolute values (pounds, inches, etc.) of the qualities:

$$\log p_{it} = a_0 + a_1x_{1it} + a_2x_{2it} + \ldots + u_{it}.\tag{2}$$

This choice was based on an inspection of the data and the convenience of this particular formulation.[5] Other forms, e.g. linear, or linear or in the logarithms, may, however, be more appropriate in a study of other commodities and qualities.

Assuming that the equation can be estimated with enough precision, it can be used to estimate the value of certain quality changes in the base period.

Moreover, one can use it to estimate the price of a new bundle of qualities which may not have been available in this period, provided that the new bundle differs only quantitatively in its "qualities" from the previously available items and does not contain some new, previously unknown or unavailable qualities. Even if the new item possesses some previously unknown qualities, the equation can be used to estimate the change in price due to changes in the subset of quantifiable qualities, and half a loaf may be better than none.

An equation of this type can be computed for each period for which we have enough observations to do it. If the results are not the same in different periods, and they are unlikely to be so, we are faced with the general index number problem of changing weights. The implicit prices we obtain will depend on the particular period or periods chosen as "weight" or reference periods, and Laspeyres' and Paasche's indexes may diverge sharply. If the periods are not too far apart and the weight pattern not too different, we can estimate the average price change directly by assuming that the equation holds well enough in both periods except for the change in the additional variable "time."

$$\log p_{it} = a_0 + a_1 x_{1it} + a_2 x_{2it} + \ldots + a_d D + u_{it} \tag{3}$$

where D is a variable that is zero in the first period and one in the second.[6] The coefficient a_d provides us with an estimate of the average percentage increase in price of these models or varieties between the two periods, holding the change in any of the measured quality dimensions constant. If we want to impose the same set of weights on more than two cross-sections, this can be achieved by specifying additional "time" or "dummy" variables, taking the value one in their reference period and the value zero in all other periods. The necessary number of such variables is one less than the number of cross-sections that are being estimated together. The resulting coefficients measure the percentage change in the average price, holding qualities constant, with the average price for the earliest cross-section being the base of measurement.

Having estimated such equations, instead of adjusting the prices or price indexes directly, we can first define an index of quality change and use that to adjust the official indexes. Consider a particular variety of a commodity, say a Plymouth Savoy four-door sedan with a six-cylinder engine, whose qualities may have changed over some time period. Then the quality change measure g is defined as

$$g_{1i}^0 = \frac{\hat{P}_{i1}}{\hat{P}_{i0}} \text{where } \hat{P}_{i0} = f_0(x_{1i0}, \ldots),$$

and $\hat{P}_{i1} = f_0(x_{1i1}, \ldots)$.[7] That is, the p's are each predicted prices for variety i on the basis of estimated equation f_0, one for the combination of qualities this variety had in period 0 and the other for the combination of qualities it has in period 1. More simply g_1^0 measures the percentage increase in price predicted by the function f_0 on the basis of the change in the level of different qualities (the x's) between the two periods. Of course, if we had used the estimated function for the second period, f_1, or a price quality function for some other period, we would have gotten a somewhat different measure. For a larger number of varieties, or models, these g's can be aggregated into a quality change index, using the same weights that are used in aggregating their prices in the price index. To get at the adjusted "real" change in prices, we would "deflate" the observed price index by the estimated quality change index.[8]

$$\text{"true price index"} = \frac{\text{observed price index}}{\text{quality change index}} = \frac{P_1}{P_0} \Big/ \frac{\hat{P}_1}{\hat{P}_0} = \frac{P_1/\hat{P}_1}{P_0/\hat{P}_0} \, .$$

Note that this "quality change" index is based only on those "qualities" for which a price is being paid or exacted, and only to the extent of the price differential. If these price differentials are "phony" or "too high" or "too low" from some omniscient point of view, the index will not take this into account. In fact, it may not take into account some aspects of "quality" which may be important, and incorporate other "imaginary" qualities such as brand names whose "superiority" over unbranded items would be denied by many people. Thus, if we observe that garments bearing one union label sell on the average at a 5 percent higher price than comparable unlabeled items, and also that garments bearing the labels of three different unions sell for 15 percent more than comparable unlabeled items, we would predict that if a similar garment were available with two union labels, it would probably sell for about 10 percent more than the unlabeled items. And we would use this in calculating our price index (or price relative) for the two-label garment, even though we are morally certain (and supported in this by extensive test laboratory findings) that there is no "real" quality difference among all these items. We would do this since we are answering only a relatively modest question: What would the price have been if it were available? And not: Would consumers be "right" in paying this particular price, or for that matter the price of any other item? Once raised, the doubt whether the evidence of the marketplace reflects adequately, if at all, the "true" marginal utility of different items or qualities to the consumer can be turned against any other price or commodity. It is not a problem peculiar to the measurement of "quality."

While it is not necessary for our purposes, it would be nice, however, if these quality indexes represented something "real" and not just the mistakes and idiosyncrasies of manufacturers' pricing policies. There are two possible sources of evidence on this point. The first, which will be explored to some extent at the end of the paper, is the evidence of second-hand markets. Do different qualities command approximately similar relative prices in the used market, a market which could be considered to be more competitive than the market for new items? If they do, this would indicate that consumers are still willing to pay these differentials even when they are not imposed by manufacturers. A second and more stringent test, which will not be pursued here, could have been made by investigating what happens to the sales of varieties or brands if their prices are too high or too low relative to their quality content. Given an estimated price–quality equation for a particular period, the estimated residual for a specific model or brand could be interpreted as a measure of over or under pricing relative to the quality content of this model. If, with the help of these residuals, we were able to predict reasonably well the market share experience of different models or brands, i.e. "over-priced" items losing and "under priced" items gaining, this would provide strong support for the correctness of our price–quality equation and its interpretation.

3 The Sample and the Variables

The analysis of price–quality relationships reported below is based on data for US passenger four-door sedans for the years 1937, 1950, and 1954 through 1960. In each of these years an attempt was made to collect price and specification data for all models and brands for which such data were easily available.[9] Since these calculations were viewed as being exploratory, no special attempts were made to assure completeness of coverage, nor were the model observations weighted by their relative importance in the market. The number of observations in each cross-section varies from a low of 50 in 1937 to a high of 103 in 1958.

The new car prices used throughout this study are factory-delivered "suggested" (list) retail prices, at approximately the beginning of the model year.[10] Unfortunately, there are no published data on actual transaction prices for a wide range of models. Discounts from list prices may have varied over time, and this will make it somewhat difficult to compare our results with the CPI, since the CPI has tried to take discounts into account, at least since 1954. Only to the extent that relative discounting is correlated with some of our quality dimensions will the use of list prices lead to any special bias in the estimates of the quality coefficients. This same difficulty would not be present if an official government agency were doing such a study. The WPI actually collects the manufacturers' wholesale price to dealers for most automobile models. Similarly, it should not prove difficult to expand the CPI sample, at least once a year, to include a wider range of models.

No adjustment was made for any changes in minor equipment items that became standard equipment at some later point in time, such as directional signals or electric clocks.[11] Major items, such as automatic transmissions, power steering, and power brakes were treated by defining independent variables that took the value of one if the item was "standard equipment" on a particular model and zero if it was not.

The major numerical "quality" variables used in this study are horsepower (advertised brake horsepower), weight (shipping), and length (wheelbase for 1937 and 1950, and overall from 1950 on). In addition, "dummy" variables, i.e. variables that take the value of one if the particular model possesses this particular "quality" and zero if it does not, are defined for the following "qualities": V-8 engine or not, hardtop or not, automatic transmission as standard equipment or not, power steering as standard equipment or not, power brakes as standard equipment or not, and for 1960 models whether a car is a "compact" or not. Note that some of these variables do not measure the consequence of having a particular item of equipment as much as they stratify and control for the type of car on which such equipment is "standard" (included in its base price). Thus, for example, the variable for power steering effectively identifies most of the large luxury cars that differ from other cars in other ways besides sheer size or the presence of power steering as standard equipment.

A variety of variables for which no convenient data are available was not included in the calculations. Most important of these are the various "performance" variables: gasoline mileage, acceleration, handling ease, durability, and styling. Scattered data already exist on some of these qualities, and I am sure that it would not prove very difficult to collect more and include such variables explicitly in a similar price–quality analysis. Variables reflecting the level of "workmanship" associated with a particular car and variables accounting for small design changes, such as the substitution of an alternator for the generator, were also omitted for lack of data. Nor were brand or manufacturer differentials taken into account. In fact, as far as the numerical qualities that are included in the analysis are concerned, they could probably all be interpreted as different aspects of one underlying quality "size" or "capacity."

The characteristics of the sample are summarized in table 6.1. Note the sharp increase in horsepower per car since 1950, due to a large extent to the introduction of the V-8 engine, and the lengthening of cars which reached its peak in 1959. The drop in the average price and specification level of cars in 1960 is due mainly to the introduction of the "compacts" and the decline in the number of high- and medium-priced models on the market.

Table 6.1 Characteristics of the cross-sections used in this study: US passenger four-door sedans – 1937, 1950, and 1954–60

Year	Number of models	Average (geometric) price ($)	Average horse-power	Average shipping weight in pounds	Average length in inches Wheelbase	Average length in inches Overall
1937	50	1,183	109	3,506	122	—
1950	72	2,113	115	3,533	122	205.7
1954	65	2,360	141	3,452	—	205.0
1955	55	2,281	166	3,429	—	205.4
1956	87	2,594	200	3,616	—	207.5
1957	95	2,785	226	3,696	—	208.9
1958	103	3,054	252	3,835	—	211.6
1959	87	3,180	251	3,907	—	213.7
1960	78	2,800	211	3,606	—	208.6

Source: See note 9 for sources of data.

4 The Regression Results

It is impossible to reproduce here the very large number of multiple regressions that were computed for different years and different combinations of years and independent variables. Due to the very high multicollinearity between the three numerical "qualities" chosen for analysis (see table 6.2) there was substantial instability in the coefficient estimates for some of the years. Usable estimates were obtained only for years in which there was some independent variation along the three numerical quality dimensions, and for combinations of years where the larger number of observations allowed us to determine the separate coefficients with greater precision.

Regression estimates for selected years are summarized in table 6.3; table 6.4 summarizes a set of regressions utilizing two adjacent annual cross-sections each and introducing an explicit variable to estimate the average price change holding quality change constant. It also presents the estimated coefficients of the overall regression for 1954–60, lumping all of the seven (1954 through 1960) cross-sections together and allowing them to differ from each other in level but not in slope.

Since our dependent variable is the logarithm of price, the resulting regression coefficients can be interpreted as the estimated percentage change in price due to a unit change in a particular "quality," holding the other qualities constant. Thus, for example, the results for the 1960 cross-section (column 1 in table 6.3) imply that the following was true, on the average, for the 1960 model cars and their list prices. An increase of 10 units in horsepower, ceteris paribus, would result on the average in a 1.2 percent increase in the price of a car (with a standard error of 0.3 percent). An increase of 100 pounds in the weight of a car was associated with a 1.4 percent increase in price. An increase of 10 inches in the length of a car, holding the other qualities constant, was associated with a 1.5 percent increase in the price of the car (but was not significantly different from zero at conventional

Table 6.2 First-order correlation coefficients: r

	Year					
Between	1960	1959	1957	1954	1950	1937
H and log P	0.89	0.85	0.85	0.89	0.84	0.88
W and log P	0.90	0.92	0.95	0.88	0.87	0.92
L and log P	0.77	0.75	0.84	0.81	0.91	0.88
H and W	0.85	0.82	0.90	0.92	0.76	0.80
H and L	0.72	0.75	0.79	0.73	0.74	0.84
W and L	0.92	0.86	0.85	0.87	0.83	0.92

H = horsepower;
N = weight;
L = length, overall, except wheelbase in 1937;
log P = logarithm of list price.

Table 6.3 Coefficients of single year cross-sectional regressions relating the logarithm of new US passenger car prices to various specifications, selected years[a]

				Model year		
				1950		
Coefficients of	1960	1959	1957	(1)	(2)	1937
H	0.119	0.118	0.117	0.365	0.585	0.867
	(0.029)	(0.029)	(0.030)	(0.110)	(0.133)	(0.181)
W	0.136	0.238	0.135	0.111	0.145	0.388
	(0.046)	(0.034)	(0.010)	(0.066)	(0.096)	(0.078)
L	0.015	−0.016	0.039	0.192	0.147	−0.009
	(0.017)	(0.015)	(0.013)	(0.026)	(0.045)	(0.078)
V	−0.039	−0.070	−0.025	−0.054	−0.091	−0.023
	(0.025)	(0.039)	(0.023)	(0.032)	(0.040)	(0.060)
T	0.058	0.027	0.028	—	—	—
	(0.016)	(0.019)	(0.012)			
A	0.003	0.063	0.114	—	—	—
	(0.040)	(0.038)	(0.025)			
P	0.225	0.188	0.078	—	—	—
	(0.037)	(0.041)	(0.030)			
B	—	—	0.159	—	—	—
	—	—	(0.026)			
C	0.048	—	—	—	—	—
	(0.039)	—	—	—	—	—
R^2	0.951	0.934	0.966	0.892	0.835	0.904

[a] While the original computations were all done with logarithms to the base 10, the results in this table are converted to natural logarithms (to the base e) as an aid to interpretation. The resulting coefficients, if multiplied by a hundred, measure the percentage impact on price of a unit change in a particular specification or "quality," holding the other qualities constant. The numbers in parentheses are the calculated standard errors of the coefficients. For 1950 regression (2) and 1937: length of wheelbase rather than overall length.
H = Advertised brake horsepower in 100s.
W = Shipping weight in thousand pounds.
L = Overall length in tens of inches.
V = 1 if the car has a V-8 engine; = 0 if it has a six-cylinder engine.
T = 1 if the car is a hardtop; = 0 if it is not.
A = 1 if automatic transmission is "standard" equipment (included in the price); = 0 if not.
P = 1 if power steering is "standard"; = 0 if not.
B = 1 if power brakes are "standard"; = 0 if not.
C = 1 if the car is designated as a "compact"; = 0 if not.
R^2 = Coefficient of multiple correlation squared.

significance levels). A V-8 engine, holding horsepower, weight, etc., constant was associated with a 4 percent lower price than a six having comparable characteristics.[12] A "hardtop" was on the average 6 percent more expensive than other comparable ("soft top"?) models. Holding other "qualities" constant, the inclusion of an automatic transmission as "standard equipment" was not associated with any significant price increase. The presence of power steering as "standard equipment" led to a 22 percent higher price over comparable models.[13] The cars designated as "compacts" were

Table 6.4 Coefficients of regressions of the logarithms of price on various "qualities": US passenger cars, two years taken together, and all the seven years, 1954 through 1960[a]

Coefficients	Model years							
	1954–60	*1959–60*	*1958–9*	*1957–8*	*1956–7*	*1955–6*	*1954–5*	*1937–50*
H	0.056	0.114	0.062	0.040	0.095	0.091	0.241	0.538
	(0.013)	(0.018)	(0.025)	(0.026)	(0.028)	(0.055)	(0.059)	(0.108)
W	0.249	0.212	0.285	0.271	0.211	0.241	0.009	0.328
	(0.021)	(0.029)	(0.034)	(0.038)	(0.039)	(0.056)	(0.060)	(0.053)
L	0.023	−0.006	−0.018	0.007	0.045	0.053	0.082	0.108
	(0.007)	(0.011)	(0.013)	(0.013)	(0.011)	(0.015)	(0.016)	(0.039)
V	0.010	−0.059	−0.026	0.005	−0.037	−0.043	−0.031	−0.093
	(0.013)	(0.023)	(0.031)	(0.026)	(0.020)	(0.031)	(0.024)	(0.035)
T	0.023	0.040	0.030	0.024	0.022	0.018	—	—
	(0.009)	(0.013)	(0.012)	(0.013)	(0.010)	(0.018)	—	—
A	0.090	0.034	0.070	0.075	0.058	0.079	0.236	—
	(0.016)	(0.027)	(0.030)	(0.026)	(0.021)	(0.028)	(0.037)	—
P	0.088	0.206	0.125	0.113	0.089	0.062	0.035	—
	(0.017)	(0.028)	(0.040)	(0.030)	(0.023)	(0.029)	(0.038)	—
B	0.109	—	0.115	0.162	0.138	0.098	−0.045	—
	(0.016)	—	(0.038)	(0.028)	(0.019)	(0.029)	(0.045)	—
C	0.157	0.052	—	—	—	—	—	—
	(0.031)	(0.031)	—	—	—	—	—	—
D	—	−0.023	0.005	0.027	0.027	0.020	−0.093	0.527
	—	(0.011)	(0.014)	(0.012)	(0.011)	(0.018)	(0.020)	(0.027)
D_1	−0.044	—	—	—	—	—	—	—
	(0.015)	—	—	—	—	—	—	—
D_2	−0.015	—	—	—	—	—	—	—
	(0.014)	—	—	—	—	—	—	—
D_3	0.019	—	—	—	—	—	—	—
	(0.015)	—	—	—	—	—	—	—
D_4	0.044	—	—	—	—	—	—	—
	(0.016)	—	—	—	—	—	—	—
D_5	0.044	—	—	—	—	—	—	—
	(0.016)	—	—	—	—	—	—	—
D_6	0.023	—	—	—	—	—	—	—
	(0.016)	—	—	—	—	—	—	—
R_2	0.922	0.943	0.915	0.929	0.945	0.924	0.904	0.916

[a] See notes to table 6.3 for the definition of most of the variables.

D = 1 in the second of two periods being estimated together; = 0 in the first. The coefficient of D can be interpreted as the percentage change (if it is multiplied by 100) in the average price of cars between the two periods, holding all the qualities constant. Thus, e.g. for 1937–50, the estimated "true" price change is approximately 53 percent.

D_1 = 1 in 1955; = 0 in other years.
D_2 = 1 in 1956; = 0 in other years.
D_3 = 1 in 1957; = 0 in other years.
D_4 = 1 in 1958; = 0 in other years.
D_5 = 1 in 1959; = 0 in other years.
D_6 = 1 in 1960; = 0 in other years.

The coefficients of these variables measure the average percentage change in price holding quality constant as of 1954. Thus for 1960, it indicates that since 1954 the average price holding quality constant increased only by about 2 percent and that, moreover, this increase is not significantly different from zero. To get the estimated percentage change between two adjacent years, one has, in this case, to take the difference between the two coefficients. Thus, e.g., the 1954 through 1960 equation estimates the average percentage change in price between 1957 and 1958 as 1.5 (4.4–1.9), against a 2.7 estimate given by the equation for 1957–8 alone.

selling for about 5 percent more than other cars, holding other "quality" differences constant, but again, this premium was not significantly different from zero.

If we look now across the rows of tables 6.3 and 6.4, several things are worth noting. The fit of these equations is quite good. With the help of a few numerical and shift variables, we manage to explain most of the time 90 or more percent of the variance of the logarithm of car prices in a particular year or set of years, even though the range of our sample extends from Ramblers to Cadillacs.[14] The coefficient of "weight" is almost always significantly different from zero, at conventional levels, and its magnitude remains relatively stable from cross-section to cross-section. The coefficient of horsepower is also statistically significant in a large fraction of cases, but varies somewhat more in magnitude around a downward trend. The coefficient of length is perhaps the most unstable of all the estimated coefficients, being very large and significant in 1950, declining rapidly in the middle fifties, and becoming insignificant and sometimes negative by 1958 and in subsequent years. This is partly the result of the generally very low variability of "length" in the sample (its coefficient of variation was only about 4 percent, on the average) and the very marked increase in the length of the lower priced cars since 1957.

Looking at the coefficients of the shift or "dummy" variables representing the presence or absence of certain "qualities," perhaps the most interesting result is the consistently negative sign attached to the coefficient of the V-8 versus six-cylinder engine variable. It is true that most of the time this coefficient is not significantly different from zero, but the consistency in sign from period to period is both surprising and instructive. While we know that a V-8 engine costs about $100 more than a six on a "comparable" car, this is not what is meant by "comparable" in the context of our equations. What the coefficient says is that if we hold horsepower and the other variables constant, a V-8 is cheaper by about 4 percent. Since the "comparable" cars are likely to differ much only in horsepower, and since there is very little overlap in the sample between the horsepower levels achieved by six-cylinder engines and the horsepower generated by the V-8s, what this coefficient is really saying is that higher horsepower levels can be achieved more cheaply if one shifts to V-8 engines than would be estimated by extrapolating the price-horsepower relationship for the six-cylinder engines alone. It is a measure of the decline in the "price" per horsepower as one shifts to V-8s even though the total expenditure on horsepower goes up.

The coefficient of the "hardtop" variable is reasonably stable over time, indicating a premium of around 3 to 4 percent for this type of car. The coefficient of the "automatic transmission included in the price" variable is always positive, but varies substantially from time to time. The coefficients of the "power steering" and "power brakes standard equipment" variables are usually very significant and relatively large in size.[15] It is quite apparent that what they measure is not so much the presence or absence of these particular equipment items, as the presence of many other "luxuriousness" attributes associated with cars on which these items are "standard equipment." In a sense, these shift variables take care of some of the nonlinearity in the relationship of the logarithm of price to numerical qualities such as weight or horsepower. Usually the high–medium and high-priced cars are priced somewhat higher than would be predicted just by extrapolating the price-horsepower (or length or weight) relationship from the lower price range. Allowing the cars having power steering, power brakes, or automatic transmission as standard equipment, to have separate constant terms, brings these cars "into line" and reduces the possible bias in the estimated price–quality relationship for the numerical qualities.

5 Price and Quality Indexes for US Automobiles

A. Hedonic Price Indexes for the Sample as a Whole

As we have noted already in our discussion of table 6.4, the results presented there provide us with an estimate of the average price change that occurred between two periods in the list prices of automobiles, holding all the specified qualities constant. This is comparable to the deflation of the change in the price of the average car in the sample by a quality index with "average" rather than base or end period weights. These "average" weights are derived from the coefficients of the regression that provides the best fit simultaneously to data for two years, a regression that imposes the same price–quality relationship (slope) on both years, but allows them to differ in level. The weights are used then to adjust for the change in the specifications of the average car in the sample that has occurred between the two periods.

The resulting price indexes are summarized in table 6.5 and compared to the Wholesale Price Index "Passenger Cars" component. The comparison with the WPI is more appropriate for two reasons. First, it is the only one of the official indexes that covers all passenger cars rather than just a few selected makes and models, and second, it is based on manufacturer prices to dealers whose relationship to the list prices used in this study has remained approximately constant over time. Unfortunately, the comparison is imperfect in the sense that the WPI is a weighted index of car prices, with weights based on the market shares of various makes (in some base period?), while our list price index is an unweighted average of all makes and models.[16] Relative to the WPI, our index gives too much weight to the high and medium priced cars.

If we disregard these reservations, or limit the implications to our sample only, the results presented in table 6.5 attest strongly to the importance of "quality" change. About one-third of the price change between 1937 and 1950, and almost all of the price increase between 1954 and 1960, is attributable to changes in a few selected specifications. If we use a chain-link index for the 1954–60 period, adjusting the 1954–5 price change by a quality index with average 1954–5 weights, adjusting the 1955–6 price change by a quality index with 1956–7 weights, and so on, we actually come to the

Table 6.5 US cars: percentage change in various price indexes, selected years

| | List prices | | | |
| | Hedonic price index based on[b] | | | |
Model year	Average car in sample[a]	Estimated adjacent two-period weights	Estimated average 1954 through 1960 weights	WPI[c]
1937–50	79.0	52.7	—	83.0
1954–5	− 3.3	− 9.3	− 4.4	2.7
1955–6	13.7	2.0	2.9	4.1
1956–7	7.7	2.7	3.4	4.7
1957–8	9.6	2.7	2.5	0.6
1958–9	3.6	0.5	0.0	5.1
1959–60	− 11.9	− 2.3	− 2.1	0.1
1954–60	18.7	− 4.2[d]	2.3	19.7

[a] Percentage change in the geometric average of all list prices in the sample.
[b] Computed from table 6.4.
[c] From various BLS releases. For 1937 and 1950 models price as of December of the previous year. For 1954 models price as of January 1954. For all subsequent model years price as of November of the preceding calendar year.
[d] Computed by multiplying all the estimated 2-year price relatives.

conclusion that the average 1960 car in our sample was cheaper than the 1954 average car, once some of the appropriate quality adjustments are made. If we use average 1954–60 weights derived from the joint multiple regression equation for all seven cross-sections, we do indicate a small price rise for the 1954–60 period (2.3 percent) but we cannot reject the hypothesis that actually there was no real change in price over the period as a whole.

B Quality and Price Indexes for the "Lower Priced Three"

Since two of the most important automobile price indexes (the automobile components of the CPI and of the Prices Paid by Farmers Index of the USDA) are based on prices for the "low-priced three" makes – Chevrolet, Ford and Plymouth – it is of some interest to develop quality and quality-adjusted price indexes that are restricted to this particular group of cars.[17] An attempt will be made to approximate a quality index appropriate to the group of cars priced by the CPI. Since it is impossible, from the published material alone, to discover all the details of the pricing and specification procedure used by the CPI, we cannot reproduce it exactly, adding only our quality adjustments.[18] In principle, however, our methods can be applied directly to the CPI data by the BLS, allowing a more firm estimate of the possible "quality bias" in the index.

The specification and list price history of the "average" Chevrolet, Ford, and Plymouth in the sample is presented in table 6.6. Some attempt is made at weighting the different makes by including only two Plymouth models in this sample versus three models each for Chevrolet and Ford cars. Also, the specification and price history of six-cylinder engine cars and V-8 engine cars

Table 6.6 Specifications and list prices of the average[a] "low-priced three" car

Year	Horsepower	Weight (pounds)	Length Wheelbase	Length Overall	Price ($)[b]
Six-cylinder engines					
1937	81	2,756	112	196.0	703
1950	94	3,099	116	196.1	1,521
1954	111	3,149	—	195.8	1,795
1955	120	3,129	—	198.7	1,839
1956	135	3,172	—	199.7	1,938
1957	139	3,255	—	203.6	2,140
1958	142	3,349	—	206.6	2,275
1959	138	3,448	—	209.6	2,415
1960	141	3,539	—	211.5	2,425
V-8 engines					
1955	163	3,185	—	198.7	1,939
1956	176	3,246	—	199.7	2,039
1957	184	3,354	—	203.6	2,240
1958	210	3,440	—	206.6	2,390
1959	202	3,525	—	209.6	2,533
1960	190	3,615	—	211.5	2,537

[a] Average for three Chevrolets, three Fords, and two (the two lower-priced series) Plymouth models, except in 1937 and 1950. The 1937 sample consists of two Chevrolets, two Plymouths, and three eight-cylinder Fords. The 1950 sample consists of four Chevrolets, two Fords, and two Plymouths. The eight-cylinder Fords in 1937 were included to raise the sample size to approximately the same levels as in the subsequent years. Since these 8s (not V-8s) had a lower list price than comparable 6s in 1937, their inclusion, if anything, will bias the quality indexes downward.
[b] Arithmetic average.

is recorded separately. Since the CPI switched over in 1956 from pricing six-cylinder cars to pricing the V-8 models of these same cars, we shall follow suit by computing separate indexes for each type of car and linking them at 1956.[19]

Table 6.7 presents some of the weights used in aggregating these "qualities." It is immediately apparent that the computed quality indexes will differ substantially depending on which sets of weights is used. To provide historical perspective, this table also records weights derived by Court in his earlier study of the same problem. The weights reproduced in this table and additional weights taken from table 6.4 are used in constructing the set of quality indexes summarized in table 6.8.

The quality indexes measure how much higher the price of the particular car (or the average price of a particular class of cars) would have been, in the weight period, if its specifications had changed by the same amount as they did between the two periods that are being compared. Using beginning period weights, we find that "quality per car" practically doubled since 1937, with most of the increase occurring since 1950. Using end period weights, the indicated increase is only about 37 percent, which is still quite substantial. Using chain-link weights, or average 1954–60 weights, produces intermediate

Table 6.7 Estimated quality weights or "prices": percentage change in the price of cars as the result of a unit change in selected "qualities," in selected years

| Years | Percentage change in price per | | |
	10-unit change in horsepower	100-pound change in weight	One-inch change in length[a]
1930–5[b]	5.5	5.7	0.31
1935–7[b]	5.3	5.8	0.01
1937–9[b]	7.1	3.0	0.15
1937[c]	8.7	3.9	−0.09
1950(2)[c]	5.8	1.5	1.47
1950(1)[c]	3.6	1.1	1.92
1957[c]	1.2	1.4	0.39
1959[c]	1.2	2.4	−0.16
1960[c]	1.2	1.4	0.15
1954 through 1960[d]	0.6	2.5	0.23

[a] Wheelbase length, 1935 through 1950(2), overall length thereafter.
[b] From Court (1939), p. 111.
[c] From table 6.3.
[d] From table 6.4.

Table 6.8 Quality indexes for the "low-priced three" (six-cylinder engines to 1956, V-8s thereafter)

| Period | Percentage change | | | |
	Beginning period weights[a]	Adjacent year weights[b]	1954 through 1960 weights[c]	End period weights[d]
1937–50	24.3	22.7	—	18.7
1950–60	61.0	—	18.7	15.1
1937–60[e]	100.1	—	—	36.6
1950–4	6.1	—	2.2	2.3
1954–5	9.3	5.7	0.7	—
1955–6	8.1	2.9	2.2	—
1956–7	12.4	4.8	4.1	—
1957–8	16.9	3.4	4.4	—
1958–9	4.3	1.4	2.3	—
1959–60	0.6	0.3	2.0	—
1954–60	51.7	20.0	16.1	12.4

[a] 1937 weights for the 1937–50 comparison and 1950(1) weights for all the subsequent comparisons. For example the 1937–50 figure is arrived at by multiplying the change in the average specifications given in table 6.6, by the 1937 weights given in table 6.7 and adding them together $(8.7 \times 1.3 + 3.9 \times 3.43 - 0.1 \times 4.0 = 24.3)$.
[b] Weights from table 6.4, i.e. the 1954–5 comparison uses average 1954–5 weights, and so on. The figure for 1954–60 is the product of all the paired year comparisons.
[c] Weights from table 6.4.
[d] 1950(2) weights for the 1937–50 comparisons and 1960 weights for the 1950–60 and 1954–60 comparisons.
[e] Derived by adding 100 each to the first two rows, multiplying, and subtracting 100.

results. Since the CPI is a Laspeyres based fixed weight index, with the latest set of weights being based on the 1950 Consumer Expenditure Survey, the "beginning period" weighted quality index is the most appropriate deflator for it. From a theoretical point of view, the chain-link index with its frequently changing weights is probably the best single measure of quality change.

Before proceeding to "deflate" the CPI by our quality indexes we have to convince ourselves that it is legitimate to do so. Since our indexes were derived from list prices, we have first to compare the CPI to an adjusted list price index for the same makes and models. Such a comparison is presented in the first two columns of table 6.9. It is apparent that the list prices and the CPI moved fairly closely together until 1954. Since 1954 the CPI has risen much less than the list prices of comparable cars (or the comparable WPI index, see table 6.5). It is not exactly clear how and why this happened, and the problem is explored in greater detail in the appendix. In part this may be due to the BLS beginning to ask for discounts in 1954; in part to absolute or relative declines in transportation costs and the cost of various attachments which were not included in the list prices. Be this as it may, unless the recent divergence between list prices and the CPI index is somehow associated with one or the other of our quality dimensions, these indexes are still appropriate deflators for the CPI. They would be inappropriate if either relative discounting were associated with some of the quality dimensions, e.g. higher horsepower cars being

Table 6.9 The "low priced three" (sixes to the 1956 model year, V-8s thereafter): percentage changes in price – list prices, the CPI and the CPI adjusted for quality change

			CPI adjusted for quality change using[c]			
Years	List prices un-adjusted[a]	CPI unadjusted[b]	Beginning period weights	Adjacent year weights	1954 through 1960 weights	End period weights
1937–50	116.0	101.3	61.2	64.1	—	69.2
1950–60	58.5	31.3	− 18.4	—	10.6	14.1
1937–60	242.4	161.3	30.6	—	—	91.3
1950–4	18.0	18.0	11.2	—	15.5	15.3
1954–5	2.5	− 1.7	− 10.0	− 8.0	− 2.4	—
1955–6	5.4	− 0.9	− 8.3	− 3.7	− 3.0	—
1956–7	9.9	5.1	− 6.5	0.3	1.0	—
1957–8	6.7	4.2	− 10.9	0.8	− 0.2	—
1959–60	0.2	0.1	− 0.2	− 0.2	− 1.9	—
1954–60	34.4	11.3	− 26.6	− 7.8	− 4.1	− 1.0

[a] Computed from table 6.7.
[b] From BLS Bulletin no. 1256 and various CPI releases. For 1937 and 1950 as of March of the same year; for 1954 as of January 1954; for subsequent years as of November of the preceding year.
[c] Computed by dividing the figures in the second column by the appropriate entry from table 6.8 (adding first 100 to each and subtracting 100 from the result).

Table 6.10 A comparison of price–quality regression coefficients of new and used cars[a]

Coefficients of	Model year									
	1960		1959		1957		1954			
									Used in	
	New	Used in 1960	New	Used in 1960	New	Used in 1958	New	1955	1956	1957
H	0.052	0.040	0.058	0.029	0.051	0.042	0.149	0.067	0.057	0.052
	(0.009)	(0.011)	(0.011)	(0.015)	(0.013)	(0.015)	(0.038)	(0.038)	(0.038)	(0.050)
W	0.063	0.069	0.090	0.112	0.059	0.053	0.084	0.126	0.122	0.118
	(0.009)	(0.011)	(0.013)	(0.017)	(0.017)	(0.020)	(0.032)	(0.032)	(0.032)	(0.042)
L	—	—	—	—	0.017	0.024	—	—	—	—
					(0.006)	(0.007)				
V	-0.017	-0.011	-0.035	-0.030	-0.011	-0.011	-0.022	0.024	0.035	0.049
	(0.010)	(0.021)	(0.015)	(0.020)	(0.010)	(0.012)	(0.015)	(0.015)	(0.015)	(0.020)
T	0.026	0.039	0.011	0.028	0.012	0.047	—	—	—	—
	(0.007)	(0.008)	(0.008)	(0.011)	(0.005)	(0.006)				
A	—	—	—	—	0.050	0.026	—	—	—	—
					(0.011)	(0.013)				
P	0.102	0.094	0.104	0.077	0.034	0.001	0.037	0.091	0.123	0.145
	(0.011)	(0.013)	(0.014)	(0.018)	(0.013)	(0.015)	(0.030)	(0.029)	(0.030)	(0.038)
B	—	—	—	—	0.069	0.095	—	—	—	—
					(0.011)	(0.014)				
R^2	0.950	0.919	0.934	0.872	0.966	0.948	0.828	0.854	0.854	0.793

[a] The results differ from those presented in table 6.3 in two ways. First, they exclude variables which turned out to be insignificant in the particular years such as length or "automatic transmission." Second, they are presented as computed, using logarithms to the base 10. To make them comparable to the results in tables 6.3 and 6.4, all the coefficients and standard errors should be divided by 0.4343 ($\log_{10}e$).

The used prices in 1960 are taken from the July issue of *Used Car Guide*. For all other years they are taken from the February issues of the *Guide*.

discounted disproportionately, or if the CPI had, in collecting its prices, linked out the particular horsepower, weight, and length increases we have used in constructing the quality indexes. Since we have no reason to believe that either is true, deflation of the CPI by these indexes appears to be warranted.

The results of deflating the changes in the CPI by the appropriate entries from table 6.8 are presented in table 6.9. For the 1937–50 period about a third of the price rise can be attributed to quality change no matter which set of weights we use.[20] In the 1950–4 period the role of quality change appears to have been minor, unless we weight it by 1950 weights. All weights point to the conclusion that "real" automobile prices fell rather than rose during 1954–60.[21] Using beginning period (1950) weights, the fall was around one-quarter. Using end period (1960) weights, the fall was very small, indicating roughly no change in "real" automobile prices. For the 1937–60 period as a whole, quality change accounted for about one-third (using end period weights) to about three-fourths (using beginning period weights) of the recorded price change in the CPI. These results are quite tentative and subject to various limitations to be discussed below. Nevertheless, if we realize that we have only scratched the surface as far as quality adjustments are concerned, considering only a very limited and narrow class of "qualities," the conclusion is inescapable that the lack of adequate quality adjustments has resulted in a very serious upward bias in the official automobile price indexes.[22]

6   Additional Tests, Limitations, and Conclusions

A   The Evidence of the Used Car Market

One of the problems associated with the use of list prices in this study is the extent to which they may just represent pricing mistakes by manufacturers at some point in time. A manufacturer may overprice or underprice a particular innovation, and there is nothing in our method that would catch it. Of course, if we had sales data broken down by makes, models, and attachments, an appropriate weighting of the original data would go a long way toward the solution of this problem. In the meantime, however, we may want to investigate the prices of these cars. The prices of used cars are not tied any more to the manufacturers' list prices and are set, presumably, more directly by the "market."

Since a used and a new car are not exactly the same commodity, we should not expect a perfect agreement between estimates of "quality prices" from these two different sets of data. In particular, as cars age, one might expect that some of the "qualities" depreciate much faster than others. Nevertheless, relatively "new" used cars should be reasonably good substitutes for new cars and their prices should reflect similar quality differentials.

Table 6.10 compares the results of using used prices instead of list prices for selected cross-sections. For the 1960 models the used prices are for approximately 6-month-old cars. For the other cross-sections they are for a little over 1-year-old cars. As can be seen by comparing the coefficients of the "new" and "used" regressions respectively, the difference between the two are relatively minor and usually well within the range of their respective standard errors. Thus, the quality weights that could be derived from the regressions using the prices of 1-year-old cars are roughly similar to those that we obtain using new car (list) prices.[23]

B Reliability

One of the advantages of the approach outlined above is the possibility of computing confidence intervals for the quality indexes or the quality adjusted price indexes. For each new combination of specifications we can compute not only its predicted price in some base period but also the "prediction interval," the probable range of the error of prediction based on the goodness of the fit of the equation and the distance of the new specifications from their mean values. Since this computation is somewhat laborious and since time was limited, no such calculations were actually performed.[24] Some insight, however, into the possible magnitude of such an interval can be obtained by examining the standard error of regression (the standard deviation of the residuals from the equation). The average error of "prediction" for any *one* particular car is quite large. It varies from about 5 percent in 1957 to about 8 percent in 1950 for single year cross-sections, from about 6 percent for the 1956–7 combined regression to about 9 percent in the 1958–9 regression, and is about 8 percent for the overall 1954 through 1960 regression. This figure is applicable if we want to predict the price of a particular make and model. We are interested, however, in predicting the *average* price for the three "low-priced" makes. In our case this is an average of eight models and the error of predicting an average goes down, approximately and under suitable conditions, as the square root of the number of items. Thus, the average residual for this group of cars as a whole is only about a third ($\sqrt{8} = 2.8$) of the individual errors quoted above. It would be even smaller if we had computed a weighted regression, since the three "low-priced" makes would probably account for about 60 percent of the weights.

C Shifting Supply Conditions and Tastes

To the extent that shifting supply conditions or changing tastes change the relative "price" of a particular quality we are back to the classical index number problem of changing weights. Not much can be done about this in practice except to shorten the timespan of comparison, compute base and end period weighted indexes, and hope that they are not too far apart. In our case, the more striking examples of such changes are the rapid decline in the "price" of horsepower, with the introduction of the V-8s and the fall in the "price" of length.

The CPI in switching to the pricing of V-8s in 1956 linked them to the previously priced six-cylinder engine cars without allowing the index to rise or fall as the result of this substitution, and we have followed suit in the calculation of our indexes. If we use contemporary weights (e.g. for 1955–6) this is about right. Our estimates of the horsepower coefficients are based on a sample that includes V-8s and thus it is not surprising that the increase in horsepower weighted by its coefficient comes close to the difference in price.[25] For the "low-priced three," if we use the horsepower and weight difference between the sizes and the V-8s in 1956 and weight them with 1955–6 quality prices, we predict that comparable V-8s should cost about 6 percent more. Actually, they were only 5.5 percent more expensive. Using the 1959 horsepower differences between these cars and 1959–60 weights we predict a 9 percent price differential against the observed 5 percent.[26] This agrees with our finding for the sample as a whole that the V-8s were about 3 or more percent cheaper than would be predicted from an extrapolation of the price–horsepower relationship for six-cylinder engine cars.

The introduction of the V-8s represented a decline of a few percentage points in the "real" price of cars that is not caught by the linking procedure. But this is only an "economies of scale" effect along a given relationship, and does not represent the total possible contribution of the V-8 engine. In fact, the appearance of the V-8 on the market in substantial quantities brought the whole level of horsepower "prices" down. Thus, if we were to value the V-8 at 1950 horsepower "prices," when there were only a few V-8 engine cars in the sample, we would estimate it to be a 15 percent "more car" (to have a 15 percent higher "quality" index) as against only a 5 percent increase in its price. The very fact of the rapid rise of the V-8 to market dominance would indicate that it was somewhat "underpriced" relative to the sixties. This is also supported by the used car price–quality regressions. In a large number of cases, the negative coefficient (discount) of the V-8 variable observed for new car prices turns into a positive coefficient (premium) once these cars get to the used car market.

Another problem is created by our use of proxy variables, of dimensions that may not be desirable per se, but which are correlated with other, more difficult to measure, but basically more desired dimensions. Weight, for example, is unlikely to be desired very much for its own sake. Rather, it is a proxy for "size." The relationship between price and weight may involve, however, other things besides "size," and the relationship between weight and the underlying desired characteristics may change over time. Our weight coefficients are derived on the basis of the difference in price between the cheap and the expensive cars, but the "large" cars may be expensive for reasons other than just "size." We have tried to control this by introducing a variety of dummy variables such as power steering and automatic transmissions which are standard equipment on the more expensive cars.[27] This prevents these cars from exerting an undue influence on the price–weight relationship for the sample as a whole. Alternatively, we could have computed separate estimates for different groups of cars; for example, the "low," "medium," and "high" priced cars. Still another approach would have been to estimate "comparable" prices for different models by subtracting from the more expensive cars the estimated "value" of most of the attachments and features not available on the lower priced cars. Since many of these are listed as "extras" for other cars, one could probably go some distance in "standardizing" prices.

The basic method would of course be seriously compromised if the relationship between any one of the measured dimensions and the more basic "real" qualities were to change from one period to the next. For example, suppose all cars were, after a given date, made of an aluminium alloy which halved their weight, but absolute and relative prices did not change. This change in weight would increase the apparent price of weight and reduce its level per car while in fact nothing may have happened except for a change in units of measurement. If we did not know what had happened, we would have mistaken this weight change for a quality change. But in practice this should not present an insuperable problem. We usually know enough about what is happening in a particular market and to a particular product to be able to make some adjustments for it. More important, such changes are unlikely to be sudden and all inclusive. Aluminium cars will probably sell for several years together with more "old-fashioned" cars, and we shall be able, by the use of dummy variables or other techniques, to detect the difference between these cars and build it into our equations.[28]

D Suggestions for Further Research and Conclusions

It is obvious that our investigation is only illustrative of a promising line of attack on the quality change problem. There are more than just a half-dozen dimensions to an automobile and they may not interact in any simple linear fashion.[29] Further work along these lines would include the intro-duction and testing of a number of additional "qualities;" an examination of the residuals from the various non-linearities; use of weighted regressions, where different cars would be weighted according to their importance in the market; division of the sample into separate subgroups to test hypotheses about the linearity of the various price–quality relationships; use of actual transaction prices instead of list prices in the analysis; and the extension of this type of analysis to a variety of other commodities such as trucks, refrigerators, and cameras.[30]

Continuous studies of the present type by the price collecting agencies should prove of great value. First, they would eventually perfect the method enough so that it could be used routinely in the computation of the official indexes. Second, they would provide them with much more information on the various dimensions of a commodity, allow the use of a more sophisticated linking procedure, and isolate the qualities or dimensions which appear to be most important. Third, the availability of such information is also likely to lead to a more useful specification of commodities for price collection purposes. And finally, such studies, if done for a wide enough range of commodities, could provide an estimate of the probable upward drift of the price indexes due to their inability to control adequately for many of the constantly occurring quality changes.

Appendix: The Official Automobile Price Indexes

There are three official automobile price indexes: The "new automobiles" component of the CPI, the "passenger cars" component of the WPI, and the automobile component of the Prices Paid Index of the US Department of Agriculture. The CPI new automobile price index is a retail price index for Chevrolet, Ford, and Plymouth sedans with V-8 engines (sixes before 1956 except Ford), automatic transmissions (since 1956), and other minor items such as extra trim, radio and heater, gasoline and antifreeze. The WPI is a wholesale (manufacturer to dealer) index of car prices, presumably covering all or most makes and models weighted by some base period production. The Agricultural Marketing Service index, which is not published separately, is based on a mail survey of prices paid by farmers for six-cylinder Chevrolets, Fords, and Plymouths, and for V-8 Chevrolets, Fords, Plymouths, and Buicks. Average prices paid for six-cylinder cars and for V-8s are published separately each quarter in *Agricultural Prices*. Again, it is not clear how the different makes and models are weighted, and what weights are used in aggregating state data into national averages.

Of the three indexes, the AMS stands alone in not specifying exactly what attachments are included in the model being priced. The CPI explicitly deals with the items that are being priced with the car and adjusts for changes in "extras." The WPI presumably prices the "standard equipment" car and adjusts for major changes in what is being considered as standard. The AMS, however, has collected prices paid by farmers for specified models and makes "together with the usual equipment bought by farmers." It has tried to control for some aspects of size by comparing similar "price lines" of each make in different years, and has priced V-8s and sixes separately, but its failure to specify other attachments allows the index to drift upward as the result of farmers shifting to the purchase of more heavily equipped cars, cars that include radios and heaters, automatic transmissions, power steering and brakes, and other extras. That this drift is serious is indicated by the fact that the difference between the average six- and eight-cylinder car priced by the AMS which stood at $200 in 1947–9 increased to $660 by November 1959. Since the price of V-8s and Buicks probably did not increase as much, percentage-wise, as the price of the "low-priced-three" sixes, most of this increase must be due to the increasing number of attachments bought with the more expensive cars.

Percentage changes in these indexes are tabulated in table 6A.1 for selected periods and are compared to changes in a list price index of the "low-priced-three" makes. Note that, in almost all of the comparisons, the AMS prices rise more than all the other indexes, including the list price one. This is another indication of the upward drift in the AMS index as the result of its relatively loose specification policy. Looking at the other indexes, we note that the movements to 1954 are roughly similar, with the WPI rising somewhat less than the CPI and the list price

Table 6A.1 A comparison of official indexes and list prices for US cars: percentage change, selected periods

Period	WPI^a	CPI^b	AMS^c	List prices Unadjusted^d	List prices Adjusted for minor equipment changes^e
1937–50	83.0	101.0	129.0	116.0	—
1947–9 to January 1954	20.6	29.7	32.7	18.0^f	16.9
January 1954 to November 1954	2.7	−1.7	2.2^g	2.5	—
November 1954 to November 1955	4.1	−0.9	3.8^h	5.4	—
November 1955 to November 1956	5.7	5.1	5.4	9.9	—
November 1956 to November 1957	0.6	4.2	3.8	6.7	—
November 1957 to November 1958	5.1	4.2	11.8	6.0	—
November 1958 to November 1959	0.1	0.1	3.0	2.0	—
January 1954 to November 1959	19.7	11.3	33.6	34.4	—
1947–9 to November 1959 (1960) models	44.3	44.3	68.4	58.6^f	—
January 1954 to November 1957	13.8	6.7	29.8	26.7	21.6

^{a} See table 6.6.
^{b} See table 6.7.
^{c} Sixes before November 1955, V-8s thereafter; the V-8s include Buick Special in addition to the "low-priced three." From various issues of *Agricultural Prices*. The 1937–50 comparison is based on an unpublished index used to deflate farmers' expenditures on automobiles.
^{d} From table 6.7: The "low-priced three." The model year is assumed to start in November of the previous calendar year.
^{e} Adjusting list prices for differences in minor equipment items included in the price, such as directional signals and electric clocks, based on data from *Administered Prices*, pp. 3548–9, 3622–6, and 3730–3. Also, including automatic transmissions in the list prices as of 1956.
^{f} Beginning with 1950 models.
^{g} January 1954 to January 1955.
^{h} January 1955 to November 1955.

index. The main divergence between these indexes comes in the 1954–8 model year period, with the CPI rising substantially less than either the WPI or list prices. It is not too surprising that the WPI rose less than the list price index for the lower priced makes. About half of its weight is given to medium and higher priced cars which have risen less percentage-wise than the lower priced makes.[31] The sharp divergence between the CPI and list prices during 1954–8 is, however, surprising and requires explanation.

A reconciliation of the two series is seriously hampered by the lack of a detailed description of how the CPI is actually computed. There is no published information on whether the index is a ratio of the average price for these makes or an average of their price relatives; what weights, if any, are used in averaging the price data for different makes and models; which models of a given make are being priced in a particular year and to what models they are being compared in the previous year; and what quality changes were "linked-in" or "out," and when and how.[32] The list price index was constructed in such a way as to approximate the CPI closely.[33] It differs from the CPI in that it does not adjust for changes in minor equipment items, it does not include transportation costs, state and local taxes, and minor accessories sold with the car, and it does not allow for changes in the discount from list prices.

It is possible to adjust the list prices for some of the minor equipment changes using more detailed price data presented in the Kefauver Hearings (*Administered Prices*). This will reduce the rise in list prices somewhat (see the last column of table 6.10), but it still leaves a very substantial difference between the CPI and list prices (or the WPI) unexplained. Some of this difference could be due to the inclusion in the CPI of various "trim" items, transportation costs, and taxes, which may have remained constant or risen less than the price of the "basic" (stripped) car. Still it could not explain it all – the actual difference is too large for that.

Another source of this difference could lie in the fact that the CPI started in 1954 to collect data on discounts offered by retailers. But even this is unlikely to explain much of the difference between the two series. Assume that before 1954 the CPI did not include discounts, that it does so since 1954, and that no linking was done to account for this. We know that list prices went up by about a third during 1954–60, that the spread between the price to the dealer and the list price remained at approximately the same percentage level (24 percent) throughout the period, and that during the same period the CPI rose only 11 percent. Consider the following arithmetic example: A representative car cost $1,350 wholesale in 1954, listed for $1,800 at retail, with the dealer's margin being $450. No discount was given in 1954. The same type of car lists for $2,400 in 1960 (a rise of 33 percent) and costs the dealer $1,800. If the actual retail price had risen only 11 percent, to $1,998, the dealer's return would have dropped from $450 to $198 per car, or from a 25 to an 8 percent margin. This seems to be too big a drop in the return to dealers in a period of rising prices to be plausible.

An additional explanation for this divergence has been suggested by John M. Blair, who was also puzzled by it (*Administered Prices*, pp. 4000–2). He has argued that since the BLS agent first asks for the list price and then separately for the magnitude of the discount, the difference between the two may not equal the actual price charged. It is said to have been common practice during 1955–8 for dealers to "pack the price," i.e. to quote a discount that was not calculated from the list price but from some higher figure. Subtracting this "unrealistic" discount figure from the list price would lead to a downward bias in the estimated price actually paid by consumers. But this should be a transitory phenomenon. Once eliminated, as it apparently has been in the most recent years, it should have led to a comparably higher rise in the CPI. This has not happened.

The final possibility is that the CPI has been much more thorough in its quality adjustments than is reflected in the published literature. That is, it could have been argued in some year, for example, that "this year's cheapest Ford model is equivalent in size, trim, and horsepower to last year's medium-priced Ford." The only detailed description of automobile price in the CPI suggests this possibility by saying:

> the automobile retail price indexes have been designed to measure solely the trend of prices paid by city workers for automobiles of as nearly fixed quality as possible. . . . Therefore, prices are collected for automobiles which are regarded as most nearly equivalent to the cars priced in the preceding year (Mack, 1955).

But then the next sentence reduces the probability of this by stating:

> Equivalent quality of new cars has been assured to a great extent by specifying as a basis of pricing the same make and body style, the *same or equivalent price series*, and the same numbers of cylinders as the car which was priced in the preceding year. (Emphasis supplied.)

Thus it appears quite unlikely that the CPI has linked out the type of horsepower, weight, and length changes used in constructing our quality indexes. If this is true, then it is quite probable that, for some unknown reason, the CPI underestimated the rise in new car prices (given its own definition) between 1954 and 1958.[34]

Notes

1 *Price Statistics*, 1961, p. 35.
2 As far as I know, this procedure was first suggested by A. T. Court (1939). A more recent exposition is given by R. Stone (1956).
3 The reader is referred to the literature on this problem, and in particular to Hofsten (1952) and to Stone (1956); see also Adelman and Griliches (1961).
4 It can always be made into a tautology by specifying enough factors or qualities.
5 If natural logarithms are used, an "a" coefficient will provide an estimate of the percentage increase in price due to a one-unit change in the particular quality, holding the level of the other qualities constant.
6 This was the procedure followed by Court (1939).
7 The designation g is borrowed from Hofsten (1952).
8 Compare this with Adelman (1960), where the quality change index is defined additively rather than multiplicatively. Ideally the varietal prices should be

deflated individually before they are aggregated into an overall price index. Only for geometrically weighted indexes will the ratio of the two equal the "true" index exactly.

9 The 1937 price and specification data for new 1937 automobile models are taken from the Sept.–Oct. 1937 issue of the *Red Book*. The 1950 model data are from the Nov. 15, 1956, issue of the *Red Book*. For 1954 through 1960 the data are taken from various issues of *Used Car Guide*. For 1955 through 1958 the data are from the February issue of the corresponding year. For 1954 models the figures are taken from the July 1959 issue; for 1959 models from the January 1959 issue; and for 1960 models from the December 1959 issue. Data on power brakes come from various issues of *Ward's Automotive Reports*.

10 Factory-advertised delivered price includes only standard equipment, federal excise tax, and dealer handling and preparation charges. Transportation and state or local taxes are not included.

11 The possible consequences of this omission are explored briefly in the appendix to this chapter.

12 There was very little overlap in horsepower between the sixes and the V-8s in the sample. What the coefficient measures, actually, is the fact that higher horsepower levels could be achieved at a price–horsepower relationship for six-cylinder engine cars. For more on this, see the text below.

13 This is more related to the "luxuriousness" of these models than to the presence of power steering *per se*.

14 This does not mean, necessarily, that we are able to predict the price for any one particular car very well. The average standard error of regression for these equations is around 8 percent.

15 The power brakes variable is not included in the years when all (or almost all) the models on which power steering is "standard equipment" also have power brakes included in their price. Note that in those years the estimated coefficient of power steering alone equals approximately the sum of the two coefficients in the other years.

16 Different makes are weighted, in a sense, by the number of models of each make included in the sample. This mitigates the problem somewhat since the more popular makes are likely to have a larger number of models on the market, but does not solve it.

17 The USDA index also includes one Buick model. The CPI will probably introduce "compact" cars into its calculations in the fall of 1960.

18 It is not clear which models within a make are being priced; what weights, if any, are attached to each model and make; whether the index averages price relatives for each model or make, or takes the relative of the average price of these models; and so forth. See also the Appendix for additional discussion of the CPI.

19 Alternatives to this linking procedure are discussed below.

20 Loosely speaking. Since the quality index is defined multiplicatively, there is no unique way of decomposing a given price change into additive "quality" and "pure" price change components. With 1937 = 100, the CPI stood at 201 in 1950, the beginning period weighted quality index at 124, and the "adjusted" CPI at 161. $1.25 \times 1.61 \approx 2.01$. The "role" of quality in change could be measured as:

$$\frac{24}{101}, \quad \text{or} \quad \frac{101-61}{101} = \frac{40}{101}, \quad \text{or as} \quad \frac{\frac{1}{2}(24+40)}{101}.$$

The last procedure leads to the "one-third" statement in the text. On this problem see the note by Levine (1960).

21 If we had deflated the list price index instead of the CPI, we would have shown some price rise for the 1954–60 period with all but the 1950 set of weights.

22 And in the CPI as a whole. Adjusting the overall CPI for quality change in only *one* commodity – automobiles (applying 1950 quality weights to the 1950–60 changes in specifications and using the 1950 weight of automobiles in the index – 3.7 percent) – results in a reduction of the index from 125.6 (in November 1959) to 123.7 (1947–9 = 100). Over 7 percent of the increase in the CPI since 1947–9 may be due just to the changing quality of one commodity.

23 There are some minor differences that foreshadow the results that would be found if we were to use prices of 3-, 4-, 5-, and 6-year-old cars in our analysis. The relative price of horsepower falls somewhat with age, while the coefficient of weight remains stable or rises somewhat. The discount on V-8s turns to a premium with age. The premium on "hardtops" rises. The "automatic transmission" premium depreciates very rapidly. In general the results for 5- and 6-year-old used cars look quite different from those reported here. They will be described elsewhere.

24 But they present no problem, in principle. See Mood (1950) and Chow (1960).

25 A V-8 engine has usually 50 more horsepower units than a comparable "six" and costs about $100 more. Since our horsepower coefficient during this period is around 1 percent per 10 horsepower units, we would predict a 5 percent higher price. But 5 percent on a $2,000 car is $100.

26 This brings out an additional problem associated with the linking procedure. The additional cost of a V-8 engine has remained approximately constant at $100 while the absolute price of cars increased. Thus a price index based on six-cylinder engine cars would rise somewhat faster than the V-8 based index. The inclusion of attachments in the pricing procedure may lead to an underestimate of the price rise of the attachmentless car if, as appears to have happened recently, attachment prices do not rise as much as the price of the "basic" unit or at all. To the extent that a substantial fraction of cars is bought without them, this could bias the index.

27 This is one reason why these estimated coefficients should not be used directly in estimating the "value" of a particular attachment. We know that power steering and brakes come to about $130, which is far from the 20 percent or so increase in price indicated by their coefficients. The main purpose of these variables is not to estimate the price of these attachments, which we know, but to reduce the possible bias in our slope estimates for the numerical qualities by allowing different groups of cars to differ in the position or intercept of these slopes.

28 The next few years will provide a good test of this assertion. One of the 1961 model year cars is already an aluminium block engine.

29 For evidence on how complicated a machine an automobile really is and for the many changes that actually occurred in it since the 1930s, see the history

of the Plymouth and its specifications summarized in *Administered Prices*, pp. 3655–65 and 3734–49.

30 A study of wheel tractor prices along these same lines is in progress.

31 Between 1954 and 1958 the prices of Buicks, Pontiacs, Mercurys, and Dodges advanced relatively less (about 15 percent) than the prices of Chevrolets, Fords, and Plymouths (which rose 23 percent). Compare also with table 6.5.

32 Many of these problems could have been settled by a consultation with BLS personnel and an examination of their records. Unfortunately, previous commitments, deadlines, and distance prevented this from being accomplished in time.

33 It differs from the CPI in that before 1956 it prices only six-cylinder engine cars (except in 1937) whereas the CPI priced eight-cylinder Fords throughout, and it does not include automatic transmissions in its price, which the CPI has done since 1956.

34 See Triplett (1971) for further discussion of this point. Much of the difference *was* apparently due to the use of discounted prices by the CPI without any linking out of the shift from list to transaction prices. [Note added in 1971.]

References

Adelman, I., 1960, "On an index of quality change". Paper given at the August 1960 meeting of the American Statistical Association, Stanford, California.

—— and Z. Griliches, 1961, "On an index of quality change," *Journal of the American Statistical Association*, 56: 535–48.

Chow, G. C., 1960, "Tests of equality between sets of coefficients in two linear regressions," *Econometrica*, 28: 591–605.

Court, A. T., 1939, "Hedonic price indexes with automotive examples," in *The Dynamics of Automobile Demand*, pp. 99–117. New York: General Motors Corporation.

Dhrymes, P. J., 1971, "Price and quality changes in consumer capital goods: an empirical study," in Z. Griliches (ed.), *Price Indexes and Quality Change*. Cambridge, MA: Harvard University Press.

Hofsten, E. von, 1952, *Price Indexes and Quality Changes*. Stockholm: Bokforlaget Forum AB.

Kravis, I. B. and R. E. Lipsey, 1971, "International price comparisons by regression methods," in Z. Griliches (ed.), *Price Indexes and Quality Change*. Cambridge, MA: Harvard University Press.

Levine, H. S., 1960, "A Small Problem in the Analysis of Growth," *Review of Economics and Statistics*, 42: 225–8.

Mack, L. J., 1955, "Automobile prices in the Consumer Price Index," *Monthly Labor Review*, 72: 5.

Mood, A., 1950, *Introduction to the Theory of Statistics*. New York: McGraw-Hill.

National Automobile Dealers Association, *Official Used Car Guide*. Washington. Various issues, 1954–60.

Red Book of Official Used Car Valuations. Chicago: National Market Reports, Inc. Various issues.

Stone, R., 1956, *Quantity and Price Indexes in National Accounts*. Paris: Organization for European Economic Co-operation.

Triplett, J. E., 1971, "Quality bias in price indexes and new methods of quality measurement," in Z. Griliches (ed.), *Price Indexes and Quality Change*. Cambridge, MA: Harvard University Press, 1971.

U.S. Department of Agriculture, *Agricultural Prices*, Washington, D.C.: Government Printing Office, various issues.

U.S. Senate, Subcommittee on Antitrust and Monopoly, 1958, *Administered Prices*, Hearings, 85th Congress, 2nd Session, Part 7: *Automobiles* (Appendix), Washington: Government Printing Office.

Ward's Automotive Reports, Detroit: Powers and Company, various issues.

7
Hedonic Price Indexes Revisited*

A decade has passed since my first attempt to revive the "hedonic" multiple regression approach to the construction of price indexes.[1] While one cannot claim that it has taken the profession by storm (it is too imperfect and difficult a tool for that), there has, in the meantime, developed a reasonably large literature on the subject with a number of rather interesting applications. It has even infiltrated into some of the official government departments responsible for the work on price statistics.[2]

Most of this work has been empirical. Automobile prices attracted most of the attention with additional work reported by Fisher, Griliches, and Kaysen (1962), Griliches (1964), Cagan (1965), Cramer (1966), Triplett (1966), and Dhrymes (1967). Tractor prices were analyzed by Fettig (1963), electric apparatus by Dean and DePodwin (1961), house prices by Bailey, Muth, and Nourse (1963), Brown (1964), Musgrave (1969), and Yoshihara, Furuya, and Suzuki (1970), diesel engines by Kravis and Lipsey (1969), refrigerators by Dhrymes (1967), and washing machines and carpets by Gavett (1967a). There are also several studies dealing with other topics which could be interpreted as implying a "hedonic" price index: Barzel (1964) on steam power generators, Knight and Barr (1966) and Chow (1967) on computers, and Hanoch (1965) on people. Unfortunately, the theoretical base of such studies has not expanded greatly since the Adelman and Griliches article (1961). The idea of a commodity as a bundle of characteristics (dimensions or qualities) has been developed further by Lancaster (1966) and Muth (1966), but it has not produced any new implications for the construction of price indexes. More important has been the development of the concept of various types of technical change (embodied, factor-augmenting, etc.) in the production function and growth literature. The implications of this literature for the construction of price indexes have been derived and extended in an important paper by Fisher and Shell (1967).

*From *Price Indexes and Quality Change: Studies in New Methods of Measurement*. Cambridge, MA: Harvard University Press, 1971.

A similar approach was used by Hall (1968) to derive measures of quality change from secondhand market data.

In this brief paper I shall first comment on regression analyses of prices as they are practiced today, then explore in some detail the promise and difficulties associated with the use of secondhand market prices for the measurement of quality change, make some observations on the current state of practice in the official price indexes, and conclude with a comment on difficulties with the concept of "quality change" itself.

The "hedonic," or, using a less value-loaded word, characteristics approach to the construction of price indexes is based on the empirical hypothesis (or research strategy) which asserts that the multitude of models and varieties of a particular commodity can be comprehended in terms of a much smaller number of characteristics or basic attributes of a commodity such as "size," "power," "trim," and "accessories," and that viewing the problem this way will reduce greatly the magnitude of the pure new commodity or "technical change" problem, since most (though not all) new "models" of commodities may be viewed as a new combination of "old" characteristics. In its parametric version, it asserts the existence of a "reasonably well-fitting" relation between the prices of different models and the level of their various but not too numerous characteristics. If one views the commodity as an *aggregate* of individual components or characteristics, there is no reason to expect that this relationship between the overall price of the bundle and the level or quantity of the various characteristics will remain constant. Both the relative and the absolute prices of the various components may change.

In practice the following questions arise:

1 What are the relevant characteristics?
2 What is the form of the relationship between prices and characteristics?
3 How does one estimate the "pure" price change from such data?

There isn't much that one can say in general about the first question – it is very much an empirical matter – except to note that most of the studies do quite well with some combination of "size" and "power" variables. I would like to warn, however, against the use of variables which are not direct characteristics of the commodity (or a transformation of them) but an outcome of the market experiment. I have in mind here such things as the use by Brown of the purchaser's income in explaining house prices, and the use by Dhrymes (in the earlier version of his work) of total quantities produced to explain relative automobile prices. The latter variable is an *outcome* of the encounter of consumers with commodities of different qualities and price. The characteristics theory would predict that models which have more "quality" per dollar will sell better, but this is a characteristic of the market, not of the commodity. I'll admit that there is an identification problem here, but I don't believe that it is relevant for the derivation of characteristic prices to be used in the construction of a "purer" price index.[3]

The Dhrymes (1971) and Triplett (1971) papers do remind us, however, of a problem overlooked in most of the previous studies, including my own. A characteristic and its price are important only to the extent that they capture some relevant fraction of the market. Most of the analyses have used unweighted data on models, specifications, and prices. But at any point of time some manufacturers may offer models with characteristics in undesirable combinations and at "unrealistic" (from the consumer's point of view) relative prices. Such models will not sell very well and hence should also not be allowed to influence our analyses greatly. There is no good argument except simplicity for the one-vote-per-model approach to regression analysis. It is true that market shares by detailed characteristics are not easy to come by, but some scattered data are available and more of them should be used.

The form of the relationship is again an empirical matter. Most of the investigators settle after some experimentation for a semi-logarithmic relationship between prices and characteristics, implying a rising supply price per characteristic unit.[4] Such experimentation, however, is usually conducted without the help of a relevant statistical framework. I would like therefore to draw attention to an article by Box and Cox (1964) which does provide the appropriate methodology for choosing between different functional forms.

There are several ways of constructing a "pure" price index from such data. The particular way chosen will depend both on the kind of price index one wants and on the type of data that one has. The first, and the one most directly in the usual price index spirit, is to use the regression equations *only* to estimate the *"prices"* of the relevant characteristics, using these in turn in the construction of a more detailed quantity-of-characteristics index and the associated price-of-characteristics index. To evade the usual Laspeyres and Paasche problem, let me proceed for a while via Divisia indexes (see Jorgenson and Griliches (1967) for an application of such indexes in a different context, and Richter (1966) for a proof of their optimality). Also, since data on characteristics are more readily and frequently available, let us start from a change in the quantity-of-characteristics index per particular model, defined as

$$\frac{dQ_i}{Q_i} = \sum_j w_j \frac{dq_j}{q_j},$$

where Q_i is the quality index for the ith model of the product, q_j is the level of the jth characteristic, and w_j is the value share of that characteristic in the total price of the (aggregate) commodity, $w_j + p_j q_j$ (and hence also $\sum w_j dq_j/q_j = \sum p_j dq_j$). The pure price change is then estimated simply as the rate of change of observed price minus the rate of the change of "quality" per commodity unit:

$$\frac{d\pi_i}{\pi_i} \equiv \frac{dP_i}{P_i} - \frac{dQ_i}{Q_i}.$$

The total pure price index for the whole class of such commodities is $d\pi/\pi = \sum v_i(d\pi_i/\pi_i)$, where v_i is the value share of the ith model in the aggregate consumption or sales of this class ($v_i = p_i n_i$, where n_i is the number of units sold or consumed). This approach calls for relatively recent and often changing "price" weights. Since such statistics come to us in discrete intervals, we are also faced with the usual Laspeyres–Paasche problem. The oftener we can change such weights, the less of a problem it will be. In practice, while one may want to use the most recent cross-section to derive the relevant price weights, such estimates may fluctuate too much for comfort as the result of multi-collinearity and sampling fluctuations. They should be smoothed in some way, either by choosing $w_i = \frac{1}{2}[w_i(t) + w_i(t+1)]$, or by using "adjacent year" regressions in estimating these weights.

This approach focuses on the estimation of quality change due to a change in a particular set of dimensions and characteristics, assuming either that the other "left out" aspects of quality are not correlated with the included ones, or, if they are, that this correlation also persists into the future. It does not pretend to accomplish everything, to adjust for all quality change. But half a loaf should be better than none.

The alternative approach, first used by Court, is to interpret the coefficient of a time dummy variable(s) in a combined two (or several) years cross-section regression of prices on specifications as a direct estimate of the pure price change. The justification for this is very simple and appealing: We allow as best we can for all of the major differences in specifications by "holding them constant" through regression techniques. That part of the average price change which is not accounted for by any of the included specifications will be reflected in the coefficient of the time dummy and represents our best estimate of the "unexplained-by-specification-change average price change." I used this approach extensively in the first half of my own early paper on this subject (included in this volume), but I am much less satisfied with it now. Besides the fact that it imposes a common set of implicit specification prices on several periods and is not well articulated with the rest of the index number literature, it is very much subject to the vagaries of sample selection. Most of the workers in this area, including myself, tried to get as large a cross-section in any year as possible, not worrying too much about the overall comparability of any two cross-sections. As a result of this, depending on the propensity for publishing such data and the changing proclivities of manufacturers to proliferate different model variants, actual cross-sections used vary significantly in sample size from year to year. This would not matter if all of the observed price variance were just due to the variables considered. Then all points would be on the line, and the different models included or excluded in a particular year would be just different versions of the same thing. But in practice there are "model effects" that persist. Consider an equation of the form

$$\log p_{it} = a + \Sigma b_k x_{kit} + v_t + e_i,$$

where x_{kit} stands for the quantity of the kth dimension in the ith model in year t, v_t is the common "year" or "pure" price change effect, and e_i is a "model" effect, the effect of other left-out qualities, assumed to be independent of calendar time and the other x's. If one uses two adjacent cross-sections and they do not contain all the same i's in the two years, even if the b's remain unchanged, the estimate of the time dummy coefficient will be unbiased *only* if Σe_n for the "new" models just equals Σe_O, the sum of the individual effects of the "old" models no longer appearing in the new sample. But this would be true only if the "included dimensions" exhausted all there was or they were perfectly correlated with the left-out variables. For example, consider a new year in which we suddenly include both the Volkswagen and Mercedes-Benz models among our standard cars. Since these are more expensive per unit of size than the regular American cars, we shall suddenly show a rise in price or a decline in quality. But what has actually happened is a change in the mix of our sample. That this may be a serious problem can be seen from the proliferation of Packard and Studebaker models just before their disappearance. The problem could be reduced somewhat if these regressions were weighted by total sales of the various models, but in any case one needs here to worry much more about the comparability of the samples than if one were only interested in the estimates of some of the slope coefficients (and not the year constants). Without further analysis of whether changing the size and composition of the various cross-sections makes much of a difference, one should not interpret the time dummy estimates presented by Triplett and me for automobiles as unbiased estimates of the pure-price change. This stricture does not apply, however, to the specific quality indexes constructed for a particular fixed sample of cars using the estimated dimensions weights, which appear later in my paper.

The time dummy approach does have the advantage, if the comparability problem can be solved, of allowing us to ignore the ever-present problem of multi-collinearity among the various dimensions. Using it, we may not care that in one year the coefficient of weight is high and horsepower is low while in another year these coefficients reverse themselves, as long as the two coefficients taken together hold the *joint* effect of weight and horsepower constant. But even here, we should use a weighted regression approach, since we are interested in an estimate of a weighted average of the pure-price change, rather than just an unweighted average over all possible models, no matter how peculiar or rare.

The idea that one ought to be able to measure quality differences with the help of prices of used items must have been in the air for quite a while. It appears first in Burstein's paper (1961), but Cagan (1965) was the first to present actual estimates based on such an approach. The basic idea is extremely simple: We can observe today both 1970 and 1969 Chevrolet cars being sold in the market. The price of the two differs because (a) the 1969 car is 1 year older than the 1970 one, and (b) because the 1970 model may be better (or worse) than the 1969 one. If we can assume that the rate of "aging" is independent of calendar time, and if we could somehow find out what it is, we could derive the implied premium of 1970 over 1969 cars (provided that this relative premium does not change with age). Consider the price of used machines or cars of model (or manufacturer) j (such as Chevrolet) in year t (say 1970) of vintage (model year) v (e.g. 1968), or equivalently of age $h = t - v$. The basic hypothesis can then be written as follows:

$$P_{tvj} = P_t Q_{tvj},$$

when P_t is the overall average price per unit of constant quality machine, and Q_{tvj} are the units (quantity) of "carness" or "machinery" still embodied in year t, in machine type j of vintage v. It is assumed that

$$Q_{(t+1)vj} = d_{hj} Q_{tvj},$$

where d is a depreciation factor (one minus the depreciation rate) which is independent of calendar time or vintage (this is what Hall calls a "stationarity" assumption) but may depend on age (h) and make (j). Moreover, we assume that we can write

$$Q_{t,v,j} = T_v Q_{t-1,v-1} e^{uvj}.$$

That is, the quantity (or quality) of a new model in year t, relative to the previous vintage (when new), is composed of an average improvement factor T over all models, and a factor special to the particular model and vintage e^{uvj}, and this relative superiority is constant and independent of age or calendar time. This is a very strong assumption about the character of technical or quality change, stating that any new version of model j can be expressed as so many units more or less than the old version of j, this premium once established being fixed and independent of everything else. This means, technically, that all quality change is of the factor- or product-augmenting type. Fisher and Shell call this the "repackaging" case. To reiterate, it implies that the relative superiority of one version of a commodity over the other is independent of market conditions, relative supplies of the two versions, and age.[5]

If we can observe a number of different vintages being sold at the same time, we can form logarithms of price ratios which given our assumptions will equal

$$\ln P_{tvj} - \ln P_{t(v-1)j} = \ln T_v + \ln d_{hj} + u_{vj} - u_{v-1,j}.$$

Given a number of vintages per year, a reasonably large number of models and makes, and several years' worth of observations, there should be enough degrees of freedom to estimate most of the parameters of interest.[6]

The appropriate approach here is via the use of dummy variables for age (d's), make, and vintage. Note, however, that for estimation purposes one must impose some constraints of the form $\Sigma \ln T_v = \Sigma \ln d_{hj} = $ constant, and hence *in general one cannot separate the effect of an average rate of quality improvement (obsolescence) from the average effect of aging (depreciation)* on the basis of such data.[7] Only if one assumes that in some period(s) there was no improvement in quality $T_v \equiv 1$, which is the procedure adopted by Cagan, can one get an unambiguous estimate of d and hence also estimates of T for other periods. But Cagan's numbers are also consistent with lower average depreciation rates and higher rates of quality change. The only thing that can be estimated unambiguously from such data is the *change in the rate of change of quality improvement*, but that may be interesting enough by itself.

The major advantage of using secondhand market prices to measure quality change lies in freeing us from the necessity of choosing and specifying a limited list of commodity characteristics and estimating their relative contributions. Such lists are never complete and such estimates are never perfect. It is bought, however, only at the cost of very specific assumptions about the nature of quality change and a fundamental identification problem. It cannot really supplant the hedonic index approach. The latter is at least needed to arbitrate the assertion that $T_v = 1$ for a particular pair of vintages allowing the identification of the rest of the parameters.[8] Moreover, by not insisting that the relative prices (weights) of the various characteristics remain constant (and independent of other variables), the hedonic index approach can adapt itself and remain valid for a much wider range of types of quality changes.

The power of the approach outlined in the previous section can be illustrated by using it to evaluate some recent official procedures. "It is standard procedure for the BLS (Bureau of Labor Statistics) to adjust for quality when measuring price changes for the Consumer Price Index and the Wholesale Price Index" (Stotz, 1966, p. 178). This statement has a somewhat strange but welcome ring to participants of the old debates on this topic. I welcome the increased attention given to quality change by the BLS, but I am concerned that by basing such adjustments largely on data furnished by manufacturers and on "producer costs" it may wind up overestimating "quality change," accepting as "improvements" expenditures which consumers may not interpret as such. For example, the reported improvement in "quality" for the basket of cars priced by the CPI was about 0.8 percent between the 1965 and 1966 models of these cars and 1.9 percent between 1966 and 1967 models (Stotz, 1966 and Commissioner Ross's statement of November 23, 1966). If this conclusion is correct, then it should be true that 1967 cars will be considered more valuable than 1966 cars were relative to the 1966 models. Using Cagan's approach, the basket of cars described by Stotz (1966), used car prices given by National Automobile Dealers' Association (for November 1967 and 1966), and rough relative weights based on model sales for 1966 from *Ward's Automotive Reports*, one must conclude that there is no evidence in the used car market that "quality change" occurred at a higher rate between 1967 and 1966 than between 1966 and 1965 (see table 7.1 for details). The overall numbers are small, such calculations are rough, and the differences may not be significant and shouldn't be taken too seriously. Nevertheless, unless the BLS presents a clearer and more detailed description of how it actually makes these adjustments, (an appeal to the confidentiality of manufacturers' data and the opinions of engineering experts doesn't really help much here), doubts about the "quality" of such quality adjustments and their objectivity will remain. I assume that the BLS is doing a good job, and I believe that it will do an even better job in the future, but I would appreciate much more detailed information on how it is actually done.

This is not the place nor is there time to go into all the intricacies of utility theory and associated price index problems, but it is worthwhile to point out that we are living in a complicated world and that it is both unrealistic and unnecessary to expect that one number, "the" price index, can summarize adequately all the changes that occur. Economists will often define a "price-of-giving" index by the question: "How much [more] income is required [relatively] *today* to make me just indifferent between facing yesterday's budget constraint [with yesterday's money income and prices] and a budget constraint defined by today's prices and the income in question?" (Fisher and Shell, 1967). Implicitly, this question assumes that any change in consumer behavior can be factored into a "real income" effect (the reciprocal of the price index) and a substitution effect, and that *this is all there is to*

Table 7.1   Prices of the CPI cars in the used car market[a]

Model and make	Model year				Ratios			Approximate weights (8)
	November 1967		November 1966					
	1967 (1)	1966 (2)	1966 (3)	1965 (4)	(1)/(2) (5)	(3)/(4) (6)	(5)/(6) (7)	
Chevelle Malibu, 2 dr. sport cp.	2340	1895	2160	1685	1.2348	1.2819	0.9633	0.08
Chevrolet Impala, Super Sport, 2 dr. hardtop	2730	2235	2580	2090	1.2215	1.2344	0.9895	0.30
Ford Mustang, 2 dr. hardtop	2350	1875	2220	1785	1.2533	1.2437	1.0077	0.12
Ford Galaxie 500, 2 dr. hardtop	2540	2060	2420	1930	1.2330	1.2539	0.9833	0.19
Plymouth Fury III, 4 dr. sedan	2440	1990	2355	1900	1.2261	1.2395	0.9892	0.06
Pontiac Catalina, 4 dr. sedan	2610	2135	2510	2075	1.2225	1.2096	1.0107	0.10
Rambler Rebel-Classic 770, 4 dr. sedan	2155	1670	1965	1595	1.2904	1.2320	1.0474	0.06
Volkswagen 113, 2 dr. sedan	1590	1345	1510	1310	1.1822	1.1527	1.0256	0.09

[a] An index of the *rate of change* of quality change: $\Sigma(7) \times (8) = 0.997$.
Source: Cols. (1)–(4): *Used Car Guide*, November 1966 and 1967 issues. Col. (8): *Ward's Automotive Reports*, 1967.
     All cars are V-8 models except for Rambler (6-cyl.) and VW (4-cyl.). Rambler Rebel in 1967, Classic 1966, 1965.

that. But many other things, all of which may affect the level of utility achieved with a given money income, may be changing at the same time. It is then a question of *definition* and research strategy whether we want to lump them all into one concept of "the" price index.

Schematically and purely definitionally, let us visualize the following set of equations summarizing the tastes of a consumer and the constraints and opportunities facing him:

$$U = U(S)$$
$$S = F(X, Z, t_s, E)$$
$$Y = W(t_y) = PX$$
$$t_s + t_y \leq T, \ Z \leq \overline{Z}$$

where $U(S)$ is the utility indicator of a stream of services S, $F(X, Z, t_s, E)$ is the "production function" of such services using purchased inputs X, non-market inputs Z, time t_s, and affected by uncontrollable environment factors E. Money income y is a function of time spent at work and in turn constrains the total value of purchased commodities PX, where P is our "price" index of goods purchased. Finally, we have the constraints on time and non-market commodities (the levels of the latter could in turn be correlated with income, as in the case of fringe benefits).[9]

Unfortunately, the distinction between the U and S function, while illuminating, is not very operational. Moreover, it can be shown that an important class of changes in S can be equally well represented by changes in the "quantities" of X. This is true for quality changes of the commodity augmenting type, the "repackaging" case, where either representation will describe the facts. If we could get measures of the changing utility efficiency of some goods, we would want to do so, because it would be an interesting piece of information of major economic significance. Whether or not we would want to incorporate them into our measure of the "price" index is then a purely definitional rather than substantive issue. It depends on what the particular "price" index purports to measure.

Most economists would agree that they would like the "price" index to be a "price-of-living or of utility" indicator. Many government statisticians in charge of producing actual price indexes will reply that they *cannot* achieve this and that therefore they should not even try, but should concentrate instead on some more "objective" index of "transaction" prices and/or allow only for those "quality" changes which are based on "production" costs. The fact that "truth" cannot be achieved doesn't mean that one shouldn't strive to do so, though I sympathize with the position that it is better to measure well something definite than to do a very poor job on a more interesting but also more nebulous concept. Nevertheless, I would deny the contention that "transaction" units or "production" costs are much more definitive concepts. In general, they too make little sense without some appeal to utility considerations.

Consider the simple example of a box of crackers of an unchanged size and price. If its contents (in terms of ounces of crackers per box) have declined, the statistician will usually record (correctly, I believe) a rise in the price of crackers. But the "transaction" unit is a "box," it is priced as a "box," and most consumers don't know or notice the exact number of ounces per box. Nevertheless, the statistician will usually decide that the relevant unit is an ounce of crackers, not the box, even if crackers are not sold by the ounce. Why? Because he believes that the ounce of crackers is a more relevant *utility unit*, that the consumer is ultimately interested in crackers and not in the package they come in.[10] Without an appeal to utility considerations, he wouldn't know what to do in this and many other cases. With this fact established, it is only fair to let the statistician haggle about its price, since bringing utility considerations into the measurement of commodities will prove to be a much harder task in some cases (such as medical service) than in others. This doesn't mean, however, that he shouldn't want to do it if he could.

Nor are "production costs" an adequate guide to quality changes without a check of their utility implications. There may be changes that cost more, such as antipollution devices for automobiles, which are "quality changes" in some sense, but not the relevant one. From the point of view of the individual consumer, if he were not willing to buy these devices on his own, their introduction by law represents a form of tax (in kind) rather than a rise in his utility. This should be recorded as a rise in price, not a fall. It may lead to externalities, possibly to an overall improvement of his environment (E), and hence to an indirect rise in his utility, which then could be perhaps represented by a decline in the "real price" of air, but that is a different matter.[11]

Nor should we ignore "costless" changes if we can measure them. If the consumer is in fact buying "horsepower," and if a design change makes it possible to deliver more horsepower from the same size and "cost" engine, then the price of horsepower to the consumer has fallen *and he is better off*.[12] There always remains the question, how do we know what the consumer is buying? What *are* the relevant units? "Hedonic" price indexes are one way of answering this question. The critical property of such price indexes is that when prices (and units) are given by such a "hedonic index for the commodities [models] within a group, all such commodities have marginal rates of transformation, vis-a-vis commodities outside the group that move in proportion to each other. Insofar as this property is substantiated by empirical evidence, [such] adjustment . . . amounts to correcting an error of aggregation" (Jorgenson and Griliches, 1967, p. 260). In simpler words, this means that we look for such units that would allow us the most *concise* and stable explanation of reality, one that is based on a smaller number of variables (than the almost infinite number of various varieties of commodities) and in terms of which the demand relations, the relations between prices and quantities purchased, are more stable, explain a larger fraction of the observed variance of prices and quantities, and require the introduction of a smaller number of ad hoc parameters, trends, or shift variables. At this level of generality, such a statement is neither a fact nor a theorem, but rather a methodological prejudice, a prejudice about what is likely to be the most fruitful way of approaching such problems and organizing our knowledge about consumer behavior and the economy at large.

Acknowledgements

This work has been supported by a grant from the National Science Foundation and the Price Statistics Committee of the Federal Reserve Board. The paper appeared earlier in slightly different form ("Hedonic Price Indexes Revisited: Some Notes on the State of the Art," 1967, *Proceedings of the Business and Economics Statistics Section* (Washington: American Statistical Association), pp. 324–32).

Notes

1 Griliches (1961) and Adelman and Griliches (1961). The major earlier references are Court (1939) and Stone (1956).

2 See the articles by Nicholson (1967) and Musgrave (1969). This is also reflected in an unpublished Bureau of Labor Statistics memorandum by Thomas W. Gavett (1967b).

3 Nor did Dhrymes use it for this purpose. His interest was in testing the homogeneity of the relationship across manufacturers.

4 See Kravis–Lipsey (1971) for a detailed discussion of this problem.

5 In addition, unless we can assume that the depreciation rate is independent of age ($d_n = d$), i.e. exponential depreciation at a fixed rate, we cannot assume that these relationships (relative model prices) are independent of the interest rate. This important point is made by Hall (1968), but I shall not pursue it further here.

6 One would, however, expect the residuals from such equations (the $u_{vj} - u_{v-1,j}$'s) to be negatively correlated, and this should be taken into account in the estimation procedure. In particular, some smoothing is in order. It is unfortunate, for example, that Cagan ends his analysis with the 1959–60 comparison and a large negative "residual." A rough computation indicates that if his data were extended to 1960–1, they'd show a positive residual for that year, largely canceling the earlier negative one and leading to a higher overall estimated rate of quality change for the whole period.

7 These difficulties of identification are explored in greater detail in Hall (1968, 1971).

8 Hall (1971) uses the hedonic approach to estimate the no-change point implicitly by defining it as no change in any of the listed characteristics. His approach, however, assumes that all of the characteristics (qualities) depreciate at the same rate. This assumption appears to be contradicted by the results of hedonic regressions using secondhand prices reported in the earlier Griliches paper (table 6.10, this volume). The implied prices (coefficients) of the various characteristics decline with age but at different rates.

 9 See Becker (1965), Lancaster (1966), and Muth (1966) for further elaboration of such framework.
10 This is not to imply that "packaging" is irrelevant to the consumer, only that in this context it is a second-order consideration.
11 Alternatively one could view it as a deterioration in his environment, the same as a colder winter that requires him to purchase more fuel to achieve a given level of satisfaction. In such a case, one may not record a rise in the "price" of living, even though there has clearly occurred a rise in the "cost" of living.
12 Gavett's discussion of this point (1967a, pp. 18–19) is confusing because he does not recognize that one could interpret such a change as a decline in price per corrected unit, and hence as a downward shift in the relevant supply function. The increase in total utility he observes is not the consequence of an increase in total consumption of the good at an unchanged price (since this couldn't happen with a fixed budget constraint) but is rather due to a fall in the relevant price.

References

Adelman, I. and Z. Griliches, 1961, "On an index of quality change," *Journal of the American Statistical Association*, 56: 535–48.

Bailey, M. J., R. F. Muth, and H. O. Nourse, 1963, "A regression method for real estate price index construction," *Journal of the American Statistical Association*, 58: 933–42.

Barzel, Y., 1964, "The production function and technical change in the steam power industry," *Journal of Political Economy*, 72: 133–50.

Becker, G. S., 1965, "A theory of the allocation of time," *Economic Journal*, 75: 493–516.

Box, G. E. and D. R. Cox, 1964, "An analysis of transformations," *Journal of the Royal Statistical Society* (b), 26(2): 211–52.

Brown, S. L., 1964, "Price variation in new houses, 1959–1961," Staff Working Paper in Economics and Statistics, no. 6. Washington: Bureau of the Census. Mimeographed.

Burstein, M. L., 1961, "Measurement of quality change in consumer durables," *Manchester School of Economics and Social Studies*, 29: 267–79.

Cagan, P., 1965, "Measuring quality changes and the purchasing power of money: an exploratory study of automobiles," *National Banking Review*, 3: 217–36. (Reprinted in Z. Griliches (ed.), *Price Indexes and Quality Change*. Cambridge: Harvard University Press, 1971.)

Chow, G. C., 1967, "Technological change and the demand for computers," *American Economic Review*, 57: 1117–30.

Court, A. T., 1939, "Hedonic price indexes with automotive examples," in: *The Dynamics of Automobile Demand*, pp. 99–117. New York: General Motors Corporation.

Cramer, J. S., 1966, "Een prijsindex van nieuwe personenauto's, 1950–1965," *Statistica Neerlandica*, 20: 215–39.

Dean, C. R. and H. J. DePodwin, 1961, "Product variation and price indexes: a case study of electrical apparatus," *Proceedings of the Business and*

Economic Statistics Section, pp. 271-9. Washington: American Statistical Association.

Dhrymes, P. L., 1967, "On the measurement of price and quality changes in some consumer capital goods," *American Economic Review*, 57: 501-18.

——, 1971, "Price and quality changes in consumer capital goods: an empirical study," in Z. Griliches (ed.), *Price Indexes and Quality Change*. Cambridge: Harvard University Press, 1971.

Fettig, Lyle P., 1963, "Adjusting farm tractor prices for quality changes, 1950-1962," *Journal of Farm Economics*, 45: 599-611.

Fisher, F. M., Z. Griliches, and C. Kaysen, 1962, "The cost of automobile model changes since 1949," *Journal of Political Economy*, 79: 433-51.

Fisher, F. M. and K. Shell, 1967, "Taste and quality change in the pure theory of the true cost-of-living index." In: *Value, Capital, and Growth: Essays in Honour of Sir John Hicks*, J. N. Wolfe (ed.). Edinburgh; University of Edinburgh Press. (Reprinted in J. Griliches (ed.), *Price Indexes and Quality Change*. Cambridge: Harvard University Press, 1971.)

Gavett, T. W., 1967a, "Quality and a pure price index," *Monthly Labor Review*, 90: 16-20.

——, 1967b, "Research on quality adjustments in price indexes," Unpublished Bureau of Labor Statistics Memorandum, Washington, DC.

Griliches, Z., 1961, "Hedonic price indexes for automobiles: an econometric analysis of quality change." In: *The Price Statistics of the Federal Government*, General Series, no. 73, pp. 137-96. New York: National Bureau of Economic Research. (Reprinted in Z. Griliches (ed.), *Price Indexes and Quality Change*. Cambridge: Harvard University Press, 1971.)

——, 1964. "Notes on the measurement of price and quality changes," in: *Models of Income Determination*, Studies in Income and Wealth, vol. 28, pp. 301-404. Princeton: National Bureau of Economic Research.

Hall, R. E., 1968, "Technical change and capital from the point of view of the dual," *Review of Economic Studies*, 35: 35-46.

——, 1971, "The measurement of quality change from vintage price data," in Z. Griliches (ed.), *Price Indexes and Quality Change*. Cambridge: Harvard University Press, 1971.

Hanoch, G., 1965, "Personal earnings and Investment in Schooling," PhD, dissertation. University of Chicago.

Jorgenson, D. W. and Z. Griliches, 1967, "The explanation of productivity change," *Review of Economic Studies*, 34: 249-83.

Knight, K. E. and J. L. Barr, 1966, "Micro measurement of technological change in the computer industry." Unpublished paper, presented at the Inter-University Committee Conference on Micro-Economics of Technological Change in Philadelphia, March 24, 1966.

Kravis, I. B. and R. E. Lipsey, 1969, "International price comparisons," *International Economic*, 10: 233-246. (Reprinted in revised form in Z. Griliches (ed.), *Price Indexes and Quality Change*. Cambridge: Harvard University Press, 1971.)

—— ——, 1971, *Price Competitiveness in World Trade*, National Bureau of Economic Research, New York: Columbia University Press.

Lancaster, K., 1966, "A new approach to consumer theory," *Journal of Political Economy*, 74: 132–57.

Musgrave, J. C., 1969, "The measurement of price changes in construction," *Journal of the American Statistical Association*, 64: 771–86.

Muth, R. F., 1966, "Household production and consumer demand functions," *Econometrica*, 34: 699–708.

National Automobile Dealers Association. *Official Used Car Guide*. Washington. Various issues. (Cited as *Red Book*.)

Nichelson, J. L., 1967, "The measurement of quality changes," *Economic Journal*, 77: 512–30.

Richter, M. K., 1966, "Invariance axioms and economic indexes," *Econometrica*, 34: 749–55.

Stone, R., 1956, *Quantity and Price Indexes in National Accounts*, Paris: Organization for European Economic Co-operation.

Stotz, M. S., 1966, "Introductory Prices of 1966 Automobile Models," *Monthly Labor Review*, 89: 178–85.

Triplett, J. E., 1966, "The Measurement of Quality Change," PhD dissertation, University of California.

Ward's Automotive Reports. Detroit: Powers and Co. Various issues.

Yoshihara, K., K. Furuya, and T. Suzuki, 1970, "The problem of accounting for productivity change in the construction index." SEAS Discussion Paper no. 5, Kyoto University. Mimeographed.

8
Postscript on Hedonics

The literature on hedonic price measurement, hedonic price indexes, and associated topics has grown rapidly since my original effort to revive this topic and my subsequent attempt, in 1971, to survey it. I will not be able to do justice to it in this brief postscript. Nevertheless, let me try to indicate what I consider to be the highlights of this literature.

There are three major issues which tend to be addressed, in different proportions, in the various papers relating to this topic. There is a range of theoretical questions: how should different "qualities," characteristics, of commodities (outputs or inputs) be modelled, entered into utility or cost functions and translated into demand and supply functions and the resulting market outcomes? Can one give a theoretically consistent interpretation to "quality-adjusted" price indexes and can one derive valid restrictions from the theory which the empirical price–characteristics regressions should satisfy? There is also a wide range of empirical problems: What are the salient characteristics of a particular commodity? Under what conditions should one expect their market valuation to remain constant? How should the regression framework be expanded; what variables should be added to it, so as to keep the resulting estimates "stable" in face of changing circumstances? And there is also a whole host of econometric methodology issues associated with the attempt to estimate a relationship that can be thought of as being the result of an interaction of both demand and supply forces, and with the use of detailed micro-data, often in the form of an unbalanced panel of data for a fixed number of manufacturers, but a different and changing number of "models" (commodity versions).

The theoretical literature tends to focus either on the demand side, Lancaster (1971), Muellbauer (1974), and Berndt (1983), among others, or the supply side (see Ohta, 1975 for an example), with very few, Rosen (1974) being a notable exemption, attempting a full general equilibrium discussion. (See also Epple (1987) for a recent discussion.) There is much finger-pointing at the restrictive assumptions required to establish the "existence" and meaning of hedonic "quality" or price indexes. (See e.g. Muellbauer, 1974, and Lucas, 1975.) While useful, I feel that this literature has misunderstood the original purpose of the hedonic suggestion. It is easy to show that, except for unique circumstances and under very stringent assumptions, it is not possible to devise a perfect price index for *any* commodity classification. With finite amounts of data, different procedures will yield (hopefully not very) different answers, and even "good" formulae, such as Divisia-type indexes, cannot be given a satisfactory theoretical interpretation except in very limiting and unrealistic circumstances. Most of the objections to attempts to construct a price index of automobiles from the consideration of their various attributes apply with the same force to the construction of a motor-vehicles price index out of the prices of cars, trucks, and motorcycles.

My own point of view is that what the hedonic approach tries to do is estimate aspects of the budget constraint facing consumers, allowing thereby the estimation of "missing" prices when quality changes. It is not in the business of estimating utility or cost functions *per se*, though it can also be very useful for these purposes. (See Cardell, 1977; McFadden, 1974, and Trajtenberg, 1983 for examples). What is being estimated is actually the locus of intersections of the demand curves of different consumers with varying tastes and the supply functions of different producers with possibly varying technologies of production. One is unlikely, therefore, to be able to recover the underlying utility and cost functions from such data alone, except in very special circumstances. Nor can theoretical derivations at the individual level really provide substantive constraints on the estimation of such "market" relations. (See the detailed discussion of many of these issues, in the context of estimating the value of urban amenities, in Bartik and Smith, 1987.) Hence my preference for the "estimation of missing prices" interpretation of this approach. Accepting that, one still faces the usual index number problems and ambiguities, but at least one is back to the "previous case." In this my views are close to those articulated by Triplett (1983a and 1986). The following passage from Ohta and Griliches (1976, p. 326) represents them reasonably well.

> Despite the theoretical proofs to the contrary, the Consumer Price Index (CPI) "exists" and is even of some use. It is thus of some value to attempt to improve it even if perfection is unattainable. What the hedonic approach attempted was to provide a tool for estimating "missing" prices, prices of particular bundles not observed in the original or later periods. It did not pretend to dispose of the question of whether various observed differentials are demand or supply determined, how the observed variety of models in the market is generated, and whether the resulting indexes have an unambiguous welfare interpretation. Its goals were modest. It offered the tool of econometrics, with all of its attendant problems, as a help to the solution of the first two issues, the detection of the relevant characteristics of a commodity and the estimation of their marginal market valuation.
>
> Because of its focus on price explanation and its purpose of "predicting" the price of unobserved variants of a commodity in particular periods, the hedonic hypothesis can be viewed as asserting the existence of a reduced-

form relationship between prices and the various characteristics of the commodity. That relationship need not be "stable" over time, but changes that occur should have some rhyme and reason to them, otherwise one would suspect that the observed results are a fluke and cannot be used in the extrapolation necessary for the derivation of missing prices. . . .

To accomplish even such limited goals, one requires much prior information on the commodity in question (econometrics is not a very good tool when wielded blindly), lots of good data, and a detailed analysis of the robustness of one's conclusions relative to the many possible alternative specifications of the model.

The theoretical developments have been useful, however, in elucidating under what conditions one might expect the hedonic price functions to be stable or shift, and which variables might be important in explaining such shifts across markets and time. My own work in this area has had more of a methodological–empirical flavor, though there were also non-negligible attempts to formulate and clarify the theory underlying such measurement techniques in Adelman and Griliches (1961), Griliches (1964), and in Ohta and Griliches (1976 and 1986). The last two papers represent also my efforts to pursue additional empirical work in this area. In the 1976 paper with Ohta we extended the earlier approach to the analysis of used automobile prices and investigated differences between performance and specification characteristics and pricing differences between manufacturers of different makes of automobiles. The 1986 paper focuses on the role of gasoline price changes in shifting the hedonic price relationships for cars, extends the theory to incorporate operating cost components, and shows that allowing for such price changes leaves the "extended" hedonic function effectively unchanged, permitting one to maintain the stability of tastes hypothesis in this market. See also Gordon (1982, 1987) and Kahn (1986) for related work.

The major recent "success" of hedonic methods has been their acceptance by the official statistical agencies after many years of resistance. Hedonic methods had been used for a long time by the Bureau of the Census to compute its index of single-family houses and much experimental work was carried on at the Bureau of Labor Statistics, but it was not until January 1986, when the *Survey of Current Business* announced a revision of the US National Income Accounts which incorporated a new price index for computers based on the hedonic methodology, that one could feel that they had received the official imprimatur. This index is described and discussed in Cole et al. (1986) and Triplett (1986).

The hedonic methodology has been applied in many different contexts. I cannot really review the vast literature on housing prices and the valuation of various amenities (see the references in Bartik and Smith, 1987). But I do want to mention the literature on earnings functions and wage differentials where similar methods have been pursued and similar problems have arisen. I myself was influenced by the hedonic work in my approach to the economics of education (see the relevant papers in this volume and Griliches, 1977).

Additional surveys of the empirical and econometric issues faced in this context can be found in Brown (1983) and Triplett (1983b).

I want to finish this incomplete review with acknowledging an original oversight: among the early pioneers of the hedonic method credit should also be given to the late Frederick Waugh, who in 1929 analyzed vegetable prices and related them to various "quality" dimensions.

9

Capital Stock in Investment Functions: Some Problems of Concept and Measurement*

This paper originated as a report on an econometric study of total farm investment in machinery and motor vehicles. For this study I needed a measure of the stock of machinery on farms, and the available "official" measures seemed to be the obvious choice. But the more I worked with the model and these data the more dissatisfied I became with the official measures and with my own understanding of the problem. Although it was not too difficult to point out the flaws in the available measures, it was not at all clear to me what the right measure should be. Thus my interest shifted from the substantive question of the determinants of farm machinery investment to the methodological question of the role of capital stock in investment functions and the quest for the right concept of capital.

In pursuing the literature on this subject, mostly in the accounting and national accounting field and not too familiar to an econometrician, I found that almost everything worth saying has already been said by somebody some place, but not, I think, loud enough or often enough.[1] In this paper I will not be able to solve all or even many of the conundrums in this field. At best, it will be a guide to some unsolved problems of concept and measurement.

The Role of Capital in Investment Functions

The first question to be asked is "why is it necessary to have a capital variable in the investment function in the first place?" Is it not possible to explain investment without getting involved in the semi-metaphysical issues of defining capital? The answer is clearly no. What we need is a behavior

*From Christ, C. et al. *Measurement in Economics*, Stanford, CA: Stanford University Press, 1963.

equation, an equation that summarizes the most important determinants of investment and allows us to interpret and, hopefully, to predict this variable. But at best, the investment function is a derived function. There is no such thing as demand for investment. Farmers or entrepreneurs do not desire a particular annual flow of purchased capital equipment *per se*. The fundamental desire or demand is for particular levels of capital, for a particular stock of capital equipment. To the extent that the desired stock does not equal the actual, there will be investment, but this investment is derived from the demand for a stock of capital – it is not an independent function.

Let me beg most of the crucial measurement and definition problems in this section and consider a very simple model in which the desired level of capital K^* is a function of a *set* of variables that I shall call X (say, the expected level of output, the rate of interest, wages, etc.).

Thus

$$K^* = f(X). \qquad (1)$$

Now, if at the end of the period actual capital always equals desired capital, i.e. all of the adjustment is completed within the unit period of analysis, *net* investment depends only on the change in $f(X)$. Thus, only if we believe that all adjustments are completed within a year (or whatever is the unit of analysis) can we disregard capital (or the past history of net investment that is summarized in the capital concept) in explaining current net investment. In this particular case net investment is a function of the changes in X and will be zero unless X changes. Thus, net investment functions that do not contain a capital variable should have all their independent variables in first-difference form or have some other provision that would reduce net investment to zero eventually if there are no continuing changes in the independent variables. Of course, if the dependent variable is gross investment, there may be no escape from the capital stock variable, since gross investment is the sum of net investment and replacement, and replacement is definitely some function of the existing level of stock and its age distribution.

The assumption that all the desired change in the stock of capital is completed within one year is clearly unrealistic. Investment takes time. There are lags in budgeting, ordering, delivering, and installing new equipment. Often one investment (e.g. a new plant) will result in a stream of investment expenditures that are spread over quite a number of years. Moreover, in a world of rapidly changing economic conditions one may discount rather heavily changes that would indicate a very different level of desired capital stock. There is little point in completing the adjustment quickly if changing economic conditions are likely to outdate this decision quickly. The best policy may be to adjust only a little at a time, arriving at the goal eventually and only if the conditions that made this level desirable persist for a long enough period. (For an extensive survey of the literature on this type of model, see Nerlove [17].)

Such considerations lead to a very simple adjustment equation,

$$\Delta K = g(K^* - K), \tag{2}$$

where K is the actual stock of capital and ΔK is net investment. Thus net investment is a function g of the difference between the desired and the actual stock of capital. For practical purposes and to simplify the exposition, I shall assume that net investment is actually a fixed fraction of this difference, though it is easy to think of reasons why this may not be so.[2] We have thus

$$K_{t+1} - K_t = \gamma(K^*_{t+1} - K_t), \tag{3}$$

where γ is the adjustment coefficient and the stock of capital is measured at the beginning of the period.

Putting equations (1) and (3) together gives us

$$K_{t+1} - K_t = \gamma f(X_t) - \gamma K_t. \tag{4}$$

We are interested, however, in gross rather than net investment. It is gross investment that is the output of the capital goods industries and the variable that we would like to predict. Also, gross investment is the datum of the economy and its explanation is a fairer test of one's model than explaining net investment, which may be largely a construct of one's (questionable) depreciation procedure. But gross investment equals net investment plus replacement:

$$I_t = K_{t+1} - K_t + R_t, \tag{5}$$

where I_t is gross investment and R_t is replacement. If in addition we assume that replacement demand is proportional to the existing stock of capital, an assumption that I shall question later on, we have

$$R_t = \delta K_t, \tag{6}$$

where δ is the implicit depreciation coefficient.[3] Putting (3), (4), and (5) together, we have

$$\begin{aligned} I_t &= \gamma f(X_t) - \gamma K_t + \delta K_t \\ &= \gamma f(X_t) + (\delta - \gamma)K_t. \end{aligned} \tag{7}$$

This form has been used widely in various investment studies [1, 11, 14]. The role of capital stock is here twofold: it has a negative influence on investment, measured by γ, owing to the damping effect of the existing stock of capital on the adjustment process, and a positive influence, measured by δ, owing to the larger replacement demand generated by a larger stock of capital. The net effect of the capital stock on investment is indeterminate; it depends on whether δ is smaller or larger than γ.

Which Concept of Capital?[4]

Besides the usual index number problems of adding cars and manure spreaders together and taking care of changing quality and price levels, which will not be discussed here, the main difficulty with capital measures arises from the most important property of capital – its durability. Most of my remarks will deal with the problem of how to compare and count new machines and 10-year-old machines.

There is a variety of possible capital measures (e.g. gross stock, adjusted gross stock, net stock, market value, depreciation or capital consumption, flow of services), and I shall discuss each of these measures briefly. The thing to remember throughout this discussion, however, is the difference between the *quantity* of capital in some physical units and its *value*. For many purposes we will be interested mainly in the first concept, whereas most of the available measures are approximations of the second concept. The reason for this distinction should be made clear: Almost all of our theorizing about investment and the desired stock of capital rests implicitly on some technological considerations and is derived from some kind of general production function. As long as we stick to the production function framework, it is clear that *quantity* rather than *value* is the relevant dimension, since the production function is defined as a relationship between the quantity of output and the quantity of various inputs. (See, e.g. Smith [18], and also chapter 5 in the very interesting book by Hood and Scott [13].) That is, if we are interested in the demand for tractors, changes in the desired level of output will imply a desire to change the *number* of machines on farms. Whether these machines are new or old is a secondary problem. Of course, if we are interested in problems of technical change, and the new machines are also "better" machines, then the distinction between old and new machines becomes important again and we might want to rephrase our model in terms of capital vintages. On this see the paper by Solow [19].

Perhaps the simplest and least ambiguous measure is the gross stock concept. A machine is carried at its original value until it is retired. This is equivalent to a one-hoss-shay assumption – the machine is as good as new until the day of its "death." There are two major variants of this measure when one comes to deal with a large number of machines. The first retires (writes off) all machines at the end of their expected (average) life span, say 20 years. This is equivalent to measuring capital stock by a moving sum of past investment expenditures, the length of the summation being determined by the average life span of the particular machine. The second variant takes into account the variance around the expected life span and retires items on the basis of survival curves, machine mortality tables, or some other information on the distribution of retirements with age. The second variant seems to be clearly preferable to the first, since the first results in the writing off of all the capital by the time its average life span is exhausted, while in fact half of the original capital stock (gross) is still "alive."

Neither of these measures, however, takes care of the case of the same machines having different degrees of durability. The best procedure would be to count machines. In this fashion, the same machines have the same weight in the total, irrespective of the differences in their durability. In practice, however, we are faced with the problem of adding together machines of different size and type, and we wind up using some measure of total investment expenditures in constant dollars. Doing this, we are likely to give a machine that lasts 10 years (approximately) twice the weight of a new machine of the same capacity and annual productivity that will last only 5 years. The thing to do in this case is to deflate gross investment for changes in the expected life span, giving the more durable machine the same weight as the standard machine but carrying it for a longer period. It may be very difficult in practice to apply this type of deflation, but it is useful to keep it in mind as one possibly ideal measure, which I shall call *adjusted gross stock*.

The net stock concept is motivated by the observed fact that the value of a capital good declines with age (and/or use). This decline is due to several factors, the main ones being the decline in the life expectancy of the asset (it has fewer work years left), the decline in the physical productivity of the asset (it has poorer work years left), and the decline in the relative market return for the productivity of this asset due to the availability of better machines and other relative price changes (its remaining work years are worth less). One may label these three major forces as exhaustion, deterioration, and obsolescence. Net stock concepts will differ, depending on which (if any) of these factors govern the rate at which the gross stock is "written down." Some depreciation patterns have very little economic justification (except accounting convenience), but most of them at least purport to approximate the decline in the economic value of the remaining services (i.e. market value). Of the various possible depreciation schemes (net stock measures), two measures seem to be of the most interest: (a) a net stock concept based on a purely physical deterioration depreciation scheme,[5] and (b) the market value of the existing stock of capital. The latter figure can be approximated by the use of depreciation rates derived from studies of used machinery prices. The first concept is of interest because it is likely to be proportional to the available *current* flow of productive services. The second concept is of interest because it represents the market valuation of the current *and* all future flows of services that will be available from a given capital stock. Since many decisions about the future are made today, the market's valuation of the available *future* services of capital should not be ignored.

Two concepts of depreciation are of special interest to us: (1) "deterioration" (the loss in physical productivity or in the capacity of the existing capital to supply *current* services)[6] and (2) devaluation (the decline in the economic value [price] of the current and all future services derivable from the given capital stock).[7] The usefulness of either of these concepts is limited mainly to computing the appropriate net stock of capital. Neither of them is a good measure of the flow of current productive capital services. Ideally, the available flow of services would be measured by machine-hours or machine-years. In a world of many different machines we would weight the different machine-hours by their respective rents.[8] Such a measure would approximate most closely the flow of productive services from a given stock of capital and would be on par with man-hours as a measure of labor input.

Although we have explored briefly an array of possible "reasonable" measures of capital, the choice of a particular capital measure to be used in the investment function is still not very clear. Each of these measures is of some interest and will be found useful in answering particular questions. It is important, however, to decide first what questions we want our measure to answer. For this purpose it is useful to distinguish between the two different roles of capital in the investment function: (a) the depressing effect of the existing stock of capital on the *rate of adjustment* to the new equilibrium, and (b) capital stock as a measure of replacement demand.

If we stick to our prejudice in favor of physical measures of the *quantity* of capital (i.e. if we derive the demand for capital from some underlying production function in which the relevant dimension is the *number* of machines), it seems quite clear that it is the existing *quantity* of machines which will affect the rate at which entrepreneurs adjust to changes in underlying conditions. It is clear that the pressure to close a given gap will be less if there are many machines, even if they are somewhat dilapidated, than if there are only a few and they are new. The desired stock of capital that is derived from production function considerations is a demand for the services of a number of machines, and the rate of adjustment of stock is a function of the difference between the desired number of machines and the existing number. These considerations lead to the following ranking of the relevant capital measures for this purpose: (1) the currently available flow of services, (2) net stock based on deterioration depreciation only, (3) adjusted gross stock (adjusted for the changing expected life span of machines), and (4) gross stock (survival curve retirement). The first measure is probably proportional to the second, and hence we may concentrate on the second and fourth measures, since we do not have the data to do an adequate job on the third. The net stock based on deterioration and the gross stock differ only in their assumption about deterioration. How the productivity of machines actually declines with age is a major empirical question to which we have very few answers. The range of possibilities extends from gross stock with the assumption of no deterioration at all to net stock based on declining balance depreciation at a relatively high rate.

The fact is that we know almost nothing about the deterioration with age in the performance of a machine. For many purposes, in particular for productivity comparisons, the one-hoss-shay assumption seems to be reasonable, or at least less extreme than some of the more commonly used assumptions (e.g. declining balance). When we turn to scattered data on hours worked by age of machine, gasoline consumption, etc., some deterioration with age is quite evident, but it is rather slow and concave to the origin. Figure 9.1 presents some evidence on the influence of age on the performance of tractors in the US [20, pp. 70, 71]. (See the curves labeled (2) and (3).) Data on the market value depreciation of tractors show, however, a strongly convex pattern (see (4) in the same figure).[9] Since market depreciation is composed of the change in interest costs, deterioration, and obsolescence, and since the change in interest costs by itself would produce a concave pattern of

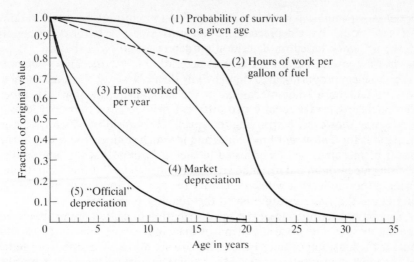

Figure 9.1 The ageing of tractors.
(1) From A. P. Brodell and A. R. Kendall, *Life of Farm Tractors*, USDA, BAE, F. M. 80. Washington, DC, Supt of Documents, June 1950. The estimates are based on purchase data and the estimated 1948 stock of tractors on farms by age.
(2) The reciprocal of an index of fuel consumption per hour of work for wheel tractors 25 horsepower and over (this index equals 1.00 for less-than-5-year-old tractors). From A. P. Brodell and A. R. Kendall, *Fuel and Motor Oil Consumption and Annual Use of Farm Tractors*, USDA, BAE, F. M. 72. Washington, DC, Supt of Documents, 1950. Data based on a 1940 survey.
(3) Average hours worked in 1940 by wheel tractors of different ages, with hours worked by less-than-5-year-old tractors equaling 1.00. From Brodell and Kendall, USDA, F. M. 72, *op. cit.*
(4) Average resale price of different-age tractors of same make and model in 1958 (new price = 1.00). Based on a sample of 23 models from the *Official Tractor and Farm Equipment Guide*. St Louis, Mo.: National Retail Farm Equipment Association, 1958.
(5) "Official" depreciation based on 18.5 percent declining balance depreciation, the rate and method used by the USDA in its estimates of farm machinery depreciation, net income, and the value of the stock of tractors on farms. See USDA, *Agriculture Handbook No. 118*, vol. 3, p. 17.

depreciation, either deterioration must be strongly convex (in spite of our scattered evidence) or, more likely, the impact of obsolescence is all-pervasive. If there is a high enough rate of obsolescence, it will produce a convex pattern in market values, even though the other two components of depreciation are concave. The fact is that we do not know what assumption is correct, and until we gather better information we should try several measures incorporating different assumptions about deterioration.

If we follow the same logic, the choice of the appropriate replacement measure is also quite clear. Replacement is the replacement of a number or units of machines; it is the replacement of productive capacity. Hence, if we assume no deterioration, replacement is just equal to retirements. If we do allow for some deterioration, it is equal to retirements plus deterioration. Thus it is equal to the depreciation in the *gross* stock, or to the depreciation in the net stock based on deterioration depreciation. Note that only if we assume a constant percentage rate of deterioration (i.e. declining balance) is replacement proportional to capital. Only then can we use the same variable to represent both roles of capital in the investment function. All other depreciation schemes require two different measures for these purposes.

We may, however, be proving too much. In a world in which the basic demand is for a *number* of machines and in which competition prevails, the prices of machines will be adjusted in such a fashion that the alternatives of using a new or an old machine are equally attractive. But there is nothing, then, in this model that would determine the age distribution of machinery, and hence the *value* as opposed to the quantity of capital.

The main determinants of the age distribution of capital are probably on the supply side. The price at which new machines can be substituted for old ones is the most important of these. But we are also abstracting from some of the most important aspects of the capital-equipment markets, from the fact that at given prices entrepreneurs differ in their preferences about different-age machines, that they differ in the value they put on reliability and in their evaluation of the risks of obsolescence. These differences of opinion and the possible different uses of the same machine make for a market in used machinery and result in the well-known downgrading in use of machinery with age.[10]

Moreover, we have up to now left out from consideration the impact of technological change or "obsolescence" on the demand for new equipment. But a theory of investment without obsolescence is like *Hamlet* without the Prince. The main reason for the continuing high levels of gross investment is probably not so much the wearing out of old machinery as the availability of better new machinery [20]. Ideally, we should have taken account of it in our definition of X, including it as one of the variables determining the desired level of capital (in constant quality units). One way of doing this would have been through a continuously declining price of new machinery per constant quality unit. In practice, however, there is very little that we can do about it, and the whole effect of obsolescence is likely to be ignored in empirical work.[11]

One way of taking some of this into account is to assume that replacement is not only replacement of retired or worn-down machines but includes also the replacement of obsolete machines, even though the obsolete machines have not been retired but only downgraded in use. The amount of obsolescence (and deterioration) can be approximated quite well by the decline in the market value of the existing capital. More generally, we may postulate that there is a set of variables which determines the desired *value* of the stock of capital (e.g. the cost of durability, speculative considerations, and transaction costs), even though our simple production-function-based theory of the demand for capital determines only the desired *quantity* of capital. If these variables remain unchanged, it would appear reasonable to assume that replacement is desired not only for the quantity of capital but also for the value of capital. If this is true, market depreciation may be the best measure of replacement, defined broadly to include not only deterioration and retirement but also the reduction in the expected life span and the loss in earning power due to obsolescence.

We still have no very strong a priori reason for choosing any one capital measure over the others. One can, however, summarize three distinct approaches to the problem. At one extreme is the position that assumes that market values reflect all the relevant calculations and that the *value* of capital and the changes in it are the only relevant aspects of capital. At the other extreme is the "technological" position, claiming that only the number of machines matters, that the *quantity* of capital is the relevant and the only relevant dimension. In this section an attempt was made to outline an intermediate position, recognizing the different questions answered by different measures of capital, and suggesting that perhaps two different measures of capital are both relevant to the explanation of gross investment. The intermediate position would come close to the quantity extreme if we were able to include the impact of the availability of better machines on the demand for new machinery explicitly among our independent variables (the X's).

Although we have expressed a preference for the intermediate approach, the evidence for it is not overpowering and some experimentation with alternative measures is in order. Moreover, even if we did decide a priori on a particular *concept* of capital, we should still experiment with alternative *measures* of capital, since empirical knowledge on the right depreciation or deterioration formula is almost nonexistent. The next section, therefore, will be devoted to a report on the experimental use of different measures of capital in explaining the fluctuations of farm gross investment in machinery and motor vehicles.

The Farm-Investment Function and Different Measures of the Stock of Machinery on Farms

The more substantive aspects of the farm-investment study will not be presented here. It will suffice to note that the major candidates for X (the set of independent variables) were the price of machinery, the prices of farm products, farm wages, prices of machinery supplies (complements) the price of land, the rate of interest, farmers' net worth, farm income, stocks of substitutes (horses and mules), and trend. After a substantial amount of experimentation the list was narrowed down to only two useful variables: the ratio of farm machinery prices to the index of prices received by farmers (machines-product price ratio) and the stock of horses and mules on farms. None of the other a priori important variables seemed to add significantly to the explanation of gross investment. The farm-wage variable and the price of machinery supplies had the right sign (positive for wages and negative for the price of machinery supplies) and were, in some cases, on the verge of being significant. The coefficient of the interest-rate variable was negative and highly significant in the absence of the horses-and-mules variable, but it became insignificantly different from zero when the horses-and-mules variable was introduced. These two variables turned out to be almost perfect substitutes for each other. The main substantive result of this study was the high significance of the real price of farm machinery in explaining the fluctuations in farm-machinery investment, most of it being accounted for, however, by the fluctuations in the denominator (farm products prices) of this variable rather than by its numerator (farm machinery prices). The coefficient of the latter variable was not significantly different from zero when the two parts of the real price were introduced separately in the regression.

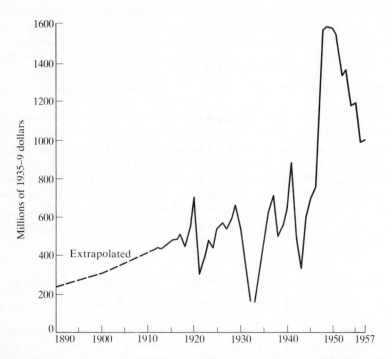

Figure 9.2   Gross farm capital expenditures on machinery and motor vehicles (for production purposes), in 1935–9 prices. Gross farm-machinery investment–gross farm investment in farm machinery, tractors, trucks and 40 percent of the gross investment in automobiles, all in 1935–9 prices. Based on unpublished data (from USDA, AMS, Farm Income Branch) underlying the official depreciation rates. The figures in current prices are published in USDA, The Farm Income Situation. All the subsequent stock estimates are based on these series. Since the USDA figures go back only to 1910, when more years were needed for the particular stock measure, the data were extrapolated backward on the basis of decade investment averages from M. W. Towne and W. D. Rasmussen, "Farm Gross Product and Gross Investment in the 19th Century" [21].

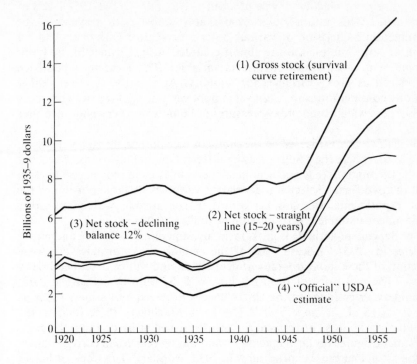

Figure 9.3 Different measures of the stock of machinery on farms.

(1) Gross stock (survival-curve retirement) – based on survival information for tractors and grain binders (tractor survival tables from A. P. Brodell and A. R. Kendall, *Life of Farm Tractors*, USDA, BAE, F. M. 80, Washington, DC, Supt of Documents, June 1950, and A. P. Brodell and R. A. Pike, *Farm Tractors: Type, Size, Age, and Life*, USDA, BAE, F. M. 30, Washington, DC, Supt of Documents, February 1942. Survival table for grain binders from E. L. Butz and O. G. Lloyd, *The Cost of Using Farm Machinery in Indiana*, Purdue University Agricultural Experiment Station, Bulletin no. 437, May 1939). It is assumed that an investment depreciates to 10 percent of its original value in 25 years and is scrapped after that. The pattern of depreciation is as follows: 0.990, 0.985, 0.975, 0.97, 0.96, 0.95, 0.93, 0.91, 0.88, 0.85, 0.81, 0.77, 0.71, 0.65, 0.59, 0.52, 0.45, 0.38, 0.31, 0.26, 0.21, 0.17, 0.13, 0.10.

(2) Net stock–straight-line depreciation: 15 years length of life before 1940, 20 years since 1940. Comparable to Goldsmith estimates [7].

(3) Net stock–declining-balance depreciation at 12 percent per year. Consistent with market depreciation figures [9].

(4) Net stock–based on unpublished USDA data underlying the official depreciation estimates, using the declining balance method at a rate of (in recent years) 18.5 percent for tractors, 22.0 percent for automobiles, 21 percent for trucks, and 14 percent for other farm machinery. Average rate used is about 17 percent (using 1950 stock values as weights). The estimates from 1940 to date are published in *The Balance Sheet of Agriculture*.

The estimated short run elasticity of investment with respect to changes in the real price of machinery was between -1.6 and -2.4 at 1957 levels of investment. This estimate was remarkably stable with respect to the introduction or deletion of various other independent variables. Another result of some interest was the almost complete insignificance of the farm-income or net-real-wealth-of-farmers variables. These variables had been introduced as proxy variables for liquidity. Apparently, however, either liquidity considerations were not very important during this period or, more likely, these widely used proxy measures of liquidity are very poor measures of it.

Since this paper is focused on the choice of the appropriate capital measure, I shall report here the results of using different capital measures to account for the fluctuations in gross farm-machinery investment only in conjunction with the two finally selected independent variables. A large portion of the same experiments was also performed in conjunction with many other independent variables, with essentially similar results.

Our dependent variable is gross farm investment in machinery and motor vehicles in 1935–9 prices.[12] All of the capital stock variables are some function of these same gross investment series, and will be described briefly below. The real price of farm machinery is the ratio of a Paasche Farm Machinery Price Index to the USDA Prices Received by Farmers Index as of March 15 of *the same year*.[13] The Farm Machinery Price Index is the "implicit farm gross investment deflator" arrived at by dividing the gross investment figures on farm machinery and motor vehicles in current dollars by the estimated gross expenditures in 1935–9 dollars. The stock of horses and mules on farms is the product of the number of horses and mules on farms times their average price in 1935–9.[14] All the stock and flow variables enter linearly into the regressions; all the others (in the reported regressions only the "real price") are defined as logarithms of the original values.[15] The period of analysis is 1920–41 and 1947–57.

Before we consider the different capital measures in detail, it is worthwhile to establish first that different measures of capital differ enough to matter. Although it is quite true that in a stationary world most measures of capital would come to the same thing, it is very misleading to apply this reasoning to the real world. Given changing rates of growth and substantial cyclical fluctuations, different measures of capital will differ both in the slope of their trends and in cyclical timing. Figure 9.2 presents the gross investment figures underlying all the various capital measures. Figure 9.3 presents a selected number of these capital measures and illustrates the resulting differences in level and movement. The differences in movement are minor, but the differences in trend are striking, particularly in the post-World-War-II period. On the whole, however, the series look enough alike to throw some doubt on the profitability of our search for "the" capital measure based on a goodness-of-fit criterion.[16]

Four capital measures were chosen for extensive analysis, and two more were added to cover more completely the range of possible assumptions.

The first measure, K_1, is the unpublished "official" USDA estimate of the net stock of machinery on farms underlying the published official estimates of depreciation and net farm income. It is based on a declining balance depreciation scheme for the different items at a rather high rate (approximately 17 percent per year). The second measure, K_2, is also a net-stock measure based on declining balance depreciation but at a lower rate (12 percent). This rate is closer to the rates of depreciation observed in the market for used machinery. The third concept of net stock, K_3, is based on straight-line depreciation (15 years of life until 1940, 20 thereafter). This type of capital measure is probably the one used most widely by economists and statisticians, and the resulting estimates are comparable and very close to Goldsmith's estimates for similar categories. The last of the first four measures, K_4, is a measure of the gross stock where depreciation is only retirement depreciation, and its form is based on survival curve data. Since for K_3 and K_4 depreciation or replacement is not proportional to the measured stock, two measures of depreciation, D_3 and D_4, were also added to the analysis. Finally, two additional net stock variables were added–K_{30} and K_{05}, each based on declining balance depreciation, but one at the very high rate of 30 percent per year and the other at the relatively low rate of 5 percent per year. We can think of K_4 as approximating the *quantity*-of-capital concept, the *number* of machines on farms, whereas K_1 and K_2 are approximations of the *value* of capital on farms. The other measures are intermediate to these two extremes, expressing different assumptions about the pattern and the rate of depreciation.

The results of using these different capital measures to explain gross farm-machinery investments in 1920–41 and 1940–57, in conjunction with the real price of farm machinery and the stock of horses on farms, are presented in table 9.1. Several interesting results emerge from these calculations: None of the measures of capital is statistical significant by itself. All the coefficients are small and most of them are smaller than their standard errors. When depreciation measures are included for those concepts which are not proportional to depreciation, K_3 and K_4, the coefficient of depreciation is significant but has the wrong sign *a priori*. Its coefficient is negative, rather than positive as we would expect if depreciation were a good measure of replacement demand. The most interesting result is represented in lines 9 and 10, where K_1 and K_4 are included together. Separately, the contribution of neither concept was statistically significant, but when they are included together, the coefficients are highly significant at conventional significance levels but *opposite in sign*. The coefficient of the approximation to market value (K_1, K_2, or K_3) is positive, whereas the coefficient of the approximation to gross stock (K_4 and D_3) is negative. The results are similar when we use K_2 or K_{30} instead of K_1. Also, the results of 4.0 using K_3 and D_3 can be interpreted similarly. Straight-line-based depreciation is proportional to gross stock (based on a fixed-life assumption), whereas net stock is closer to market value than to anything else. Thus, the results support our hypothesis that to explain gross investment adequately we may need *two* capital concepts:

Table 9.1 Coefficients of different concepts of the stock of machinery on farms in gross farm-machinery investment regressions

	K_1	K_2	K_3	D_3	K_4	D_4	K_{30}	K_{05}	R
				Coefficients					
1	0.007 (0.027)								0.954
2		−0.005 (0.025)							0.954
3			−0.013 (0.016)						0.955
4					−0.021 (0.016)				0.956
5							−0.060 (0.040)		0.957
6								−0.020 (0.022)	0.955
7			0.118 (0.053)	−2.78 (1.08)					0.964
8					0.012 (0.013)	−4.46 (0.085)			0.978
9	0.326 (0.047)				−2.205 (0.028)				0.984
10		0.381 (0.063)			−0.259 (0.041)				0.981
11					−0.110 (0.017)		+0.281 (0.042)		0.984

The dependent variable is gross farm investment in machinery (farm machinery, tractors, trucks, and 40 percent of automobiles) in billions of dollars at 1935–9 prices. Each of the regressions contains two additional variables. The logarithm of the real price of farm machinery (index of farm-machinery prices divided by the index of prices received by farmers), and the stock of horses and mules on farms in billions of dollars at 1935–9 prices. The coefficients of these two variables are always significantly different from zero at conventional significance levels in all the regressions and vary between −1.7 and −2.4 for the coefficient of real price and between −0.5 and −1.0 for the coefficient of the stock of horses and mules on farms.

The numbers in parentheses are the estimated standard errors of the coefficients, and R is the coefficient of multiple correlation.

K_1 = USDA estimate of the stock of machinery on farms based on declining-balance depreciation scheme for the individual items at an average 17 percent rate.

K_2 = net stock of machinery on farms based on 12 percent declining-balance depreciation per year.

K_3 = net stock based on straight-line depreciation (15 years life before 1940, 20 thereafter).

$D_{3t} = K_{3t} - K_{3t+1} + I_t$.

K_4 is the gross stock of machinery on farms based on survival-curve depreciation. The assumed pattern of depreciation is as follows: 0.990, 0.985, 0.975, 0.97, 0.96, 0.95, 0.93, 0.91, 0.88, 0.81, 0.77, 0.71, 0.65, 0.59, 0.52, 0.45, 0.38, 0.31, 0.26, 0.21, 0.17, 0.13, 0.10, 0.00.

$D_{4t} = K_{4t} - K_{4t+1} + I_t$.

K_{30} = net stock based on 30 percent declining-balance depreciation.

K_{05} = net stock based on 5 percent declining-balance depreciation.

(1) a quantity or number-of-machines concept for the adjustment mechanism, and (2) a value concept for replacement demand. As we have seen, a single capital concept by itself does not work, and when two concepts are included together not only do they contribute significantly to the explanation of gross investment, but also the concepts that approximate market value most closely always have a positive sign, whereas the approximations to the quantity of capital emerge with a negative sign.[17]

Before we take these results too seriously, we should discuss some alternative hypotheses that might explain them. A very simple statistical explanation of these results has been suggested to me by Milton Friedman. Consider the possibility that there is serial correlation in gross investment, which our model does not capture via the few independent variables that it uses. For various reasons, when investment is high in one year it is also more likely to be high in the next year and vice versa. But when investment has been relatively high in the recent past, the *value* concept of capital will be higher relative to the *quantity* concept and will contribute positively to the prediction of next year's investment. When investment has been below the average for some time, the *quantity* concept will be higher relative to the *value* concept and will contribute negatively to the prediction of next year's investment. That is, for various reasons we should have included lagged gross investment in our model. Since we have not done this, the use of these two capital measures approximates the use of lagged investment, and this also explains the different signs that we get for the two measures.

This result can be derived rigorously from a model in which the desired stock of capital depends not only on X but on expected X^*, and the expectations are formed by the function

$$X_t^* - X_{t-1}^* = \beta(X_i - X_{t-1}^*);$$

i.e. expectations are revised by some fraction of the difference between the realized X and the previously expected X^*. This assumption, together with the adjustment equation and the stock-flow identities, leads to the following model:

$$S_{t+1}^* = aX_t^*,$$
$$S_{t+1} - S_t = \gamma(S_{t+1}^* - S_t),$$
$$X_t^* - X_{t-1}^* = \beta(X_t - X_{t-1}^*),$$
$$I_t = S_{t+1} - S_t + \delta S_t.$$

All these equations, substituted into each other and lagged twice, result in the following estimating equation:

$$I_t = \gamma\beta aX + \frac{(\delta - \gamma)\,(\beta - \delta)}{(1 - \delta)}\,S_t + \frac{(1 - \beta)\,(1 - \gamma)}{(1 - \delta)}I_{t-1}.$$

Thus, the introduction of an expected X equation on top of the adjustment equation transforms the resulting estimating equation from a first-order difference equation to a second-order difference equation (in S) and introduces I_{t-1} explicitly into it.

Table 9.2   Farm-machinery investment regressions: different measures of stock versus lagged investment

	Coefficients			
	K_1	K_4	I_{t-1}	R
1	0.326	−0.205		0.984
	(0.047)	(0.028)		
2			0.416	0.976
			(0.082)	
3	−0.068		0.571	0.984
	(0.020)		(0.084)	
4		−0.414	0.496	0.985
		(0.0099)	(0.068)	
5	0.162	−0.125	0.291	0.987
	(0.080)	(0.042)	(0.120)	

Note: The same comments apply as in table 9.1. The new variable I_{t-1} lagged one year.

We can test this explanation of our results by including lagged investment explicitly in our model. Table 9.2 presents the results of substituting lagged gross investment for the various capital measures and the results of including it together with two measures of capital. It is clear from this table that there is something to the charge that we may have ignored the serial correlation in investment, but it does not invalidate our previous conclusions. Although lagged investment contributes significantly to the explanation of current investment both by itself and in conjunction with other capital measures, it does not supersede the contribution of either of the capital measures nor does it change their sign. It works in adddition to, rather than instead of, the two capital measures. Both K_1 and K_4 retain their statistical significance and their opposite signs in the presence of I_{t-1}.

Another possible explanation of these results is that, using two extreme-assumption capital measures together, we allow, in effect, for an intermediate measure of capital based on a more complicated depreciation scheme. Let us then explore briefly the depreciation pattern implied by our results. Take, for example, line 10 in table 9.1, where the coefficients of K_2 and K_4 are $+0.381$ and -0.259, respectively. We know that

$$K_{2t+1} = \sum_{i=0}^{\infty} I_{t-i}(0.88)^i \quad \text{and} \quad K_{4t+1} = \sum_{i=0}^{25} I_{t-1}W_i,$$

where the W's are the weights given in the notes to figure 9.2. Thus

$$
\begin{aligned}
0.381K_{2t+1} - 0.259K_{4t+1} = &+0.125I_t \quad +0.080I_{t-1} +0.040I_{t-2} \\
&+0.008I_{t-3} -0.020I_{t-4} -0.044I_{t-5} \\
&-0.062I_{t-6} -0.080I_{t-7} -0.091I_{t-8} \\
&-0.098I_{t-9} -0.103I_{t-10} -0.104I_{t-11} \\
&-0.100I_{t-12} -0.096I_{t-13} -0.088I_{t-14} \\
&-0.078I_{t-15} -0.067I_{t-16} - \cdots \cdots .
\end{aligned}
$$

We can see that the net weights have different signs in different years, and no one depreciation scheme could produce these results. Figure 9.4 presents a graphic picture of the estimated influence of past gross investment on current gross investment. It is evident that the impact of past gross investment on current gross investment is positive for the first four years, becomes negative in the fifth year, and reaches a peak of negative influence at 12 years, after which the influence of past investment declines slowly to about zero at age 25. The positive impact of the previous few years is due either to serial correlation in the left-out variables or to the fact that replacement demand is strongly correlated in adjacent years. After the first few years this influence wears off, and the contribution of past investment becomes strongly negative, reflecting the depressing effect of the existing stock of capital on new

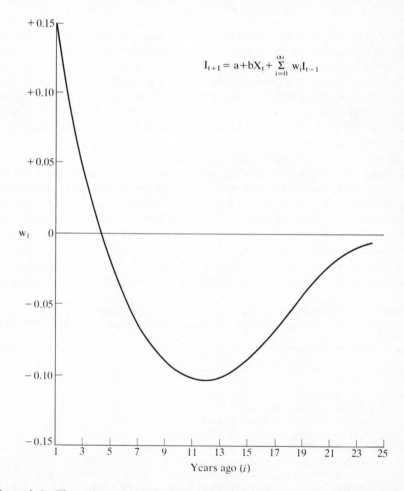

$$I_{t+1} = a + bX_t + \sum_{i=0}^{\infty} w_i I_{t-1}$$

Years ago (i)

Figure 9.4 The estimated impact of investment i years ago on current investment (from line 10, table 9.1).

investment. The fact that the impact of past investment is, on balance, strongly negative should not surprise us. Given our simple model, we would expect the stock of capital to have a negative net effect on gross investment, since the depreciation coefficient should not exceed 0.2, whereas the adjustment coefficients are likely to be higher than that, on the order of 0.3 to 0.5. Be that as it may, it is clear that the resulting pattern of the effects of lagged investment cannot be rationalized on the basis of some intermediate depreciation scheme. It seems preferable, and much easier, to interpret this as the result of the working of two distinct measures of capital, each having a different influence on gross investment.

Summary

In the previous sections I have argued that it is not obvious what kind of measure of capital we want to use in an investment function, but that most of the argument favors the use of two different measures of capital simultaneously, one approximating the idea of capital as a quantity (or number) of machines and the other measure approximating the idea of capital as the current value of the existing stock of machines. I have also argued that I would expect these two different measures to affect gross investment in opposite directions.

In spite of the very high multi-colinearity among the different measures of capital (the simple intercorrelation coefficients are presented in table 9.3), our empirical results tend to support this position. The results are not very clear-cut and are subject to different interpretations, but they still suggest that two measures of capital are better than one. To gain more insight into these problems we shall have to improve our empirical knowledge about what happens to machines as they age. We have very little information about the expected life of different machines, about the rate of deterioration in their physical productivity with age, or about the factors that determine the relative

Table 9.3 Simple correlation coefficient between farm machinery investment, lagged investment, and different measures of the stock of machinery on farms, 1920–41, 1947–57[a]

	I_i	K_1	K_2	K_3	K_3	K_4	D_4	K_{30}	K_{05}
K_1	0.741								
K_2	0.734	0.989							
K_3	0.693	0.986	0.994						
D_3	0.587	0.945	0.968	0.980					
K_4	0.672	0.969	0.991	0.996	0.988				
D_4	0.606	0.783	0.865	0.856	0.886	0.897			
K_{30}	0.809	0.985	0.978	0.957	0.906	0.940	0.779		
K_{05}	0.742	0.948	0.984	0.975	0.959	0.985	0.936	0.944	
I_{t-1}	0.917	0.857	0.841	0.798	0.717	0.773	0.642	0.921	0.814

[a] For the definition of variables see table 9.1.

prices of different age machines in the used-machinery markets. Without this knowledge we don't know how to measure any kind of capital. With this knowledge we will still need different measures of capital to answer different questions.

Notes

1 Even though I do not agree with his main conclusions, the best discussion of many of these problems is to be found in Denison's paper [3]. See also Kuznets [15, 16].

2 Adjustments may not be symmetric: it may take longer to disinvest than to invest, the adjustment path may be nonlinear, and so on. The rate of adjustment may depend on other variables, e.g. "liquidity." It seems quite clear that the usual liquidity variables have no place in the desired-capital function. What they affect is not the desired level of capital, but only the rate at which these desires can be satisfied. Thus, such variables, if included at all in investment functions, should be entered in a form in which their influence will die out eventually.

3 Note that this is equivalent to assuming a declining balance depreciation scheme (the capital stock "disappears" at a constant percentage rate), and it immediately implies one particular way of measuring K – as a weighted average of all past investments, with weights declining geometrically.

4 I am greatly indebted to Trygve Haavelmo, in whose seminar on investment theory (Winter 1958, at the University of Chicago) I participated, for some of the ideas in this section. See also his monograph on investment theory [12]. He, of course, is not responsible for my interpretation and mistakes.

5 Deterioration would be subtracted from the adjusted gross stock. Ideally, this net stock would equal the number of existing machines times one minus the percentage decline in their average physical productivity.

6 I am well aware of the fact that it is extremely difficult in practice to distinguish between physical and economic aspects of a factor of production and that only economically relevant physical aspects are of interest to us. Nevertheless, it seems to me both important and useful to keep these distinctions in mind, even if there is very little that we can do about them in practice.

7 In Fabricant's language [4] the latter would include both capital consumption and capital adjustment.

8 In a perfectly competitive world the annual rent of a machine would equal the marginal product of its services. The rent itself would be determined by the interest costs on the investment, the deterioration in the future productivity of the machine due to current use, and the expected change in the price of the machine (obsolescence). To the extent that the world deviates from the competitive model, this weighting scheme may not be fully appropriate.

9 For an interesting discussion of a particular depreciation model, see [2].

10 It is really outside the scope of this paper to pursue these interesting problems further. For an outline of what a theory of used-machinery markets should contain, see Fox [6]. Also, see Farrell [5].

11 Since this was first written I have become somewhat more optimistic about the possibility of adjusting machinery price indexes for quality change. For an example of such an attempt, see [10].

12 Actually, it is equal to farmers' expenditures on farm machinery, tractors, trucks, and *40 percent* of the expenditures on automobiles, since only 40 percent of the total automobile expenditures are considered by the USDA as a *productive* investment, the rest being classified as consumption expenditures. In retrospect, I feel that it might have been better either to include all automobile purchases or to exclude all. At the time this study was begun, however, I wanted to cover all productive investment excluding construction, and it seemed easiest to follow the USDA definition. I am indebted to the Farm Income Branch, AMS, USDA, for making available to me these unpublished estimates of gross investment in 1935–9 prices.

13 Wherever I use the word "machinery" alone it is a shorthand for all farm machinery, tractors, trucks, and 40 percent of automobiles.

14 Weighting horses and mules separately according to age groups did not lead to appreciably different results.

15 This form was chosen to allow for the linear identities between stocks and flows and for some nonlinearity in the behavior relation. The result is a semi-logarithmic desired-demand function with a linear adjustment equation and identities.

16 From the point of view of productivity analysis, these same differences would loom much larger, implying substantially different rates of growth in "total factor productivity" in the post-World-War-II period. (See [9].)

17 Similar results were obtained in the tractor-demand study. The two concepts of capital used there were the USDA estimated net value of tractors on farms (based on 18.5 percent declining-balance depreciation) and the actual *number* of tractors on farms. In this case each of the variables was also significant separately. Together, however, they contributed distinctly more to the explanation of gross investment with the coefficient of net stock having a positive sign, whereas the coefficient of the *number* of tractors on farms had a negative sign. For details of this study see [8].

References

[1] Chenery, Hollis, Overcapacity and the acceleration principle, *Econometrica*, 20 (January 1952), 1–28.

[2] Cramer, J. S., The depreciation and mortality of motor cars, *J. Roy. Stat. Soc.*, Ser. A, **121**, Part I (1958).

[3] Denison, Edward F., Theoretical aspects of quality change, capital consumption, and net capital formation. In *Problems of Capital Formation: Concepts, Measurement, and Controlling Factors* (Conference on Research in Income and Wealth 19). New York: National Bureau of Economic Research (distr. Princeton University Press), 1957, pp. 215–60.

[4] Fabricant, Solomon, *Capital Consumption and Adjustment*. New York: National Bureau of Economic Research, 1938.

[5] Farrell, M. J., The demand for motor cars in the United States, *J. Roy. Stat. Soc.*, **117** (1954), 171–93.

[6] Fox, Arthur, H., A theory of second-hand markets, *Economica*, **24** (May 1957), 99–115.

[7] Goldsmith, Raymond W., *Study of Saving in the United States*, vols. I–III. Princeton, NJ: Princeton University Press, 1955–56.

[8] Griliches, Zvi, The demand for a durable input: U.S. farm tractors, 1921–1957. In A. C. Harberger (ed.), *The Demand for Durable Goods*. Chicago: University of Chicago Press, 1960.

[9] Griliches, Zvi, Measuring inputs in agriculture: a critical survey, *J. Farm Econ.*, XLII (December 1960), 1411–27.

[10] Griliches, Zvi, Hedonic price indexes for automobiles: an econometric analysis of quality change. In *The Price Statistics of the Federal Government* (Gen. Ser. 73). New York: National Bureau of Economic Research (distr. Princeton University Press), 1961, pp. 173–96.

[11] Grunfeld, Yehuda, The determinants of corporate investment. In A. C. Harberger (ed.), *The Demand for Durable Goods*. Chicago: University of Chicago Press, 1960.

[12] Haavelmo, Trygve, *A Study in the Theory of Investment*. Chicago: University of Chicago Press, 1960.

[13] Hood, Wm. C. and Antony Scott, *Output, Labour and Capital in the Canadian Economy*, Royal Commission on Canada's Economic Prospects, February 1957 (Queen's Printer: Hull).

[14] Kuh, Edwin, The validity of cross-sectionally estimated behavior equations in time series applications, *Econometrica*, **27** (April 1959), 197–214.

[15] Kuznets, Simon, Comment, *Studies in Income and Wealth*, vol. XIV (Conference on Research in Income and Wealth 14). New York: National Bureau of Economic Research (distr. Princeton University Press), 1951.

[16] Kuznets, Simon, Comment. In *Problems of Capital Formation: Concepts, Measurement, and Controlling Factors* (Conference on Research in Income and Wealth 19). New York: National Bureau of Economic Research (distr. Princeton University Press), 1957.

[17] Nerlove, Marc, *Distributed Lags and Demand Analysis of Agricultural and Other Commodities*, USDA Agriculture Handbook no. 141, Washington, DC, Supt of Documents, June 1958.

[18] Smith, Vernon L., The theory of investment and production, *Quart. J. Econ.*, LXXIII (February 1959), 67–87.

[19] Solow, R. M., Investment and technical progress. In Kenneth J. Arrow, Samuel Karlin, and Patrick Suppes (eds), *Mathematical Methods in the Social Sciences, 1959*. Stanford: Stanford University Press, 1960, pp. 93–104.

[20] Terborgh, George, *Dynamic Equipment Policy*. New York: McGraw–Hill, 1949.

[21] Towne, M. W., and W. D. Rasmussen, Farm gross product and gross investment in the 19th century. In *Trends in the American Economy in the 19th Century* (Conference on Research in Income and Wealth 24). New York: National Bureau of Economic Research (distr. Princeton University Press), 1960.

Part II

Education and
Productivity

10
Notes on the Role of Education in Production Functions and Growth Accounting*

Introduction

This paper started out as a survey of the uses of "education" variables in aggregate production functions and of the problems associated with the measurement of such variables and with the specification and estimation of models that use them. It soon became clear that some of the issues to be investigated (e.g. the relative contributions of ability and schooling to a labor quality index) were very complex and possessed a literature of such magnitude that any "quick" survey of it would be both superficial and inadvisable. This paper, therefore, is in the form of a progress report on this survey, containing also a list of questions which this literature and future work may help eventually to elucidate. Not all of the interesting questions will be asked, however, nor all of the possible problems raised. I have limited myself to those areas which seem to require the most immediate attention as we proceed beyond the work already accomplished.

As it currently stands this paper first recapitulates and brings up to date the construction of a "quality of labor" index based on the changing distribution of the US labor force by years of school completed. It then surveys several attempts to "validate" such an index through the estimation of aggregate production functions and reviews some alternative approaches suggested in the literature. Next, the question of how many "dimensions" of labor it is useful to distinguish is raised and explored briefly. The puzzle of the apparent constancy of rates of return to education and of skilled–unskilled wage differentials in the past two decades provides a unifying thread through the latter parts of this paper as the discussion turns to the implications of the ability-education-income inter-relationships for the assessment of the

*From: *Education, Income, and Human Capital*, W. Lee Hansen (ed.), *Studies in Income and Wealth*, NBER, vol. 35. New York: Columbia University Press, 1970.

contribution of education to growth, the possible sources of the differential growth in the demand for educated versus uneducated labor, and the possible complementarities between the accumulation of physical and human capital. While many questions are raised, only a few are answered.

The Quality of Labor and Growth Accounting

One of the earliest responses to the appearance of a large "residual" in the works of Schmookler [50], Kendrick [39], Solow [56] and others was to point to the improving quality of the labor force as one of its major sources. More or less independently, calculations of the possible magnitude of this course of economic growth were made by Schultz [53, 54] based on the human capital approach and by Griliches [22] and Denison [16] based on a standardization of the labor force for "mix-changes." Both approaches used the changing distribution of school years completed in the labor force as the major quality dimension, weighting it either by human capital based on "production costs" times an estimated rate of return, or by weights derived from income-by-education data.[1]

At the simplest level, the issue of the quality of labor is the issue of the measurement of labor input in constant prices and a question of correct aggregation. It is standard national-income accounting practice to distinguish classes of items, even within the same commodity class, if they differ in value per unit. Thus, it is agreed (rightly or wrongly) that an increase of 100 units in the production of bulldozers will increase "real income" (GNP in "constant" prices) by more than a similar numeric increase in the production of garden tractors. Similarly, as long as plumbers are paid more than clergymen, an increase in the number of plumbers results in a larger increase in total "real" labor input than a similar increase in the number of clergymen. We can illustrate the construction of such indexes by the following highly simplified example:

| | Number | | Base period |
Labor category	Period 1	Period 2	wage
Unskilled	10	10	1
Skilled	10	20	2
Total	20	30	

The index of the unweighted number of workers in period 2 is just $N_2 = 30/20 = 1.5$. The "correct" (weighted) index of labor input is

$$L_2 = \frac{10 + 2 \times 20}{10 + 2 \times 10} = \frac{50}{30} = 1.67$$

The index of the average quality of labor per worker can be defined either as the ratio of the second to the first measure or equivalently as the "predicted" index of the average wage rate, based on the second period's labor mix and base period wages:

$$\bar{w}_1 = 1.5, \bar{w}_2^* = \frac{10 + 2 \times 20}{30} = 1.67, E_2 = \frac{1.67}{1.5} = L_2/N_2 \doteq 1.113.$$

Note that we have said nothing about what happened to actual relative wages in the second period. If they changed, then we could have also constructed indexes of the Paasche type which would have told a similar but not numerically equivalent story. It is then more convenient, however, and more appropriate to use a (chain-linked) Divisia total-labor-input index based on a weighted average of the rates of growth of different categories of labor, using the relative shares in total labor compensation as weights.[2] To represent such an index of total labor input, let L_l be the quantity of input of the lth labor service, measured in man-hours. The rate of growth of the index of total labor input, say L, is:

$$\frac{\dot{L}}{L} = \Sigma v_l \frac{\dot{L}}{L_l}$$

where v_l is the relative share of the lth category of labor in the total value of labor input.[3] The number of man-hours for each labor service is the product of the number of men, say n_l, and hours per man, say h_l; using this notation the index of total labor input may be rewritten:

$$\frac{\dot{L}}{L} = \Sigma v_l \frac{\dot{n}_l}{n_l} + \Sigma v_l \frac{\dot{h}_l}{h_l}.$$

The index of labor input can be separated into three components – change in the total number of men, change in hours per man, and change in the average quality of labor input per man (or man-hour). Assuming that the relative change in the number of hours per man is the same for all categories of labor services, say $\dot{H}/H$,[4] and letting N represent the total number of men and e_l the proportion of the workers in the lth category of labor services, one may write the index of the total labor input in the form:

$$\frac{\dot{L}}{L} = \frac{\dot{H}}{H} + \frac{\dot{N}}{N} + \Sigma v_l \frac{\dot{e}_l}{e_l}.$$

Thus, to eliminate errors of aggregation one must correct the rate of growth of man-hours as conventionally measured by adding to it an index of the quality of labor input per man. The third term in the above expression for total input provides such a correction. Calling this quality index E, we have

$$\frac{\dot{E}}{E} = \Sigma v_l \frac{\dot{e}_l}{e_l}.$$

For computational purposes it is convenient to note that this index may be written as follows:

$$\frac{\dot{E}}{E} = \Sigma \frac{p_l}{\Sigma p_l e_l} \dot{e}_l = \Sigma p'_l \dot{e}_l,$$

Table 10.1 Civilian labor force, males 18–64 years old, percentage distribution by years of school completed

School year completed	1940	1948	1952		1957	1959	1959[a]	1962[a]	1965[a]	1967[a]
Elementary 0–4	10.2	7.9	7.6		6.3	5.5	5.9	5.1	4.3	3.6
5–6 or 5–7[b]	10.2	7.1	6.6	11.6	11.4	10.4	10.7	9.8	8.3	7.8
7–8 or 8[b]	33.7	26.9	25.1	16.8	16.8	15.6	15.8	13.9	12.7	11.6
High school 1–3	18.3	20.7	19.4		20.1	20.7	19.8	19.2	18.9	18.5
4	16.6	23.6	24.6		27.2	28.1	27.5	29.1	32.3	33.1
College 1–3	5.7	7.1	8.3		8.5	9.2	9.4	10.6	10.6	11.9
4+ or 4	5.4	6.7	8.3		9.6	10.5	6.3	7.3	7.5	8.0
5+	—	—	—		—	—	4.7	5.0	5.4	5.5

[a] Employed, 18 years and over.

[b] 5–6 and 7–8 for 1940, 1948 and the first part of 1952, 5–7 and 8 thereafter.

Source: The basic data for columns 1, 3, 4, 5, and 6 are taken from US department of Labor, *Special Labor Force Report*, No. 1 "Educational Attainment of Workers, 1959." The 5–8 years class is broken down into the 5–7 and 8 (5–6 and 7–8 for 1940, 1948, and 1952) on the basis of data provided in *Current Population Report*, Series p–50, nos. 14, 49, and 78. The 1940 data were broken down using the 1940 *Census of Population*, vol. III, Part 1, table 13. For 1952 the division of the 5–7 class into 5–6 and 7 was based on the educational attainment of all males by single years of school completed from the 1950 *Census of Population*. The 1962, 1965, and 1967 data are taken from Special Labor Force Reports Nos. 30, 65, and 92 respectively.

where p_l is the price of the lth category of labor services and p'_l is its relative price. The relative price is the ratio of the price of the lth category of labor services to the average price of labor services, $\Sigma p_l e_l$.

In principle, it would be desirable to distinguish as many categories of labor as possible, cross-classified by sex, number of school years completed, type and quality of schooling, occupation, age, native ability (if one could measure it independently), and so on. In practice, this is a job of such magnitude that it hasn't yet been tackled in its full generality by anybody, as far as I know. Actually, it is only worthwhile to distinguish those categories in which the relative numbers have changed significantly.[5] Since our interest is centered on the contributions of "education," I shall present the necessary data and construct such an index of input quality labor for the United States, for the period 1940–67, based on a classification by years of school completed of the *male* labor force only. These numbers are taken from the Jorgenson–Griliches [37] paper, but have been extended to 1967.

Table 10.1 presents the basic data on the distribution of the male labor force by years of school completed. Note, for example, the sharp drop in the percentage of the labor force having no high school education (from 54 percent in 1940 to 23 percent in 1967) and the sharp rise in the percentage completing high school and more (from 28 in 1940 to 58 in 1967). Table 10.2 presents data on mean income of males by school years completed, and table 10.3 uses these data together with table 10.1 to derive an estimate of the implied rate of growth of labor input (quality) per worker.[6] The columns in table 10.3 come in pairs (for example, the columns headed 1939 and 1940–8). The first column gives the estimated relative wage (income) of a particular class and is derived by expressing the corresponding numbers in table 10.2 as ratios to their average (the average being computed using the corresponding entries of table 10.1 weights). The second column of each pair is derived as the difference between two corresponding columns of table 10.1. It gives the change in percentages of the labor force accounted for by different educational classes. The estimated rate of growth of labor quality during a particular period is then derived simply as the sum of the products of the two columns, and is converted to per annum units.[7]

For the period as a whole, the quality of the labor force so computed grew at approximately 0.8 percent per year. Since the total share of labor compensation in GNP during this period was about 0.7, about 0.6 percent per year of aggregate growth can be associated with this variable, accounting for about one-third of the measured "residual." A comparison and review of similar estimates for other countries can be found in Selowsky's [52] dissertation and Denison [18].

Note that in these computations no adjustment was made to the relative weights for the possible influence of "ability" on these differentials. Also, while a portion of observed growth can be attributed to the changing educational *composition* of the labor force, it should not be interpreted to imply that all of it has been produced by or can be attributed to the educational *system*. I shall elaborate on both of these points later on in this paper.

Table 10.2 Mean annual earnings of males, 25 years and over by school years completed, selected years ($)

School year completed	1939	1949	1956	1958	1959	1963	1966
Elementary 0–4	665	1,724	2,127	2,046	2,935	2,465	2,816
5–6 or 5–7	900	2,268	2,927	2,829	4,058	3,409	3,886
7–8 or 8	1,188	2,693 2,829	3,732	3,769	4,725	4,432	4,896
High school 1–3	1,379	3,226	4,480	4,618	5,379	5,370	6,315
4	1,661	3,784	5,439	5,567	6,132	6,588	7,626
College 1–3	1,931	4,423	6,363	6,966	7,401	7,693	9,058
4+ or 4	2,607	6,179	8,490	9,206	9,255	9,523	11,602
5+			—	—	11,136	10,487	13,221

Note: Earnings in 1939 and 1959; total income in 1949, 1958, 1963 and 1966.

Source: Columns 1, 2, 3, 4, H. P. Miller [42, table 1, p. 966]. Column 5 from 1960 *Census of Population* PC(2)–7B, "Occupation by Earnings and Education." Columns 6 and 7 computed from *Current Population Reports*, Series p-60, no. 43 and 53, tables 22 and 4 respectively, using midpoints of class intervals and $44,000 for the over $25,000 class. The total elementary figure in 1940 broken down on the basis of data from the 1940 *Census of Population*. The "less than 8 years" figure in 1949 split on the basis of data given in H. S. Houthakker [34]. In 1956, 1958, 1959, 1963, and 1966, split on the basis of data on earnings of males 25–64 from the 1959 1-in-a-1000 Census sample. We are indebted to G. Hanoch [31] for providing us with this tabulation.

It is important to note that by using a Divisia type of index with shifting weights, one can to a large extent escape the criticism of using "average" instead of "marginal" rates (or products) to weight the various education categories. If the return to a particular type of education is declining, such indexes will pick it up with not too great a lag and readjust its weights accordingly. Also, note that I have not elaborated on the alternative of using the growth in "human capital" to construct similar indexes. For productivity measurement purposes, we want indexes based on "rental" rather than "stock" values as weights. It can be shown (see Selowsky [52]), that if similar data are used consistently, there is no operational difference between the quality index described above and a "human capital times rate of return" approach, provided the capital valuation is made at "market prices" (i.e., based on observed rentals) rather than at production costs. For my purposes, the construction of "human capital" series would only add to the "round-aboutness" of the calculations. Such calculations (or at least the calculation of the rates of return associated with them) are, of course, required for discussions of optimal investment in education programs.

Education as a Variable in Aggregate Production Functions

Much of the criticism of the use of such education per man indexes as measures of the quality of the labor force is summarized by two related questions: (1) Does education "really" affect productivity? (2) Is "education" and its contribution measured correctly for the purpose at hand? After all, the measures I have presented are not much more than accounting conventions. Evidence (in some casual sense) has yet to be presented that "education" explains productivity differentials and that, moreover, the particular form of this variable suggested above does it best. There is, of course, a great deal of evidence that differences in schooling are a major determinant of differences in wages and income, even holding many other things constant.[8] Also, rational behavior on the part of employers would lead to the allocation of the labor force in such a way that the value of the marginal product of the different types of labor will be roughly proportional to their relative wages. Still, a more satisfactory way of really nailing down this point, at least for me, is to examine the role of such variables in econometric aggregate production function studies. Such studies can provide us with a procedure for "validating" the various suggested quality adjustments, and possibly also a way of discriminating between alternative forms and measures of "education."

Consider a very simple Cobb-Douglas type of aggregate production function:

$$Y = AK^{\alpha}L^{\beta}$$

where Y is output, K is a measure of capital services, and L is a measure of labor input in "constant quality units." Let the correct labor input measure be defined as

Table 10.3 Relative prices,[a] changes in distribution of the labor force, and indexes of labor input per man, US males, civilian labor force, 1940–64

I. Relative prices and changes in the distribution of the labor force

School years completed	p' 1939	ė 1940–8	p' 1949	ė 1948–52	p' 1956	ė 1952–7	p' 1958	ė 1957–9	p' 1958	ė 1959–62	p' 1963	ė 1962–5	p' 1966	ė 1965–7
Elementary														
0–4	0.497	−2.3	0.521	−0.3	0.452	−1.3	0.409	−0.8	0.498	−0.8	0.407	−0.8	0.380	−0.7
5–6 or 5–7	0.672	−3.1	0.685	−0.5	0.624	−0.2	0.565	−1.0	0.688	−0.9	0.562	−1.5	0.525	−0.5
7–8 or 8	0.887	−6.8	0.813	−1.8	0.796	−3.3	0.753	−1.2	0.801	−1.9	0.731	−1.2	0.661	−1.1
High school														
1–3	1.030	2.4	0.974	−1.3	0.955	0.7	0.923	0.6	0.912	−0.6	0.886	−0.3	0.861	−0.4
4	1.241	7.0	1.143	1.0	1.159	2.6	1.113	0.9	1.039	1.6	1.087	3.2	1.030	+0.8
College														
1–3	1.442	1.4	1.336	1.2	1.356	0.2	1.392	0.7	1.255	1.3	1.269	—	1.223	+1.3
4+ or 4	1.947	1.3	1.866	1.6	1.810	1.3	1.840	0.9	1.569	1.0	1.571	0.2	1.566	+0.5
5+	—	—	—	—	—	—	—	—	1.888	0.3	1.730	0.4	1.785	+0.1

II. Labor input per man: percentage change

	1940–8	1948–52	1952–7	1957–9	1959–62	1962–5	1965–7
Total	6.45	2.50	2.97	2.39	2.36	2.13	1.77
Annual	0.78	0.62	0.59	1.20	0.79	0.72	0.88

[a]The relative prices are computed using the appropriate beginning period distribution of the labor force as weights.
Source: Derived from tables 10.1 and 10.2.

Table 10.4 Education and skill variables in aggregate production function studies

Industry, unit of observation, period and sample size	Labor coefficient	Education or skill variable coefficient	R^2
1. US agriculture, 68 regions, 1949	(a) 0.45 (0.07)		0.977
	(b) 0.52 (0.08)	0.43 (0.18)	0.979
2. US agriculture, 39 "states," 1949–54–9	(a) 0.43 (0.05)		0.980
	(b) 0.51 (0.06)	0.41 (0.16)	0.981
3. US manufacturing, states and two-digit industries, $N = 417$, 1958	(a) 0.67 (0.01)		0.547
	(b) 0.69 (0.01)	0.95 (0.07)	0.665
4. US manufacturing, states and two-digit industries, $N = 783$, 1954–7–63	(a) 0.71 (0.01)		0.623
	(b) 0.75 (0.01)	0.96 (0.06)	0.757
	(c) 0.85 (0.01)	0.56 (0.16)	0.884

Note: All the variables (except for state industry, or time dummy variables) are in the form of logarithms of original values. The numbers in parentheses are the calculated standard errors of the respective coefficients.

Sources: 1. Griliches [23], table 1. Dependent variable: sales, home consumption, inventory change, and government payments. Labor: full-time equivalent man-years. "Education" – average education of the rural farm population weighted by average income by education class-weights for the US as a whole, per man. Other variables included in the regression: livestock inputs, machinery inputs, land, buildings, and other current inputs. All variables (except education) are averages per commercial farm in a region. 2. Griliches [24], table 2. Dependent variable: same as in (1) but deflated for price change. Labor: total man-days, with downward adjustments for operators over 65 and unpaid family workers. Education: similar to (1). Other variables: Machinery inputs, Land and buildings, Fertilizer, "Other", and time dummies. All of the variables (except education and the time dummies) are per farm state averages. 3. Griliches [25], table 5. Dependent variable: Value added per man-hour. Labor: total man-hours. Skill: Occupational mix-annual average income predicted for the particular labor force on the basis of its occupational mix and national average incomes by occupation. Other variable: Capital services. All variables in per-establishment units. 4. Griliches [27], table 3. Dependent, labor, and skill variables same as above. Other variables: (a) and (b) Capital based on estimated gross-book-value of fixed assets; (c) also includes 18 industry and 20 regional dummy variables.

$$L = E \cdot N,$$

where N is the "unweighted" number of workers and E is an index of the quality of the labor force. Substituting EN for L in the production function, we have

$$Y = AK^{\alpha}E^{\beta}N^{\beta},$$

providing us with a way of testing the relevance of any particular candidate for the role of E. At this level of approximation, if our index of quality is correct and relevant, when the aggregate production function is estimated using N and E as separate variables, the coefficient of quality (E) should both be "significant" in some statistical sense and of the same order of magnitude as the coefficient of the number of workers (N).[9] It is this type of reasoning which led me, among other things, to embark on a series of econometric production function studies using regional data for US agriculture and manufacturing industries. The results of these studies, as far as they relate to the quality of labor variables, are summarized in table 10.4.[10]

In general they support the relevance of such "quality" variables fairly well. The education or skill variables are "significant" at conventional statistical levels and their coefficients are, in general, of the same order of magnitude (not "significantly" different from) as the coefficients of the conventional labor input measures. It is only fair to note that the inclusion of education variables in the agricultural studies does not increase greatly the explained variance of output per farm at the cross-sectional level, while the expected equality of the coefficients of E and N is only very approximate in the manufacturing studies. Nevertheless, this is about the only direct and reasonably strong evidence on the aggregate productivity of "education" known to me, and I interpret it as supporting both the relevance of labor quality so measured and the particular way of measuring it.[11]

There have been a few attempts to introduce education variables in a different way. Hildebrand and Liu [33] considered the possibility that an education variable may modify the exponent of a conventional measure of labor in a Cobb–Douglas type production function. Their empirical results, however, did not provide any support for such a hypothesis, partly because of lack of relevant data. They used the education of the total labor force in a state for the measurement of the quality of the labor force of individual industries within the same state. But the difficulty of estimating interaction terms of the form $E \log L$ implied by their hypothesis, arises mostly, I believe, because there is no good theoretical reason to expect this particular hypothesis (that education affects the share of labor in total production) to be true. Brown and Conrad [13] have proposed the more general (and hence to some extent emptier) hypothesis that education affects *all* the parameters of the production function. They did not, however, estimate a production function directly, including instead a measure of the median years of schooling in ACMS type of time series regressions of value added per worker on wage rates and other variables. Their results are hard to interpret, in part because their education variables are fundamentally trends (having been interpolated between the observed 1950 and 1960 values), and because the same final equation is implied by the very much simpler errors-in-the-measurement-of-labor model. Nelson and Phelps [46] have suggested that education may affect the rate of diffusion of new techniques more than their level. This would imply in cross-sectional data that education affects the overall efficiency parameter instead of serving as a modifier of the labor variable. Nelson and Phelps do not present any empirical estimates of their model. Without further detailed specification of their hypothesis, it is not operationally different from the quality of labor view of education in a Cobb–Douglas world, since any multiplicative variable can always be viewed as modifying the constant instead of one of the other variables.[12]

No studies, as far as I know, have used a human capital variable as an alternative to the labor-augmenting quality index in estimating production functions. While at the national accounting level it need not make any difference which variable is used, the two approaches used in a Cobb–Douglas framework would imply different elasticities of substitution between different types or components of labor. Consider two alternative aggregate production function models.

$$Y = AK^\alpha L^\beta = AK^\alpha N^\beta E^\beta$$

where $E = \Sigma_i r_i N_i / N$ and the r_i's are some base period rentals (wages) for the different categories of labor, and

$$Y = BK^a N^b H^c$$

where H is a measure of "human capital." To be consistent with the E measure it would have to be based on a capitalization of the wage differentials over and above the returns to "raw," unskilled, or uneducated labor (r_0).[13] Thus, approximately

$$H = \delta \Sigma_i \ (r_i - r_0) N_i$$

where δ is a capitalization ratio on the order of one over the discount rate. Note, that given our definitions we can rewrite H as

$$H = \delta(EN - r_0 N) = \delta N(E - r_0)$$

and substituting it into the human capital version of the production function we get

$$Y = CK^a N^{b+c} E^c \left(1 - \frac{r_0}{E} \right)^c.$$

Thus, the H version implies that the production function written in terms of E is not homothetic with respect to E. Moreover, it implies that the elasticity of substitution between H and N is unity, while the E version assumes (for fixed r's) that the elasticity of substitution between different types of labor (the N_i) is infinite, at least in the neighborhood of the observed price ratios.

While such different assumptions are not operationally equivalent, it is probably impossible to discriminate between them on the basis of the type and

Table 10.5   Various education measures in an aggregate agricultural production function (68 regions, US, 1949)

Education variable	Coefficients of		R^2
	X_6 (man-years)	Education variable	
S	0.539	0.0165 (0.0065)	0.9789
log S	0.536	0.297 (0.119)	0.9789
E	0.524	0.431 (0.181)	0.9787
E_2	0.520	0.455 (0.203)	0.9785

S–Mean school years completed of the rural farm population (25 years old and over). E–Logarithm of the school years completed distribution of the rural farm population weighted by mean income of all US males, 25 years and over in 1949. Mean incomes from H. Houthakker [34]. E_2–Same as E except that the weights are mean wage and salary income of native white males (over 25) in 1939. Mean incomes by school years completed computed from the 1940 *Census of Population, Education*, Washington, 1947, pp. 147 and 190. Other variables are the same as in row 1 of table 10.4.
Source: Unpublished mimeographed appendix to Griliches [23].

amounts of data currently available to us. Consider the last equation; it differs from the straight E version by having a different coefficient on E than on N. If we estimate the E equation in an H world, we shall be leaving out the variable $\log(1 - r_0/E)$ with a c coefficient in front of it. But $\log(1 - r_0/E)$ is approximately equal to $-r_0/E$, since $r_0/E < 1$, and the regression coefficient of the left out variable, in the form of $1/E$ on the included variable $\log E$, will be on the order of one, for not too large variations in E. Hence, the estimated coefficient of E in an H world will be on the order of $2c$, which is not likely to be too different from the coefficient of N^{b+c}.

More generally, it is probably impossible to distinguish between various different but similar hypotheses about how the index E should be measured, at least on the basis of the kind of data I have had access to. Whether one uses "specific" or national income weights, or just simply the average number of school years completed, one has variables that are very highly correlated with each other. This is illustrated by the results reported in table 10.5, based on an unpublished appendix to my 1963 study. Our data are just not good enough to discriminate between "fine" hypotheses about the form (curvature) of the relationship or the way in which such a variable is to be measured.

Aggregation

Obviously, in constructing such indexes of "quality" (or human capital) we are engaged in a great deal of aggregation. There are many different types

Table 10.6 Ratios of mean incomes for US males by schooling categories

Year	High school graduates to elementary school graduates		College graduates to high school graduates	
1939	1.40[a]		1.57[c]	
1949	1.41	1.34[b]	1.63	
1958		1.48	1.65	
1959		1.30		1.51[d]
1963		1.49		1.45
1966		1.56		1.52

[a] Elementary 7–8 years
[b] Elementary 8 years
[c] College 4 + years
[d] College 4 years
Source: From table 10.3.

and qualities of "education" and much of the richness and the mystery of the world is lost when all are lumped into one index or number. Nevertheless, as long as we are dealing with aggregate data and asking overall questions, the relevant consideration is not whether the underlying world is really more complex than we are depicting it, but rather whether that matters for the purpose of our analysis. And even if we decide that one index of E hides more than it reveals, our response will surely not be "therefore let's look at 23 or 119 separate labor or education categories," but rather what kind of two-, three-, or four-way disaggregation of E will give us the most insight into the problem.

From a formal point of view, we can appeal either to the Hicks composite-good or to the Leontief separability theorems to guide us in the quest for correct aggregation. If relative prices (rentals or wages) of labor with different schooling or skill levels have remained constant, then we lose little in aggregating them into one composite input measure. A glance at the "relative prices" for different educational classes reported for the United States in table 10.3 does not reveal any drastic changes in them. Thus, it is unlikely that at this level of aggregation much violence is done to the data by putting them further together into one L or E index. Similar results can be gleaned from a variety of occupational and skill differential data (see tables 10.6 and 10.7). In general they have remained remarkably stable in the face of very large changes in relative numbers and other aspects of the economy.[14] In fact the apparent constancy of such numbers constitutes a major economic puzzle to which I shall come back later.

When we abandon the notion of one aggregate labor input and are faced with a list of eight major occupations, eight schooling classes, several regions, two sexes, at least two races, and an even longer list of detailed occupations, there doesn't seem to be much point in trying to distinguish all these aspects of the labor force simultaneously. The next small step is obviously not in the direction of a very large number of types of labor but rather toward the

Table 10.7 Ratios of mean incomes of US employed and salaried males: professional and technical workers to operatives and kindred

Year	Ratio
1947	1.67
1950	1.58
1953	1.55
1959	1.67
1964	1.63

Source: From US Bureau of Census, *Trends in Income of Families and Persons in the US: 1947–1964*, Technical Paper no. 17, Washington, 1967, table 38.

question of whether there are a few underlying relevant "dimensions" of "labor" which could explain, satisfactorily, the observed diversity in the wages paid to different "kinds" of labor. The obvious analogy here is to the hedonic or characteristics approach to the analysis of quality change in consumer goods, where an attempt is made to reduce the observed diversity of "models" to a smaller set of relevant characteristics such as size, power, durability, and so forth.[15] One can identify the "human capital" approach as a one-dimensional version of such an approach.[16] Each person is thought of as consisting of one unit of raw labor and some particular level of embodied human capital. Hence, the wage received by such a person can be viewed as the combination of the market price of "bodies" and the rental value of units of human capital attached to (embodied in) that body:

$$w_i = w_0 + rH_i + u_i$$

where u_i stands for all other relevant characteristics (either included explicitly as variables, controlled by selecting an appropriate sub-class, or assumed to be random and hence uncorrelated with H_i). If direct estimates of H are available, this type of framework can be used to estimate r. If proxy variables are used for H, such as years of schooling, age, or "experience," one can proceed to the estimation of income-generating functions as did Hanoch [31] and Thurow [59] which, in turn, can be interpreted as "hedonic" regressions for people. Alternatively, if one is willing to assume that the implicit prices (w_0 and r) are constant, and one has repeated observations for a given i, one can use such a framework to estimate the unobserved "latent" H_i variable. Consider, for example, a sample of wages by occupation for different industries: If one assumes that occupations differ only by the amount of human capital embodied per capita, and that the price of "bodies" and of "skill" is equalized across industries, then this is just a one-factor analysis model, and it can be used to estimate the implied relative levels of H_i, for different occupations. Of course, having gone so far one need not stop at one factor, or only one underlying skill dimension. The question can be pushed further to how many latent factors or dimensions are necessary or adequate for an explanation of the observed differences in wages across occupations and schooling classes?

This is, in fact, the approach pursued by Mitchell [44] in analyzing the variation of the average wage in manufacturing industries by states. He concludes that one "quality" dimension is enough for his purposes. He does, however, make the very stringent assumption that the implied relative price ratio of bodies to human capital, or of skilled and unskilled wages (w_0/r), is constant across states and countries. This is a very strong assumption, one that is unlikely to be true for data cross-classified by schooling. Studies of US data (see e.g. Welch [62] and Schwartz [51]) have in general found significantly more regional variation in the price of unskilled or uneducated labor than in the price of skilled or highly educated labor, implying the non-constancy of skill differentials across regions (and presumably also countries).

In another paper, Welch [63] outlines a several dimensions model of the general form

$$w_{ij} = w_{0j} + r_{1j}S_{1i} + r_{2j}S_{2i}$$

where i is the index for the level of school years completed, j is the index for states, S_1 and S_2 are two unobserved underlying skill components associated with different educational levels. This is not strictly a factor-analysis model any longer, both because the r's are assumed to vary across states and because no orthogonality assumptions are made about the two latent skill levels. With a few additional assumptions, Welch shows that if the model is correct one should be able to explain the wage of a particular educational or skill level by a linear combination of wages for other skill levels and by no more than three such wages (since there are only three prices here: two "skills" and one "body"). The linearity arises from the implicit assumption that at given prices any unit of S_1 or S_2 (and "body") is a perfect substitute for another. Thus, even though different types of labor are made up of a smaller number of different qualities which may not be perfectly substitutable for each other, because the whole bundle is defined linearly, one can find linear combinations of several types of labor which will be perfectly substitutable for another type of labor. For example, while college and high school graduates may not be perfect substitutes, one college graduate plus one elementary school graduate may be perfect substitutes for two high school graduates. Welch analyzes incomes by education by states and concludes that in general one doesn't need more than three underlying dimensions to explain eight observed wage levels, and that often two are enough. It is not clear whether Welch is using the best possible and most parsimonious normalization, or whether a generalization of the factor-analytic approach with oblique factors could not be adapted to this problem, but clearly this is a very interesting and promising line of analysis.

The approximate constancy of relative labor prices by type, the implicit linearity of the Welch model, and some scattered estimates of rather high elasticities of substitution between different kinds of labor or education levels (e.g. Bowles [8]) all imply that we will lose little by aggregating all the different types of labor into one overall index as long as our interest is not primarily in the behavior of these components and their relative prices.

Ability

This is a very difficult topic with a large literature and very little data. What little relevant data there are have been recently surveyed by Becker [2] and Denison [17]. It has been widely suggested that the usual income-by-education figures overestimate the "pure" contribution of education because of the observed correlation between measured ability and years of school completed. On the basis of scattered evidence both Becker and Denison decide to adjust downward the observed income-by-education differentials, Denison suggesting that all differentials should be reduced by about one-third.

It is useful, at this point, to set up a little model to help clarify the issues. Assume that the true relation in cross-sectional data is

$$Y_i = \beta_0 + \beta_1 S_i + \beta_2 A_i + u$$

where Y is income, S is schooling and A is ability, however measured. The usual calculation of an income–schooling relation alone leads to an estimate of a schooling coefficient (b_{ya}) whose expected value is higher than the true "net" coefficient of schooling (β_1), as long as the correlation between schooling and the left out ability variable is positive. The exact bias is given by the following formula:

$$Eb_{ya} = \beta_1 + \beta_2 b_{AS}$$

where b_{AS} is the regression coefficient in the (auxiliary) regression of the left out variable A on S, the included one. Moving to time series now, and still assuming that the underlying parameters (β_1 and β_2) do not change we have the relationship

$$\bar{Y} = \beta_0 + \beta_1 \bar{S} + \beta_2 \bar{A} + u$$

where the bars stand for averages in a particular year. Now if b_{YS} is derived from cross-sectional data and is used in conjunction with the change in the average schooling level to predict (or explain) changes in $\bar{Y}$ over time, it will overpredict them (give too high a weight to $\bar{S}$) unless $\bar{A}$ changes *pari passu*. But it is assumed that the distribution of A, innate ability, is fixed over time and hence, its mean ($\bar{A}$) does not change. This, therefore, is the rationale for considering the cross-sectional income-education weights with some suspicion and for adjusting them downward for the bias caused by the correlation of schooling with ability.

I should like to question these downward adjustments on three related grounds: (1) Much of measured ability is the product of "learning," even if it is not all a product of "schooling." Often what passes for "ability" is actually some measure of "achievement," and the argument could be made that it in turn is determined by a relation of the form

$$A = \alpha_0 + \alpha_1 S + \alpha_2 QS + \alpha_3 LH + \alpha_4 G + v$$

where S is the level of schooling, QS is the quality of schooling, LH are the learning inputs at "home," and G is the original genetic endowment. If one were to substitute this equation into the original relation for Y one would find that the "total" coefficient of the schooling variables is given by

$$\beta_1 + \beta_2 \alpha_1 + \beta_2 \alpha_2 b_{QS \cdot S}$$

where $b_{QS \cdot S}$ is the relation between the quality and quantity of schooling in the cross-sectional data, and the "total" coefficient associated with changes in total "reproducible" human capital (including that produced at home) by

$$\beta_1 + \beta_2 |\alpha_1 + \alpha_2 b_{QS \cdot S} + \alpha_3 b_{LH \cdot S}|$$

where $b_{LH \cdot S}$ summarizes the relation between learning at home and at school in cross-section. Now while the simple coefficient of income and schooling b_{YS} may overestimate the partial effect of schooling (β_1) holding achievement constant, it may not overestimate that much, if at all, the "total" effect of schooling. (2) The estimated downward adjustments for ability may be overdone particularly in the light of strong interaction of "ability" and schooling as they affect earnings. That is, since the relation between A and Y holding S constant is strong only at higher S levels, b_{AS} may be quite low, and the bias in the estimated b_{YS} may not be all that large. (3) Moreover, the whole issue hinges on whether or not $\overline{A}$ as measured has really remained constant over time. To the extent that proxies such as father's education are used in lieu of "ability," it can be shown that at least their levels did not remain constant.

It is probably best, at this point, to confess ignorance. "Ability," "intelligence," and "learning" are all very slippery concepts. Nor do we know much about the technology of schooling or education. What are the important inputs and outputs, what is the production function of education, how do the various inputs interact? Some work on this is in progress (see Bowles [91]) and perhaps we will know more about it in the future. We do know, however, the following things: (1) Intelligence is not a fixed datum independent of schooling and other environmental influences. (2) It can be affected by schooling.[17] (3) It in turn affects the amount of learning achieved in a given schooling situation. (4) Because the scale in which it is measured is arbitrary, it is not clear whether the relative distribution of "intelligence" or "learning abilities" has remained constant over time.

> The doctrine that intelligence is a unitary something that is established for each person by heredity and that stays fixed through life should be summarily banished. There is abundant proof that greater intelligence is associated with increased education. . . . On the basis of present information it would be best to regard each intellectual ability of a person as a

somewhat generalized skill that has developed through the circumstances of experience, within a certain culture, and that can be further developed by means of the right kind of exercise. There may be limits to ability set by heredity, but it is probably safe to say that very rarely does an individual really test such limits.[18]

Actually, IQ and achievement tests are so intimately intertwined with education that we may never be successful in disentangling all their separate contributions. IQ tests were originally designed to determine which children could not learn at "normal" rates. Consequently, children with above average IQ are expected to learn at above normal rates. The effect of intelligence on learning is presumably twofold (or are these two sides of the same coin?): Higher IQ children know more to start with and this "knowing more" makes it easier to learn a given new subject (since knowing more implies that it is less "new" that it would otherwise be), and higher IQ children are "quicker." They absorb more for a similar length of exposure, and hence know more at the end of a given period. Since schools try, in a sense, to maximize the students' "achievement," and since achievement and IQ tests are highly enough correlated for us to treat them interchangeably, one might venture to define the gross output of the schooling system as ability. That is, schools use the time of teachers and students and their respective abilities to increase the abilities of the students. From this point of view, the student's ability is both the raw material that he brings to the schooling process, which will determine how much he will get out of it, and the final output that he takes away from it. Hence, at least part of the apparent returns to "ability" should be imputed to the schooling system.[19] How much depends on what is the bottleneck in the production of educated people – the educational system or the limited number of "able" people that can benefit from it. If, as I believe may be the case, ability constraints have not been really binding, very little, if any, of the gross return to education should be imputed to the not very scarce resource of innate ability.

Actually, the little data we have shows a surprisingly poor relation between earnings and "ability" measures when formal schooling is held constant. Wolfle summarized the conclusions of such studies as of 1960:

> High school grades, intelligence-test scores, and father's occupation were all correlated with the salaries being earned fifteen to twenty years after graduation from high school, but the amount of education beyond high school was more clearly, more distinctly related to the salaries being earned.
>
> There is another conclusion from the data, one of perhaps greater importance. It is this: the differences in income were greatest for those of highest ability. It is of some financial advantage for a mediocre student to attend college, but it is of greater financial advantage for a highly superior student to do so.[20]

Examining the tables from Wolfle's studies reproduced in Becker and Denison, one is struck both by the importance of interaction, and by the very limited effect of IQ on earnings except for those within the upper tail of the educational distribution.[21] In fact, the IQ adjustment constitutes only a very small portion of Denison's total "ability" adjustment. One of his major adjustments is based on a cross-classification of earnings-by-education by father's occupation. It is not clear at all why this is an "ability" dimension.[22] Higher-income and -status fathers will provide both more schooling at home and *buy* better quality schooling in the market. To the extent that these differences reflect the latter rather than the former, it does not seem reasonable to adjust for them at all.

In most studies that use IQ or achievement tests, these tests are taken at the end of the secondary school period. As we have noted, such test scores are to some unknown degree themselves the product of the educational system (at the high school and elementary level). To separate the "value added" component of schools one would like to have such scores at a much younger age, upon entry into the schooling system. I have come across only one set of data, for the city of Malmö in Sweden (from Husen [35]), which provides a distribution of earnings at age 35 by formal schooling and by IQ at age *10 years*.[23] They are reproduced in table 10.8. One of the important aspects of this particular sample is that it does cover the whole range of both the schooling and IQ distribution. We can use these data to investigate how much change there is in the income–education coefficient when IQ is introduced as an explicit variable.

After some experimentation with scaling and the algebraic form of the relationship, the following weighted regressions were computed for these data (19 observations) with $N/(SD)^2$ as weights:[24]

$$\log Y = 9.317 + 0.053S \qquad\qquad R^2 = 0.589$$
$$ (0.011)$$

and

$$\log Y = 8.938 + 0.051S + 0.0042A \quad R^2 = 0.836$$
$$ (0.007) \quad (0.0009)$$

where Y is income, S is years of school completed and A is the IQ score. In these data "original" IQ is an important variable, "explaining" an additional 30 percent of the variance in the logarithm of incomes, but its introduction does almost nothing to change the coefficient of schooling. There would have been little bias from ignoring it.[25]

Similar results, but based on much more tenuous evidence, can also be had for the United States. For the United States we do not have yet any data on earnings by education and ability on a large scale, but we do have a large body of income-by-education data from the 1960 census of population, and a distribution of "ability" (Armed Forces Qualification Test) by school years completed based on tabulation of army induction tests of youths in 1964–5 (Karpinos, [38]).[26] Since the two bodies of data are not for the same population or time periods, what follows is very much an approximation, prompted by the desire to see whether these data could be of some use after all. Consider the equation

$$\log Y_{ij} = \alpha_j + \beta S_{ij} + \gamma A_{ij}$$

Table 10.8 Taxed income at 35 by number of years of formal schooling and IQ at age 10, males, Malmö, Sweden

IQ at age 10	Number of years of formal schooling											
	Less than 8 years			8–10 years			11–14 years			14 years and more		
	M	SD	N	M	SD	N	M	SD	N	M	SD	N
115+	17,450	4,260	20	21,943	7,363	35	33,750	35,238	32	43,158	19,219	19
108–114	16,625	5,165	32	19,538	7,793	26	19,429	12,893	14	41,000	18,267	13
93–107	15,266	5,270	109	18,176	8,118	68	21,735	7,477	34	31,400	26,567	5
86–92	17,744	10,306	39	17,462	5,955	13	20,500	7,527	10	—	—	—
–85	14,548	4,041	73	14,929	4,611	28	17,750	4,763	4	35,533	6,182	3

Source: From Husén [35] table 16.
Note: M = mean income in kroner; SD = standard deviation; N = number of cases.

Table 10.9 Regression coefficients of the logarithm of income on schooling and of "ability" on schooling by regions

	$b_{(\log Y)S}$	b_{AS}
Northwest, total	0.0663	0.426
Northcentral, total	0.0702	0.470
South, total	0.1011	0.424
South, nonwhite	0.0726	0.343
West, total	0.0760	0.475

Source: *Income by schooling*, data from 1960 Census of Population, median income for males age 35–44; schooling estimated as the midpoint of the class intervals and 18 for 5 + years of college category. "*Ability*" *by schooling*, based on the estimated distribution of the Army Forces Qualification Test for youths aged 19–21 in 1960; from Karpinos [38]. AFQT percentiles scaled by the approximate average score (probit) associated with the particular percentile range (-5, -2, 0, 2, 5 for less than 9, 10–35, 36–64, 65–92, and 93–100, respectively).

On the basis of the above numbers, the implied β and γ in the equation log $Y_{ij} = \alpha_j + \beta S_{ij} + \gamma A_{ij}$ and 0.112 and -0.07, respectively.

where the index i goes over schooling classes and j over regions. We cannot compute this relationship directly, since we do not have the covariances of A with log Y, but we do have information on the relationship of A to S. If we ignore A and estimate a truncated relationship of log Y on S for each region separately, we would get as our coefficient of S in each region (using the left-out-variable formula):

$$Eb_j = \beta + \gamma \cdot b_{A\ Sj}$$

where the term b_{AS} is the regression coefficient in the auxiliary regression connecting the left-out variable A with the included variable S. Since we have such b_j's and b_{ASj} for several regions, we can compute the implied β and γ by another round of least squares. Table 10.9 summarizes such a computation based on data for five regions of the United States. Note that implicit in this computation is the assumption that regional differences in the observed slope of the income-education relationship must be due to regional differences in the association between schooling and ability.

The figures reported in table 10.9 actually imply a negative γ. That is, if an adjustment were made for ability, it would *increase* the estimated influence of schooling on income. This is largely the result of the fact that the only major difference in the income-schooling slope is observed for the South (total, white and nonwhite), while the observed increase in ability with education in the South is only average or even lower.[27] Given the quality of these data, the inherent arbitrariness in the scaling of A, and the many tenuous assumptions required, these results should not be taken seriously. But they too do not come up with any strong evidence for the overwhelming importance of "ability" as distinct from "schooling."[28]

There are two more points to be made. First, to the extent that measured ability is an important determinant of earnings *only* at the higher education levels, it is not correct to "reduce" the education coefficient, or the weights attached to the higher education classes in the construction of quality indexes, unless there is evidence that the observed increases in educational attainment have been associated with a lowering in the average ability of the educated. I know of no evidence which points in this direction. There is no evidence that the growth in educational attainment has been restricted by the drawing down of the "pool of ability."[29] There is a large body of evidence pointing to the existence of many high-ability lower-class children who do not go on to college or finish high school for a variety of economic and social reasons.[30] In spite of the tremendous increase in the number of college graduates in this country, the distribution of college students by social origin (father's occupation) has not changed significantly or adversely in the last thirty years.[31] Also, if ability were a major constraint one might have expected that the observed income differentials would narrow, as poorer-quality people were getting more education.[32] This does not appear to have happened, however. One might even conjecture that, as education spread, the selection processes were actually improved, and hence that there may be a higher correlation of ability with education today than was true 30 years ago.[33] It is also possible that our children being taller and healthier than the previous generations may also be more intelligent.[34]

The second point to be made is that much of what is used as a proxy for "ability" is not really an innate ability and need not and has not remained constant over time. Denison in examining the Wolfle–Smith data concludes that about 6 percent of the observed college-high school income differential can be attributed to the "rank in high school" and about 3 percent to IQ. These, of course, are not independent, but in any case, at most 10 percent of the differential can be ascribed on the basis of the internal evidence of the Wolfle–Smith data to something that could be a reflection of innate ability. An additional 7 percent of the differential is ascribed to differences in father's occupation. The rest, about *half* of the one-third adjustment, is based on the difference between the size of the overall differential as reported in these data (for Illinois, Minnesota, and Rochester men) and the national average differential. At best, therefore, this is not an adjustment for "ability" but for *regional* differences in income and the regional correlation between average incomes and levels of education.[35] Thus, less than one-third of the "one-third" adjustment is related conceptually to ability *per se*.

Even if one allows that the underlying IQ distribution has not changed over time, the other proxy variables, such as father's occupation and regional distribution have changed. Consider, for example, the simple model where father's education is used as a control variable. Then in cross-section we have

$$Y_s = \beta_0 + \beta_1 S_s + \beta_2 S_F$$

where the subscripts s and F stand for sons and fathers respectively. If one ignored the education of fathers variable, one would estimate

$$Eb_{YS} = \beta_1 + \beta_2 b_{S_F S}.$$

where $b_{S_F S_s}$ is probably less than one (the slope of the relationship between the schooling of fathers and sons). In time series, however, we would have

$$\bar{Y}_s = \beta_0 + \beta_1 \bar{S}_s + \beta_2 \bar{S}_F.$$

But the average schooling of fathers has been growing at approximately the same rate as that of sons! Hence, the total effect of schooling should be measured by $\beta_1 + \beta_2$, and the unadjusted b_{YS} is closer to that than the "net" β_1!

Similarly, the average level of education in the North grew at about the same rate as that in the South. Since the North had more education to start with and a higher average wage associated with it, this would lead to a growth in the *share* of the North in "total" labor quality. Hence, if one holds region constant in deriving the educational weights, one should, on the other hand, also adjust the labor input upward for the fact that the share of the higher quality regions grew at the same time. Thus, the one-third downward adjustment suggested by Denison may be a serious overadjustment if what we are interested in is an estimate of the rate of growth in the *total* quality of the labor force. One should recognize, however, that not all of the growth attributed to changing educational attainment is the *net* product of the education system *per se*.

The Puzzle of the Constancy of Differentials

The main evidence on the relative constancy of educational and skill differentials in the post-World War II period in the US is summarized in tables 10.3, 10.6, and 10.7 and has been alluded to before. Becker [2] (table 14, p. 128) reaches similar conclusions about the behavior of rates of return to higher education over time. The puzzling thing is that these differentials and rates of return should change so little in the face of very large shifts in the relative numbers of educated workers. Between 1952 and 1966 the ratio of males between the ages of 18 and 64 in the US civilian labor force with high school education and more to those with elementary education and less changed from about 1 to 1 to about 2.5 to 1, and still their relative incomes did not change greatly.

There appear to be four possible explanations of the phenomenon, three on the demand side and one on the supply side.[36] On the demand side we can conceive three sources of increased demand for skilled workers which could have counterbalanced the depressing effect of the increase in their supply: (1) It may just happen that goods that have an income elasticity higher than one have on average a higher skill content embedded in them than do goods whose income elasticity is less than one. (2) It may be that for some reason not yet clear technical change has been on the average "skill using" and "unskilled labor saving." (3) It is possible, and plausible, that physical capital is more complementary with skilled than with unskilled labor. Since physical capital has been growing at a higher rate than the labor force, this would imply also a growth in the relative demand for educated labor. (4) Finally, it may be that all of this is essentially a reflection of the nature of the supply of skills. The most important factor in the production of skill is the labor of students and teachers. If the production function (time requirements) of skills does not change much over time, the prices of skilled and unskilled labor must move roughly in proportion to each other, since skilled labor can be "manufactured" from unskilled labor in a roughly unchanging way, using resources whose price is proportional both to the input and output price of this process. The existence of such a relation does not, of course, contradict the various demand hypotheses, but makes it very much harder to distinguish among them.

There is very little empirical evidence on any of these points. Nor is it obvious that a priori they are all plausible. It is easy, for example, to think of some commodities such as "food away from home" that have a high income elasticity and a rather low skill content. A crude check on the overall demand hypothesis can be made by investigating whether changes in employment between 1950 and 1960 by industries have any association with the average educational attainment of the labor force in each of the industries. Using data from the 1960 Census of Population for 149 industries we get a correlation coefficient of about 0.33 for the relationship between the percentage change in the employment of males between 1950 and 1960 and the logarithm of mean school years completed by industry in 1960.[37] This is a statistically significant but not very strong relationship. A similar calculation for females yields no relationship at all ($r^2 = -0.07$). While such a relation could be due to several causes, there does appear to be something in the demand hypothesis which may warrant further exploration.

Since we do not know how to measure neutral technical change very well, the probability of measuring the "skill-bias" of technical change in a nontautological fashion is even lower, and I shall not pursue this further here. There remains yet the possibility of capital–skill complementarity which will be explored in the next section.

Are Physical and Human Capital Complements?

To investigate this question we have to start with a three-input production-function output depending on capital and two types of labor (or "bodies" and "skill"). We shall write it in the form

$$Y = F(K, L, S)$$

with the hypothesis to be investigated being that (in the Allen sense):

$$0 < \sigma_{LK} > \sigma_{SK} \lessgtr 0$$

where the σ_{ij}'s are the respective partial elasticities of substitution. It is not clear where one could get some evidence on this. At the aggregate level things are much too collinear to be of much help (moreover, we can't really measure anything more than the trends in K, L, and S with any degree of accuracy). At the micro level, one usually does not have data on S and K at the same time or place, and what is even worse one rarely has any relevant input price data and the price data one has (such as wages) are subject to significant biases precisely because of the existence of the third variable S, significant differences in the quality of the labor force. The following model may, however, give some hope of success:[38] If one starts with inputs defined per unit of output (assuming for this purpose constant returns to scale), and measures everything in logarithms of the variables, one can write (as an approximation), the demand functions for inputs as

$$x_i = a_i + \Sigma \eta_{ij} p_i = a_i + \Sigma v_i \sigma_{ij} p_j$$

where x_i is the logarithm of the ith input per unit of output, p_j is the logarithm of the "real" price of the jth input, η_{ij}'s are the respective price elasticities ($\Sigma_i \eta_{ij} = 0$), v_j is the share of the jth factor in total cost, and the σ_{ij}'s are the Allen-Uzawa partial elasticities of substitution ($\sigma_{ij} = \sigma_{ji}$, and $\sigma_{ii} < 0$). Consider now the special case of three inputs: L–labor, K–capital, and S–skill or schooling, with the corresponding (rental) prices W, R, and Z. Then, using the homogeneity condition, we can write (using lower case letters to denote the logarithms of the corresponding variables):

$$l = \eta_{ll}(w - z) + \eta_{lk}(r - z) + a_l$$

$$k = \eta_{kl}(w - z) + \eta_{kk}(r - z) + a_k$$

$$s = \eta_{sl}(w - z) + \eta_{sk}(r - z) + a_s$$

and subtracting the first two equations from the third

$$s - l = (\eta_{sl} - \eta_{ll})(w - z) + (\eta_{sk} - \eta_{lk})(r - z) + (a_s - a_l)$$

$$s - k = (\eta_{sl} - \eta_{kl})(w - z) + (\eta_{sk} - \eta_{kk})(r - z) + (a_s - a_k)$$

or

$$(s - l) = v_l(\sigma_{sl} - \sigma_{ll})(w - z) + v_k(\sigma_{sk} - \sigma_{lk})(r - z) + c_1$$

$$(s - k) = v_l(\sigma_{sl} - \sigma_k)(w - z) + v_k(\sigma_{sk} - \sigma_{kk})(r - z) + c_2$$

The hypothesis that skill or education is more complementary with physical capital than is physical (or unskilled, or unschooled) labor would imply that the coefficient of $(r - z)$ in the first equation is negative ($\sigma_{lk} > \sigma_{sk}$) and that the coefficient of $(w - z)$ in the second equation is also negative ($\sigma_{kl} > \sigma_{sl}$). At the same time, one would expect the other two coefficients to be positive.[39]

If data were available on the relevant prices, one could estimate either version, in one case assuming that the approximation is better when one assumes the demand elasticities to be constant over the observed range, or alternatively, making the same assumption about the elasticities of substitution.

There is an additional set of assumptions that may allow us to estimate these questions almost without any price data. If one has data by state and industry for the two-digit manufacturing industries in the United States (and assumes that σ's are approximately the same for all these industries, though not the v's), one may hypothesize (a) that the true rental price of capital r does not differ among states but may differ as between industries (because of different depreciation, obsolescence, and risk rates); (b) that the real price of skilled labor has been effectively equalized by migration and the unions, and hence that z is a constant at a point of time; and (c) that the price of pure physical labor does differ between states (not having been equalized by migration) but is essentially the same for all industries within the same state. The coefficients of $(w-z)$ could then be effectively estimated by state dummy variables (or more correctly cross-dummies, if we allow also the v's to vary, which we'll have to, to achieve identification) and the coefficients of $(r-z)$ by industry dummies. The expected sign relations could then be checked by computing the ratio of the respective coefficients in the two equations (e.g. $\overline{v_l(\sigma_{sl} - \sigma_{kl})(w-z)} / \overline{v_l(\sigma_{sl} - \sigma_{ll})(w-z)}$ should be negative).

Alternatively, at a more aggregate time-series level, one may assume that factor prices are changing in the same way for all industries. Then, using an alternative but equivalent set of two equations, we have

$$(s-l) = v_l(\sigma_{sl} - \sigma_{ll})(w-z) + v_k(\sigma_{sk} - \sigma_{lk})(r-z)$$
$$(k-l) = v_l(\sigma_{kl} - \sigma_{ll})(w-z) + v_k(\sigma_{kk} - \sigma_{lk})(r-z).$$

In time series we expect that the rate growth of $r-z$ will be negative, that the rate of growth of $(w-z)$ may be positive (due to the larger increase in S over time) but close to zero, which should lead to a positive correlation between the change in $(s-l)$ and $(k-l)$ under our hypotheses, with the regression coefficient of $(s-l)$ on $(k-l)$ being less than one. This is implied by our hunch that $\sigma_{sl} < \sigma_{kl} > 0$ and $\sigma_{kk} < \sigma_{sk} < \sigma_{lk} > 0$.

A preliminary and crude foray into data for twenty-eight "two-digit" industries in the United States in 1949 and 1963 yielded some not very strong support for the hypothesis outlined here.[40] There is a positive relation between *capital* per unskilled worker and *skilled worker* per unskilled worker across these industries. The simple weighted correlation coefficient between the logarithms of these variables is 0.48 in 1949, 0.50 in 1963, and 0.47 for the change in these variables between these two years.[41] Assuming that the omission of the capital–rental variable does not significantly bias the other results (this is equivalent to assuming that it is uncorrelated with the unskilled labor wage rate across industries), we get the "right" signs in the regressions $s-l$ and $s-k$ on w. The estimated coefficients in the two equations are respectively 4.4 (0.9) and $-2.0(1.8)$ for 1949 and 2.6(0.5) and $-0.8(0.8)$ for 1963. The second of the four coefficients is the one we are most interested in, as it is proportional to $\sigma_{sl} - \sigma_{kl}$ and is negative, as expected, but this finding, however, is not statistically "significant" at conventional levels. Similarly, the regression coefficient of $\Delta(s-l)$ on $\Delta(k-l)$ is positive and less than one (0.47 with an estimated standard error of 0.17), but again this result should not be taken too seriously. It could be due to common errors in the measurement of l and the spuriousness arising out of the appearance of L

in the denominator of both variables. Better and more extensive data for testing such hypotheses is being assembled, but their analysis is only in its earliest stages.

Tentative Summary

There are a large number of important topics which have not been even touched upon in this survey. I have neglected the very important one of the interaction of on-the-job training, schooling, experience, obsolescence, and aging. I have also said nothing about different types of education and the measurement of the quality of education. Nor have I discussed models of optimal investment in human capital or the correct treatment of the educational sector and the investment in human capital in a more comprehensive set of national accounts. Hopefully, many of these topics will be dealt with by other participants in this conference.[42]

It would seem to me that the overall state of the measurement of the contribution of education is in reasonably good shape and has been validated by econometric studies. What needs more work is the elucidation of the processes of production of human capital and the determinants of the rates of return to different types of educational investment.

Acknowledgements

The work on this paper has been supported by National Science Foundation Grants nos. GS 712 and GS 2026X. I am indebted to C. A. Anderson, Mary Jean Bowman, E. F. Denison, R. J. Gordon, and T. W. Schultz for comments and suggestions.

Notes

1 Kendrick [39] had a similar "mix" adjustment based on the distribution of the labor force by industries. Bowman [10] provides a very good review and comparison of the Denison and Schultz approaches.

2 See Jorgenson and Griliches [37], from which the following paragraph is taken almost verbatim, for more detail on the construction of such indexes, and Richter [48] for a list of axioms for such indexes and a proof that they are satisfied only by such indexes.

3 Where the $\dot{x}$ notation stands for dx/dt, and $\dot{x}/x$ represents the relative rate of growth of x per unit of time; and $v_j = p_j L_j / \Sigma_i p_i L_i$. In practice one never has continuous data and so the Laspeyres-Paasche problem is raised again, albeit in attenuated form. Substituting $\Delta L = L_t - L_{t-1}$ for L, one should also substitute $v_{jt} = \frac{1}{2} (v_{jt} + v_{jt-1})$ for v_{jt} in these formulae. This is only approximated below by trying to choose the p_t's in the middle of the various periods defined by the respective Δe_t's.

4 This assumption of proportionality in the change in the hours worked of different men, allows us to talk interchangeably about the "quality" of men and the quality of man-hours. If this assumption is too restrictive, one should add another "quality" term to the expression below, $\Sigma v_i \dot{m}_i / m_i$, where $m_i = h_i / H$ is the relative employment intensity (per year) of the ith category of labor.

5 To adjust for changes in the age distribution, one would need to know more about the rate of "time depreciation" of human capital services and distinguish it from declines with age due to "obsolescence," which are not relevant for a "constant price" accounting. See Hall [29] for more details on this problem.

6 These income figures are deficient in several respects; among others: they are not standardized for age, and the use of a common $44,000 figure for the "over $25,000" class probably results in an underestimation of educational earnings differentials. I am indebted to E. F. Denison for pointing this out to me.

7 The percentage change so calculated between any two dates, is the same as would be obtained by weighting the two educational distributions by the base (weight) period i earnings, aggregating and computing the percentage change.

8 See Blaug [6] and Schultz [55] for extensive bibliographies on this subject.

9 The E measure as used here is equivalent to the "labor-augmenting technical change" discussed in much of recent growth literature. I prefer, however, to interpret it as an approximation to a more general production function based on a number of different types of labor inputs. Allowing changing weights in the construction of such an E index implicitly allows for a very general production function (at least over the subset of different L types) and imposes very few restrictions on it. An interpretation of E as an index of embodied quality in different types and vintages of labor, fixed once and for all and independent of levels of K, would be very restrictive and is not necessary at this level of aggregation.

10 The data sources and many caveats are described in detail in the original articles cited in table 10.4 and will not be reproduced here. Note that for manufacturing, the quality variable is based on an occupation-by-industry rather than education-by-industry distribution, since the latter was not available at the state level. On the other hand, the first manufacturing study (Griliches [25]) also explored the influence of age, sex, and race differences on productivity, topics which will not be pursued further here.

11 Somewhat similar results have also been reported by Besen [5].

12 Data from the 1964 Census of Agriculture may allow a test of the Nelson–Phelps hypothesis. These data provide separate information on the education of the farm *operator* as distinct from that of the rest of the farm labor force. The Nelson–Phelps hypothesis implies that the education of entrepreneurs is a more crucial, in some sense, determinant of productivity than the education of the rest of the labor force.

13 An H index based on costs (income forgone and the direct costs of schooling) would be similar to the one described in the text only if all rates of return to different levels of education were equal to each other and to the rate used in the construction of the human capital estimate.

14 The constancy of relative differentials implies a rise in absolute differential and a rise in the incentive to individuals to invest more in their education.

15 See Griliches [26] and Lancaster [41] for a recent survey and exposition of such an approach.

16 Actually, it could be thought of as a two-dimensional or factors model, body and skill, but since each person is taken to have only one unit of body (even a Marilyn Monroe), the B dimension becomes a numeraire and for practical purposes this reduces itself to a one-factor model.

17 See e.g. the studies of separated identical twins summarized in Bloom [7].

18 Guilford [28], p. 619.

19 Consider two extreme worlds. In one, the only product of the school systems is "ability" or "achievement." In this world, school years completed are just a poor measure of the product of schools. If correct measures of "ability" were available, they would dominate any earnings–education–ability regressions and imply zero coefficients of the school years completed variable. Nevertheless, almost all of the observed "ability" differential would be the product of "education." A second world is one in which the educational system does nothing more than select people for "ability," by putting them through finer and finer sieves, without adding anything to their innate ability in the process. Again, an earnings–education–ability correlation would come out with zero coefficients to education net of ability. Still, in an uncertain world with significant costs of information, there is a significant social product even in the operation of grading and sorting schemes. Even in such a world there is a net value added produced by the educational system, though it may be very hard to measure it. See Zusman [66] for the beginning of an economic analysis of sorting phenomena.

20 Wolfle [65], p. 178.

21 This is also supported by the greater role of "ability" at the lower end of the educational distribution found by Hansen, Weisbrod, and Scanlon [32]. IQ tests, however, are not very good discriminators at the very extremes of the distribution. For a sample of Woodrow Wilson Fellowship holders, Ashenfelter and Mooney [1] found that: "The inclusion of an ability variable affected the estimate of the other education-related variables only in a very marginal fashion. . . . The misspecifications caused by the absence of an ability variable seem to be quite small indeed" (for samples of highly educated people).

22 ". . . it is what the parents *do* in the home rather than their *status* characteristics which are the powerful determinants of home environment" (Bloom [7], p. 124).

23 I am indebted to C. A. Anderson for drawing my attention to these data.

24 The scaling chosen was 6, 9, 13, and 17 and 73, 89, 100, 111, 127 for the schooling and IQ categories respectively.

25 The results were essentially the same for the linear and log–log forms. The semilog forms reported in the text fit the data best on the "standard error in comparable units" criterion. The results are also similar for unweighted regressions, except that the coefficient of schooling is significantly higher.

26 The AFQT is primarily an achievement rather than an innate ability test: "The examinee's score on the tests depends on several factors: on the level of his educational attainment; on the quality of his education (quality of the school facilities); and other knowledge he gained from his educational training or otherwise, in and outside of the school. These are interrelated factors, which obviously vary with the youth's socio-economic and cultural environment, in addition to his innate ability to learn–commonly understood as IQ, nor are they to be translated in terms of IQ." From Karpinos, [38]. Thus, it is probably inappropriate to use these data to get at a pure "net" schooling effect.

27 A more detailed analysis using the AFQT-schooling distribution for ten rather than five regions and mean income by schooling data for males aged forty-five–fifty-four, yielded very similar results and will not be reported here. I am indebted to F. Welch for providing me with the adjusted state data on mean income by schooling.

28 The Carroll and Ihnnen [15] study of a group of North Carolina technical school graduates can also be interpreted to support this view.

29 See Halsey [30] for a discussion and criticism of this metaphor.

30 See, e.g. Telser [58], and Folger and Nam [21].

31 The following table adapted from the US Bureau of the Census (*Current Population Reports*, Census p. 20, no. 132, 1964, "Educational Changes in a Generation," table 1) sheds some light on this question:

Distribution of males by years of school completed, by age, and by father's occupation and education, US, March 1962 (percentages)

Age and cohort	College graduates		Some college	
	Father white collar	Father white collar, some college education	Father white collar	Father white collar, some college education
20–24				
1938–42	—	—	88	31
25–34				
1928–37	64	27	60	23
35–44				
1918–27	62	26	55	19
45–54				
1908–17	62	25	55	19
55–64				
1898–1907	60	14	56	15

Similar implications can also be read into the British data reported by Floud and Halsey [20].

32 Unless, of course, the quality of high school graduates deteriorated more than that of college graduates. Given the relative size of the two groups and the observed minor effect of "ability" on earnings for high school graduates, this is an unlikely event, at least as measured by conventional IQ scores. But the widening of the differential between elementary and high school graduates may indicate that those who do not get past elementary education may today be much more affected by ability constraints and other handicaps than used to be the case in the past.

33 Something like this is implied in the slightly higher regression and correlation coefficient for the relationship of ability to schooling in the North Central States than in the South, reported in table 10.8. This is also supported by the following table taken from a recent article by Turnbull [60], p. 1426; the data are derived from Wolfle's study and from the Project TALENT survey:

Percentages of high school graduates going on to college, by Ability Group

Ability group	Wolfle 1953	TALENT 1960
Lowest (fourth) quarter	20	19
Third quarter	32	32
Second quarter	38	54
Top (first) quarter	48	80

34 The Educational Testing Service recently notified a large number of students who took the Graduate Record Examination that instead of having scored in the, say, ninety-eighth percentile, as previously announced, upon restandardization for the more recent experience, their scores were more accurately described as being in the ninety-fourth percentile. The mean score of men in the 1964–7 norm group was 5.5 percent higher on the verbal ability test and 10.7 percent higher on the quantitative ability test than that of the 1952 basic reference group. (See Educational Testing Service [19].)

35 This adjustment is also probably overdone, since we know that education differentials in the North Central States were lower than in most of the rest of the US. In that sense, the Wolfle–Smith figures are not representative.

36 This section and also parts of section 4 have been inspired and owe a great deal to my reading of Welch's [64] unpublished paper on this topic.

37 The figures were taken from the 1960 Census of Population, vol. I, part 1, table 211 and vol. II, part 7F, "Industrial Characteristics," table 21. The results of using median years of school completed, a weighted E index, and the logarithm of the E index were almost identical.

38 This model is based on unpublished notes by H. G. Lewis. See also Mundlak [45].

39 This model could be "simplified" further by noting the relationship between input and output prices and using it to solve one of the input prices, substituting the output price throughout. But unless one has good data on output prices by states and industries, or is willing to assume that they are constant, there is little to be gained from such a substitution.

40 The limits of this set of data are based on the availability of time series on gross capital stocks in constant prices at the two-digit level. These are derived from a forthcoming book by M. Gort. (I am indebted to Michael Gort for making these yet unpublished data available to me.) The list of industries runs through ten manufacturing industries from "food" to "other transportation equipment," and nine other large industries: mining, railroads, electric ulitities, gas utilities, telephone, contract construction, wholesale trade, and retail trade. The capital stocks are as of beginning of 1949 and 1963. Total employment for these industries for 1948 and 1963 is taken from BLS Bulletin 1312–5; the percentage of the male labor force that consists of professional and technical workers is taken from the "Occupation by Industry" volumes of the 1950 and 1960 Census of Population; and the wage of unskilled labor is identified with the median earnings of laborers who worked 50–52 weeks in these industries and is taken from the respective 1950 and 1960 Census volumes on occupational characteristics. Denoting total employment by N and professionals as a fraction of the labor force by P, S is given by $N \cdot P$ and L, the level of the "unskilled" labor force by $N(1 - P)$. All the regressions and correlations were computed using total employment in 1963 as weights.

41 While such correlation coefficients are "significant" at conventional levels, the overall fit is quite poor and there are a number of notable outliers. The chemical industry has a high capital-labor and a high skill ratio, but the electric machinery industry has a high skill ratio and a relatively low capital-labor ratio, while the utility industries have very high capital-labor ratios but only average skill ratios. Similarly, the highest rates of growth in capital-per-man occurred in this period in mining and construction. Mining had also probably the highest rate of growth in the relative number of highly skilled workers, while construction had one of the lowest. There are no easy answers.

42 On the issue of training and experience see Mincer [43] and Thurow [59] among others; see Denison [16], Welch [62], and the Coleman report [14] for very different ways of estimating the relative quality of schooling; see Ben-Porath [4] for a model of human capital investment and Bowman [12] and Kendrick [40] for discussions of the treatment of education in national accounts.

References

[Note: This list includes several important works not referred to explicitly in the text.]
[1] Ashenfelter, O. and J. D. Mooney, "Graduate education ability, and earnings," *Review of Economics and Statistics*, XLX(1), 1968, 78–86.
[2] Becker, G. S., *Human Capital*, New York, NBER, 1964.
[3] ——, *Human Capital and the Personal Distribution of Income*, Woytinsky Lecture no. 1, University of Michigan, 1967.
[4] Ben-Porath, Y., "The production of human capital and the life cycle of earnings," *Journal of Political Economy*, 75(4), 1967, 352–65.
[5] Besen, S. M., "Education and productivity in U.S. manufacturing: some cross-section evidence," *Journal of Political Economy*, 76(3), 1968.

[6] Blaug, M., *Economics of Education*, A Selected Annotated Bibliography, Oxford, 1966.

[7] Bloom, B. S., *Stability and Change in Human Characteristics*, New York, 1964.

[8] Bowles, S., "Aggregation of labor inputs in economics of growth and planning," *Journal of Political Economy*, January–February 1970.

[9] ——, "Towards an educational production function," in W. Lee Hansen (ed.), *Education, Income and Human Capital*, NBER, *Studies in Income and Wealth*, 1970, no. 35. pp. 71–105. New York: Columbia University Press.

[10] Bowman, M. J., "Schultz, Denison, and the contribution of 'Eds' to national income growth," *Journal of Political Economy*, 72(5), 1964, 450–64.

[11] Bowman, M. J. and C. A. Anderson, "Distributed effects of educational programs," in *Income Distribution Analysis*, North Carolina State University, 1966.

[12] Bowman, M. J., "Principles in the valuation of human capital," *Review of Income and Wealth*, 14(3), 1964, 217–46.

[13] Brown, M. and A. Conrad, "The influence of research and education on CES production relations," in M. Brown (ed.), *The Theory and Empirical Analysis of Production*, New York, NBER, 1967, 341–71.

[14] Coleman, J. S., et al., *Equality of Educational Opportunity*, US Dept. of Health, Education and Welfare, Washington, DC, 1966.

[15] Carroll, A. B. and L. A. Ihnnen, "Costs and returns for two years of post secondary technical schooling: a pilot study," *Journal of Political Economy*, 75(6), 1967, 862–73.

[16] Denison, E. F., *The Sources of Economic Growth in the United States and the Alternatives Before Us*, Supplementary Paper no. 13, New York, Committee for Economic Development, 1962.

[17] ——, "Measuring the contribution of education," in *The Residual Factor and Economic Growth*, OECD, 1964, pp. 13–55, 77–102.

[18] ——, *Why Growth Rates Differ*, Brookings Institution, 1967.

[19] Educational Testing Service, *Handbook for the Interpretation of GRE Scores, 1967–68*, Princeton, 1967.

[20] Floud, J. and A. H. Halsey, "Social class, intelligence tests, and selection for secondary schols," in Halsey, Floud and Anderson, (eds), *Education, Economy and Society*, Glencoe, Ill., 1961, pp. 209–15.

[21] Folger, J. K. and C. B. Nam, *Education of the American Population*, A 1960 Census Monograph, Washing, DC, 1967.

[22] Griliches, Z., "Measuring inputs in agriculture: a critical survey," *Journal of Farm Economics*, XLII(5), 1960, 1411–27.

[23] ——, "Estimates of the aggregate agricultural production function from cross-sectional data," *Journal of Farm Economics*, XLV(2), 1963.

[24] ——, "Research expenditures, education, and the aggregate agricultural production function," *The American Economic Review*, LIV(6), 1964, 961–74.

[25] ——, "Production functions in manufacturing: some preliminary results," in M. Brown, (ed.), *The Theory and Empirical Analysis of Production*, Studies in Income and Wealth, vol. 31, New York, NBER, 1967.

[26] ——, "Hedonic price indexes revisited: some notes on the state of the art," in Amer. Stat. Assoc., *1967 Proceedings of the Business and Economics Section*, 1967b.

[27] ——, "Production functions in manufacturing: some additional results," *Southern Economic Journal*, 35(2), 1968, 151–56.

[28] Guilford, J. P., "Intelligence has three facets," *Science*, 160(3828), 1968, 615–20.

[29] Hall, R. E., "Technical change and capital from the point of view of the dual," *Review of Economic Studies*, 1968.

[30] Halsey, A. H. (ed.), *Ability and Educational Opportunity*, OECD, 1961.

[31] Hanoch, G., "An economic analysis of earnings and schooling," *Journal of Human Resources*, II(3), 1967, 310–29. Also unpublished PhD dissertation, "Personal earnings and investment in schooling," Chicago, 1965.

[32] Hansen, W. L., B. A. Weisbrod and W. J. Scanlon, "Schooling and earnings of low achievers," *American Economic Review*, 60(3), June 1970.

[33] Hildebrand, G. H. and T. C. Liu, *Manufacturing Production Functions in the U.S., 1957*, Cornell University Studies in Industrial and Labor Relations, vol. XV, 1965.

[34] Houthakker, H. S., "Education and income," *Review of Economics and Statistics*, 41(1), February 1959, 24–28.

[35] Husen, T., "Talent, opportunity, and career: a 26 year follow-up," *School Review*, 76(4), December 1968, University of Chicago.

[36] Johnson, H. G., "Comment," in *The Residual Factor and Economic Growth*, OECD, 1964, pp. 219–27.

[37] Jorgenson, D. W. and Z. Griliches, "The explanation of productivity change," *Review of Economic Studies*, XXXIV(3), no. 99, 1967.

[38] Karpinos, B. D., "The mental qualifications of American youths for military service and its relationship to educational attainment," *1966 Proceedings of the Social Statistics Section of the American Statistical Association*, 1966, pp. 92–111.

[39] Kendrick, J. W., *Productivity Trends in the United States*, Princeton, 1961.

[40] ——, "Investment expenditure and imputed income in gross national product," in *47th Annual Report of the NBER*, New York, 1967.

[41] Lancaster, K., "A new approach to consumer theory," *Journal of Political Economy*, 74(2), 1966, 132–57.

[42] Miller, H. P., "Annual and lifetime income in relation to education," *The American Economic Review*, (5), December 1960, 962–86.

[43] Mincer, J., "On-the-job training: costs, returns, and some implications," *Journal of Political Economy*, 70(5), 1962, Part 2, Supplement: *Investment in Human Beings*.

[44] Mitchell, E. J., "Explaining the international pattern of labor productivity and wages: a production model with two labor inputs," *Review of Economics and Statistics*, L(4), 1968, 461–9.

[45] Mundlak, Y., "Elasticities of substitution and the theory of derived demand," *Review of Economic Studies*, 1968.

[46] Nelson, R. R. and E. S. Phelps, "Investment in humans, technological diffusion, and economic growth," *The American Economic Review*, LVI(2), 1966.

[47] Nerlove, M., "Embodiment and all that: a note," unpublished mimeographed paper, 1965.

[48] Richter, M. K., "Invariance axioms and economic indexes," *Econometrica*, October 1966.

[49] Rosen, S., "Short-run employment variation on class-I railroads in the US, 1947–63," *Econometrica*, 36(3–4), 1968, 511–29.

[50] Schmookler, J., "The changing efficiency of the American economy, 1869–1938," *Review of Economics and Statistics*, 34(3), August 1952, 214–31.

[51] Schwartz, A., "Migration and life span earnings in the US," unpublished PhD dissertation, University of Chicago, 1968.

[52] Selowsky, M., "Education and economic growth: some international comparisons," unpublished PhD dissertation, University of Chicago, 1967.

[53] Schultz, T. W., "Capital formation by education," *Journal of Political Economy*, 1960, 571–83.

[54] ——, "Education and economic growth," in N. B. Henry (ed.), *Social Forces Influencing American Education*, Chicago, 1961.

[55] ——, *The Economic Value of Education*, New York, 1963.

[56] Solow, R. M., "Technical change and the aggregate production function," *Review of Economics and Statistics*, 39(3), August 1957, 312–20.

[57] Strumlin, S. G., "The economic significance of national education," 1925, translated from Russian and reprinted in Robinson and Vaizey (eds), *The Economics of Education*, New York, 1966, pp. 276–323.

[58] Telser, L. G., "Some economic aspects of college education," Economics of Education Research Paper no. 59–5, University of Chicago, unpublished paper, 1959.

[59] Thurow, L. C., *Poverty and Discrimination*, Brookings Institution, Washington, DC, 1969, chapter V.

[60] Turnbull, W. W., "Relevance in testing," *Science*, 160, 1968, 1424–9.

[61] US Bureau of the Census, *Trends in the Income of Families and Persons in the U.S.: 1947–64*, Technical Paper no. 17, Washington, DC, 1967.

[62] Welch, F., "The measurement of the quality of schooling," *The American Economic Review*, XLVI(2), 1966, 379–92.

[63] ——, "Linear synthesis of skill distributions," *Journal of Human Resources*, IV(3), Summer 1969, 311–27.

[64] ——, "Education in production," *Journal of Political Economy*, January–February 1970.

[65] Wolfle, D., "Economies and educational values," in S. E. Harris (ed.), *Higher Education in the U.S.: The Economic Problems*, Cambridge, Mass., 1960.

[66] Zusman, P., "A theoretical basis for determination of grading and sorting schemes," *Journal of Farm Economics*, 49(1), 1967, 89–106.

[67] Griliches, Z., "Capital–skill complementarity," *Review of Economics and Statistics*, 51(40), November 1969, 465–68.

[68] Jensen, Arthur R., "How much can we boost IQ and scholastic achievement?" *Harvard Educational Review*, 39(1), Winter 1969, 1–123.

[69] Schrader, W. B., "Test data as social indicators," Educational Testing Service, SR-68-77, Princeton, NJ, September 1968.

[70] Taubman, P. and Wales, T. J., "Mental ability and higher educational attainment since 1910," University of Pennsylvania Discussion Paper no. 139, October 1969.

11
Education, Income, and Ability*

(With W. M. Mason)

1 Introduction

Current estimates of the contribution of education to economic growth have been questioned because they ignore the correlation of education with ability. Whether the neglect of ability differences results in estimates of the contribution of education to income that are too high was considered in an earlier paper by one of the authors (Griliches, 1970) and a negative answer was conjectured. In this chapter we pursue this question a bit further, using a new and larger body of data. A definitive answer to this question remains elusive because of the vagueness and elasticity of "education" and "ability" as analytical concepts and because of the lack of data on early (preschooling) intelligence.

The data examined in this paper are based on a 1964 sample of US military veterans. The variables measured include scores on a mental ability test, indicators of parental status, region of residence while growing up, school years completed before service, and school years completed during or after service. This allows us to inquire into the separate effects of parental background, intelligence, and schooling.

The basic problems and analytical framework can be set out very simply. Let income (or its logarithm) be a linear function of education and ability:

$$Y = \beta_1 E + \beta_2 G + u,$$

where Y is income, E is education, G is ability, u represents other factors affecting income, assumed to be random and uncorrelated with E and G, and we have suppressed the constant term for notational convenience. The relation is presumed to hold true for cross-sectional data. If education and ability are positively associated, then a measure of the contribution of

*From: *Journal of Political Economy*, 1972. Reprinted from the slightly corrected version in *Structural Equation Models in the Social Sciences*, A. S. Goldberger and O. D. Duncan, (eds). New York: Seminar Press, 1973.

education to income that ignores the ability variable (most commonly, the simple regression coefficient of Y on E) will be biased upward from β_1 by the amount $\beta_2 b_{GE}$, where b_{GE} is the regression coefficient of ability on education in the sample. The first substantive section of this paper (section 3) investigates the magnitude of this bias, via the estimation of income equations which contain measures of both education and ability.[1]

In our data the output of the educational process is measured by the number of school grades completed in the formal education system, while ability is measured by the performance on a test at an age when most of the schooling has already been completed. Both of these measures are far from ideal for our purposes. Consider the education variable: What we would like to have is a measure of education achieved (E); what we actually have is years of schooling completed (S), without reference to the conditions under which individuals obtained their formal schooling and the kinds of schooling pursued. We write $E = S + Q$, where the discrepancy Q ($=$ quality of schooling) is assumed to be uncorrelated with the quantity of schooling (S).[2] The quality of schooling is likely to be correlated with ability, because there is some correlation between socioeconomic status and ability, because more able students are more likely to get into better schools, and because performance on intelligence tests taken at age 18 or so also reflects in part differences in both the quantity and quality of education.

Allowing for differences in the quality of education makes the assessment of the bias somewhat more complicated. The true income-generating equation becomes

$$Y = \beta_1 E + \beta_2 G + u = \beta_1 S + \beta_1 Q + \beta_2 G + u.$$

In this framework, ignoring not only G but also Q leads to the same result as before, namely $b_{YS} = \beta_1 + \beta_2 b_{GS}$, since b_{QS} (the regression coefficient of quality on quantity of schooling) is zero by assumption. But when Y is regressed on S and G, the estimated education coefficient becomes

$$b_{YS.G} = \beta_1 b_{QS.G},$$

where $b_{QS.G}$ is the partial regression coefficient of quality on quantity of schooling holding ability constant.[3] Given our assumptions it can be shown (see the Appendix) that

$$b_{QS.G} = b_{QG} \cdot b_{GS}/(1 - r_{GS}^2),$$

where r_{GS}^2 is the squared correlation coefficient between the quantity of schooling and ability. Since we expect both b_{QG} (the regression coefficient of educational quality on ability) and b_{GS} (the regression coefficient of ability on educational quantity) to be positive, $b_{QS.G}$ will be negative. Substituting this expression for $b_{QS.G}$ back into the expression for $b_{YS.G}$ gives

$$b_{YS.G} = \beta_1 - \beta_1 b_{QG} \cdot b_{GS}/(1 - r_{GS}^2).$$

Comparing this with

$$b_{YS} = \beta_1 + \beta_2 b_{GS},$$

it is clear that by going from b_{YS} to $b_{YS.G}$ we reduce the coefficient of schooling for two reasons: First, we eliminate the upward bias due to the earlier omission of ability. Second, however, we *introduce* another bias due to the correlation of ability with the omitted quality variable. This new bias is partly a function of the correlation between ability and quantity of schooling.

In this chapter we solve the problem of the second bias by concentrating our attention on that part of schooling which occurred during or after military service. This "schooling increment" (SI) turns out to be virtually uncorrelated with our measure of ability (implying $b_{G(SI)} \doteq 0$) and hence is not subject to this type of bias.

The schooling increment variable helps us also to solve another vexing problem – how to disentangle causality when the available measure of ability may itself be in part the result of schooling. Since the intelligence test reported in our data is administered prior to entering service, performance on it cannot be affected by the schooling increment. Thus, because our measure of ability is causally prior to SI, and because using SI reduces the bias problem in estimating the effects of education on income, we shall put most of the stress on the results obtained for only a *part* of schooling, namely SI.

We have already noted that our ability measure is not ideal, because it is obtained after much of the formal schooling has already been completed. What we would like is a measure of ability obtained before the major effects of the school system have been felt. Although it is possible using data such as ours to construct models that incorporate estimates of the effects of *early* ability (see Duncan, 1968; Bowles, 1972), we have chosen to work exclusively with our measure of *late* ability. Even so, our ability variable is still not ideal for our purposes. It is possible that our measure of ability, *taken as a measure of late ability*, still has errors in it. To the extent that the errors are random, a direct application of least squares in their presence may understate the effect of ability on income and consequently bias the estimated education coefficient upward. To circumvent this, we devise, in section 4, a model of income determination that contains an unobserved ability variable in place of measured ability. Manipulation of this model permits an estimate of the effect of ability freed of random errors.

The last part of the paper, section 5, summarizes our results and compares them to previous work in this field.

2   The Sample and the Variables

Our analysis is based on a sample of post-World War II veterans of the US military, contacted in a 1964 Current Population Survey (CPS) of the Bureau of the Census. The population consisted of men who were then in the age range 16 to 34, essentially the ages of draft eligibility. The sample includes about 3,000 veterans, for whom supplementary information from individual military records was collated with the CPS questionnaire responses.[4] For a substantial proportion of the veterans, their military records contain individual scores on the Armed Forces Qualification Test (AFQT), which we use here in lieu of standard civilian mental ability (IQ) tests.

The men who serve in the United States military do not represent any recent cohort of draft-age men, since those at either extreme of the ability and socioeconomic distributions are less likely to serve than those in the middle.[5] Thus, conclusions based on our analysis of these data apply only to the veterans population. But, since this population is sizable, the data are of interest despite their obvious limitation. Moreover, this is one of the few relatively large sets of data combining information on income, education, demographic characteristics, mental test scores, and socioeconomic background, the latter three being important as controls in estimating the income–education relationship.

Table 11.1 Means and standard deviations of variables for veterans age 21–34 in the 1964 CPS[a]

Variable	Mean or proportion	Standard deviation	Symbol in subsequent tables	Group label
Personal background:				
Age (years)	29.0	3.5	AGE	
Color (white)	0.96	0.20	COLOR	
Schooling before service (years)	11.5	2.3	SB	
Schooling increment (years)	0.8	1.4	SI	
Total schooling (years)	12.3	2.5	ST	
AFQT (percentile)	54.6	24.8	AFQT	
Length of active military service (months)	30.7	16.9	AMS	
Father's schooling (years)	8.7	3.2	FS	} Fa. Stat.
Father's occupational SES	29.0	20.6	FO	
Grew up in South	0.29	0.45	ROS	}
Grew up in large city	0.22	0.42	POC	} Reg. Bef.
Grew up in suburb of large city	0.05	0.22	POS	}
Current location:				
Now living in the South	0.27	0.44	RNS	}
Now living in the West	0.15	0.35	RNW	} Reg. Now
Now living in an SMSA	0.68	0.47	SMSA	}
Current achievement:				
Length of time in current job (months)	54.3	42.8	LCJ	}
Never married	0.14	0.35	NM	} Curr. Exp.
Current occupational SES	39.2	22.7	—	
Log current occupational SES	3.47	0.68	LOSES	
Income (weekly, dollars)	122.5	52.4	—	
Log income	4.73	0.40	LINC	

[a]$N = 1454$, for this and subsequent tables based on the 1964 CPS.

Within the veterans sample, the individuals on whom we base our conclusions are 1454 men who were employed full-time when contacted by the CPS, who were between the ages of 21 and 34, not then enrolled in school, who were either white or black, who provided complete information about their current occupation, income, education, family background, and for whom AFQT scores were available.

The major characteristics of our sample and the variables we use are summarized in table 11.1. The definition and measurement of most of the variables are standard and we shall comment here only on a few of the more important ones.

Income is gross weekly earnings in dollars. It is an answer to the question: "Give your usual earnings on this job before taxes and other deductions." The data also provide another concept of income: "earnings expected from all jobs in 1964." We experimented at some length with both concepts of income, getting somewhat better (more stable) results for the actual income measure. Since the major results were similar, we shall report here only those for the actual income measure. We also experimented a bit with functional form before settling on the semilog form for the "income-generating" function, leading to the use of the logarithm of income (LINC) as our main dependent variable.

Education is measured in years of school (highest grade) completed and is recorded at two points in time: before entry into military service and at the time of the survey. Taking the difference between total grades of school completed (ST) and grades of school completed before military service (SB) gives us a measure of the increment in schooling (SI) acquired during or after military service.[6] The minimum value of this variable is zero (no increment in schooling) and the maximum is six grades. This measure of incremental education is central to our analysis both because it occurs after the time at which ability was measured and because it is so little correlated with our measure of ability.

Performance on the AFQT is scaled as a percentile score estimated from eight grouped categories.[7] This test includes questions on vocabulary, arithmetic, and spatial relations, but also contains a section on tool knowledge. The AFQT has been treated by other investigators (e.g. Duncan, 1968; Jensen, 1969) as an intelligence test, so that we are following in the footsteps of others in this regard.[8]

It is clear that *some* error may arise from using the AFQT as an intelligence test, in addition to the kinds of errors which could be present in using one of the standard civilian IQ tests.[9] Another difficulty with the use of the AFQT in our analysis, a difficulty which is inherent in the use of *any* global IQ test for purposes such as ours, is that IQ by definition is an aggregation of several different traits (e.g. verbal ability, mathematical ability) sampled from some larger population of traits. The weights used in combining these traits to obtain a global IQ score are not necessarily those that would maximize the contribution of each trait to some other variable (e.g. income). Therefore, the use of AFQT instead of the separate traits which comprise it, and the use of only those traits, may lead to attenuation in our estimate of the effect of ability on income. This explains our interest in the errors-in-variables approach to be taken up in section 4.

The long list of other variables considered can be divided, somewhat imperfectly, into personal background, and current location and achievement variables. Among the first group, we have the usual variables for age (in years), color (dummy: white = 1, black = 0), and region and place of origin dummies (these are in terms of places "you lived most until age 15") which record growing up in the South, in a large city (over 100,000 in population), or in a suburb of such a city. In addition to these, we have also two measures of parental status: father's schooling (in years of school completed) and father's occupation (coded according to Duncan's 1961 SES scale).[10]

The usual rationale for including an age variable is that older men are likely to have had more training on the job and more opportunity to find the better jobs that are appropriate to their training. This, however, is probably measured better not by calendar time but by the actual time spent in the civilian labor force accumulating work "experience."[11] We can estimate this roughly by defining:

Potential experience = age − 18 − (education before service − 12)
    − education after service
    − (total months in service)/12.

Since this measure is a linear function of variables that we include anyway (age and schooling), there is no need to compute it explicitly. It does provide, however, an interpretation for the role of time spent in military service (AMS), when the latter variable is introduced separately.[12]

The "current location" and "current achievement" variables are represented by regional dummy variables for current location in the South and West (RNS and RNW), a dummy variable for current residence in a Standard Metropolitan Statistical Area (SMSA), a measure of the length of time on current job (LCJ), a dummy variable for never married (NM), and a measure of the socioeconomic status of the individual's current occupation (LOSES, the logarithm of Duncan's occupational SES index).

In our model each of these current location and achievement factors intervenes between education and income, and helps to explain the relationship between those two variables. For example, more education may lead to greater interpersonal competence and other socially desirable characteristics, which in turn may lead to a greater likelihood of being married. Married individuals may be expected to have the incentive of responsibility for others, and this may in turn lead to higher income. Although we present some results that take into account factors intervening between education and income, they are not of central interest to us. We shall, therefore, concentrate on the contribution to income of education and ability in the presence of background factors alone.

Table 11.1 presents means and standard deviations for the variables to be used. Note that this group of veterans is young, and hence will not exhibit differentials in income by education as large as those occurring in later, peak-earning, years. Also, because the number of blacks is quite small, white-black income differences will be characterized only by the coefficient for the color dummy variable. Although there are interactions between the color dummy variable and some of the other variables in the income-generating equation (Duncan, 1969), there are too few blacks in our sample to estimate the coefficients of the interaction terms reliably.

Observe, finally, that the average increment in schooling for our sample is nearly one complete grade (0.8). Actually, 68 percent of the group did not return to school after service, so that those with additional schooling must have completed on average more than one additional grade. Since the grades completed range from high school to graduate school, the incremental schooling variable is not limited to a particular range of schooling, and its coefficient may thus be taken as representing the general effect of additional schooling.

In table 11.2 we list the simple correlation coefficients among the major variables of our sample. Note that there is very little correlation between the increment in schooling and various personal background variables such as color, father's schooling and occupation, and the respondent's AFQT score. We have in this variable something as close to a well-designed experimental situation as we are likely to get in social science statistics.

3   Direct Results

The causal model we use to guide our assessment of the relationships between income, education, ability, and other variables at our disposal can be stated as follows, using the variable labels given in table 11.1:

$$SB = F(\text{Fa. Stat., Reg. Bef., COLOR}) \tag{1}$$

$$AFQT = G(\text{Fa. Stat., Reg. Bef., COLOR, SB}) \tag{2}$$

$$AMS = H(\text{Fa. Stat., Reg. Bef., COLOR, AGE, SB, AFQT}) \tag{3}$$

$$SI = J(\text{Fa. Stat., Reg. Bef., COLOR, AGE, SB, AFQT, AMS}) \tag{4}$$

$$LINC = K(\text{Fa. Stat., Reg. Bef., COLOR, AGE, SB, AFQT, AMS,} \\ SI), \tag{5}$$

where each of these functions is a linear structural equation. Figure 11.1 provides a slightly more globally stated graphical equivalent to (1)–(5). As it stands, the model is given by a set of recursive equations. Including other functional relationships linking current achievement and location variables to income and other factors would complicate the model unnecessarily for our purposes. Thus, since we are primarily interested in the *total* effects of schooling and ability *net* of potential labor force experience and background factors, we will not report on all the structural equations that inclusion of occupational SES, marital status, and other variables would entail. For the

Table 11.2 Correlations between selected variables in the 1964 CPS subsample

Variables	Variables									
	(1)	*(2)*	*(3)*	*(4)*	*(5)*	*(6)*	*(7)*	*(8)*	*(9)*	*(10)*
(1) AGE	1.000	−0.055	−0.010	0.109	0.052	−0.056	−0.093	−0.004	0.120	0.216
(2) COLOR		1.000	0.011	−0.028	−0.006	0.174	0.004	0.089	0.031	0.116
(3) SB			1.000	−0.170	0.832	0.469	0.283	0.307	0.397	0.264
(4) SI				1.000	0.405	0.098	0.103	0.085	0.216	0.149
(5) ST					1.000	0.490	0.321	0.333	0.490	0.329
(6) AFQT						1.000	0.229	0.242	0.311	0.235
(7) FS							1.000	0.431	0.250	0.114
(8) FO								1.000	0.266	0.229
(9) LOSES									1.000	0.338
(10) LINC										1.000

same reason, we concentrate in this section on the income equation, using the actual estimates for the rest of the causal model only to obtain a few secondary results.

The organization for the rest of this section is as follows. First we describe the sensitivity of the education coefficients in eq. (5) to inclusion of ability and personal background characteristics. We also appraise the contribution of education to income more generally. Next we describe the contribution of the other explanatory variables in eq. (5), and to some extent their contributions in the model taken as a whole. Finally, we summarize some of the relationships among variables other than income.

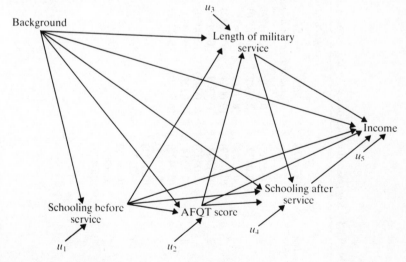

Figure 11.1 Basic causal model for income determination (random disturbances are denoted by the u values).

Table 11.3 Regression equations with log income as dependent variable

Reg. No.	COLOR	SB	SI	ST	AFQT	Other sets of variables	R^2
		Coefficient of (Standard error in parentheses)					
1	0.2548 (0.0472)	0.0520 (0.0042)	0.0528 (0.0070)			AGE, AMS	0.1666
2	0.2225 (0.0479)	0.0418 (0.0049)	0.0475 (0.0072)		0.00154 (0.00045)	AGE, AMS	0.1732
3	0.1904 (0.0473)	0.0379 (0.0045)	0.0496 (0.0070)			AGE, AMS, Fa. Stat., Reg. Bef.	0.2129
4	0.1714 (0.0479)	0.0328 (0.0050)	0.0462 (0.0071)		0.00105 (0.00045)	AGE, AMS, Fa, Stat., Reg. Bef.	0.2159
5	0.2544 (0.0471)			0.0508 (0.0039)		AGE, AMS	0.1665
6	0.2224 (0.0479)			0.0433 (0.0044)	0.00150 (0.00045)	AGE, AMS	0.1729
7	0.1907 (0.0473)			0.0408 (0.0041)		AGE, AMS, Fa. Stat., Reg. Bef.	0.2115

8	0.00097 (0.00044)	0.0365 (0.0046)			0.1732 (0.0479)	AGE, AMS, Fa. Stat., Reg. Bef.	0.2141
9	0.00252 (0.00041)				0.1335 (0.0487)	AGE, AMS, Fa. Stat., Reg. Bef.	0.1794
10					0.1742 (0.0488)	AGE, AMS, Fa. Stat., Reg. Bef.	0.1578
11	0.00115 (0.00045)		0.0445 (0.0068)	0.0320 (0.0048)	0.2052 (0.0456)	AGE, AMS, Fa. Stat., Reg. Bef., Curr. Exp.	0.2979
12	0.00129 (0.00043)		0.0468 (0.0068)	0.0372 (0.0046)	0.2240 (0.0449)	AGE, AMS, Reg. Now, Curr. Exp.	0.2851
13	0.00095 (0.00042)		0.0352 (0.0100)	0.0244 (0.0050)	0.1970 (0.0452)	AGE, AMS, Fa. Stat., Reg. Bef., Curr. Exp., LOSES	0.3114
14						AGE, AMS, Reg. Bef.	0.1178
15					0.1994 (0.0494)	AGE, AMS, Reg. Bef.	0.1277
16	0.00252 (0.00041)				0.1335 (0.0486)	AGE, AMS, Fa. Stat., Reg. Bef	0.1794

A   Education

There are several ways of measuring the bias in the education coefficient due to omission of ability. The bias can be assessed before, or after, the inclusion of personal background factors. Also, we use two schooling variables, so that there are two schooling coefficients to examine for each assessment of bias. We derive the needed bias figures by regressing the logarithm of income on (1) education, (2) education and ability, (3) education and personal background factors, (4) education, personal background factors, and ability. Comparisons made among these four regressions provide the necessary figures for assessing bias.

Table 11.3 presents a number of regression results relating the logarithm of income to selected variables. All of the regressions include age, length of military service, and color, so that the education, ability, and background effects are all net of color and the potential experience variable defined earlier. Regressions 1–4 use the two schooling variables separately; regressions 5–8 use total schooling instead.

Regression 1 provides the baseline estimates of the two schooling coefficients, estimates that do not allow for the effects of ability, father's status, and region of origin. Regressions 2 and 3, respectively, add AFQT and personal background factors to the baseline regression. Regression 4 includes both AFQT and personal background factors. Using these results, an estimate of the proportional bias in the schooling coefficients due to the omission of a relevant variable can be computed as one minus the ratio of the schooling coefficient after the inclusion of the relevant variable to the corresponding schooling coefficient in the equation that excludes this variable. Table 11.4 presents estimates, based on regressions 1–4 and 5–8 in table 11.3, of the proportional bias in the schooling coefficients due to the exclusion of AFQT and personal background variables.

Table 11.4   Estimates of proportional bias in schooling coefficients due to omission of ability and background factors

Proportional bias in the coefficient of

SB	SI	ST	Variables omitted
0.17	0.10	0.15	AFQT (Compare regressions 1 and 2, 5, and 6)
0.25	0.06	0.20	Fa. Stat., Reg. Bef. (Compare regressions 1 and 3, 5, and 7)
0.35	0.12	0.28	AFQT, Fa. Stat., Reg. Bef. (Compare regressions 1 and 4, 5, and 8)
0.13	0.07	0.11	AFQT omitted; Fa. Stat., Reg. Bef. included (Compare regressions 3 and 4, 7, and 8)
0.22	0.03	0.16	Fa. Stat., Reg. Bef. omitted; AFQT included (Compare regressions 2 and 4, 6, and 8)

Looking first at the figures for SB and SI, AFQT accounts for a drop of 7-10 percent in the coefficient of SI, and 13-17 percent in the coefficient of SB. On the other hand, when the personal-background factors are included, the drop in the SB coefficient (22-25 percent) is much greater than in the SI coefficient (3-6 percent). Moreover, the decline in the SB coefficient when both AFQT and personal background variables are included is nearly three times (35 percent) the decline in the *SI* coefficient (12 percent). These results were to be expected. Education before service is more highly correlated than education after service to personal background factors and ability, and more likely to be biased downward because of the absence of a measure of school quality.[13] This is why we prefer the coefficient of SI as an estimate of the potential effects of changes in schooling levels. But even using Total Schooling, the decline (28 percent) is not all that great. Of the total decline in the coefficient for ST, 11-15 percent can be attributed to the introduction of the AFQT variable, the rest being due to parental background and region and size of city of origin, variables that are likely to be closely related to the omitted school quality dimension.

For analyses of the contribution of education to economic growth, the most appropriate estimate is that given by the coefficient of incremental schooling in regression 4, a regression which includes background and ability measures, but does not contain any later current experience and success variables. The value of this coefficient is 0.0462, and we have already observed that this is only 12 percent lower than the 0.0528 given by first regression, which includes no background or ability measures. Thus, while the usual estimates of the contribution of education may be biased upward due to the omission of such variables, this bias does not appear to be large, and is much smaller than the 40 percent originally suggested by Denison (1962).

In regression 9 both schooling variables are excluded. Comparing the results of this equation with those of regression 4, we can see that education does in fact provide a significant independent contribution to the explanation of income. Comparison of regressions 4 and 8 indicates that even though the two schooling variables are acquired at different times and under different circumstances, their effects on income are similar. In fact, the difference between the two schooling coefficients in regression 4 is not statistically significant at the conventional 5 percent level, although this difference is significant at about the 8 percent level (which the computed $F = 3.2$ satisfies). We would expect the difference to be more highly significant with a larger sample, and we would also expect the inclusion of a school quality measure to eliminate it completely.

B   AFQT

The performance of AFQT is more modest than we anticipated, in view of the current emphasis on the role of intelligence in the achievement process, and the common use of the AFQT as a measure of IQ. While AFQT is relatively highly intercorrelated with schooling before military service and with the other personal background variables, its own *net* contribution to the explanation of the variance in the income of individuals is very small. For example, introducing AFQT into regression 2 increases the R^2 by only 0.007 relative to regression 1; introducing it into regression 4 increases the R^2 by 0.003 relative to regression 3. Even if one attributed all of the joint schooling–intelligence effects (including schooling before service and hence before the date of these tests) to the AFQT variable, one would raise its contribution to the R^2 to only 0.022 (regression 9 versus 10).

Another way to look at the relation between income and AFQT is to decompose the correlation between them into components using path coefficients. Doing so is equivalent to a repeated application of the excluded-variable formula, with all the variables scaled to have mean zero and a unit standard deviation. The advantage of such a decomposition is that it is additive, whereas a decomposition in terms of changes in R^2 is not.

The path-coefficient decomposition presupposes a causal ordering, which our model supplies. Labeling and grouping our variables into AFQT (T), SI (S), AMS (M), and Other (O), calling income Y, and using the excluded-variables formula repeatedly, we get the path-coefficient decomposition of:

$$r_{YT} = \beta_{YT.MSO} + \beta_{YM.SOT} \cdot \beta_{MT.O} + \beta_{YS.TOM}(\beta_{ST.MO} + \beta_{SM.OT} \cdot \beta_{MT.O})$$
$$+ r_{OT}[\beta_{YO.TSM} + \beta_{YS.TOM}(\beta_{SO.TM} + \beta_{SM.TO} \cdot \beta_{MO.T}]$$
$$+ \beta_{YM.SOT} \cdot \beta_{MO.T}],$$

where the betas are the standardized partial regression coefficients and $\beta_{ij} = r_{ij}$. The first term on the right-hand side is the net effect of T on Y, the second and third terms together give the effect of T via M and S, and the last term gives the effect of T which is "due to" or "joint with" the other variables O.

The decomposition of r_{YT} via path coefficients yields the conclusion that more than half of the observed simple correlation between income and AFQT is "due to" or "joint with" the logically prior variables of COLOR, Fa. Stat., Reg. Bef., SB, and Age. The estimates for eqs. (1)–(5) of our model imply that $r_{\text{LINC,AFQT}} = 0.2355 = (0.0657 \text{ net}) + (0.0361 \text{ through SI and AMS}) + (0.1337 \text{ joint with, or due to, other factors}) = (0.102 \text{ attributable to AFQT net of prior factors}) + (0.133 \text{ attributable to correlations between AFQT and prior factors}).$

In terms of the model used here, over half of the initial correlation between income and AFQT is explained by factors in the model that are prior to AFQT. And, even if schooling before service and the background variables were not taken as predetermined with respect to AFQT, it would still be the case that over half of the zero-order correlation would be allocated to *joint* influence with these other independent variables. Note, moreover, that for $r = 0.1$ (the approximate role of AFQT net of prior factors), $r^2 = 0.01$. The fraction of the variance in income accounted for by this component of AFQT is minute.

The literature on the residual factor in economic growth (e.g. Denison, 1964) has frequently adjusted, rather arbitrarily, the observed income distributions for variation presumed to be due to differences in the unobserved genetic endowment of individuals. Relevant variation in this latent variable is usually held to be measured best by variation in performances on intelligence tests, and to some extent by variation in parental social status. Since in this paper we have measures of these variables, we are in a position to question how much they contribute to the explanation of income differences. With our data, adding AFQT and Fa. Stat. to a regression of income on Age, AMS, COLOR, and Reg. Bef. increases the R^2 by only 0.052 (compare regressions 15 and 16 in table 11.3). Adding COLOR, AFQT, and Fa. Stat. to a regression of income on Age, AMS, and Reg. Bef. increases the R^2 by only 0.061 (compare regressions 14 and 16 in table 11.3). The increment in explained variance due to these "heredity"-associated variables is thus only about a fifth of the total "explainable" variance in income (the maximal R^2 in predicting income is given by regression 13 as 0.31). And, this calculation makes no allowance for the effects of quality of schooling and discrimination that are confounded with color, regional origin, and parental status variables.

Thus, the measurable effects of genetic diversity on income appear to be much smaller than is usually implied in debates on this subject. It follows, therefore, that since most of the effects of heredity are indirect, there is little bias in an estimate of a schooling coefficient which does not take heredity into account. Heredity will affect the distribution of schooling attained, but the estimated schooling coefficient measures its contribution to income correctly whatever the source of a change in schooling.

C Additional Details and Relationships

In table 11.5 we display all the coefficients of regression 13. This regression includes almost all of the variables available to us, and accounts for about a third of the observed variance in logarithmic income. It is clear that the bulk of the variance in individual income is not accounted for by our equations, even when using a rather long list of variables, a result that is common to most other similar studies based on observations on individuals (e.g. see Hanoch, 1967).

The regression coefficients displayed in table 11.5 provide some more information on our results. Since the dependent variable is the logarithm of income, these coefficients (times 100) give the percentage effect on income of a unit change in the explanatory variables. The more interesting findings here are: (1) The nonsignificance of the father's schooling variable in the presence of father's occupational SES score. This is true also in most of the other regressions. (2) The relative importance of current location (being in an SMSA and in the West). (3) The rather surprising strong negative effect of never having married. (4) The negative effect of time spent in the military and the implied positive effect of potential experience in the labor force on income.[14]

In table 11.6 we gather some auxiliary results on regressions relating other variables in our model. Among the more interesting of these are the highly

Table 11.5 Regression of log income on all available relevant variables

Variable	Coefficient	t-ratio
AGE	0.0126	(4.3)
COLOR	0.1970	(4.4)
FO	0.0016	(3.2)
FS	−0.0038	(−1.2)
POC	0.0325	(1.4)
POS	0.0971	(2.4)
ROS	−0.0238	(−0.7)
SB	0.0244	(4.9)
AFQT	0.00095	(2.2)
SI	0.0352	(4.8)
RNS	−0.0751	(−2.3)
RNW	0.1173	(4.5)
SMSA	0.1365	(6.7)
LCJ	0.0013	(5.7)
NM	−0.1496	(−5.7)
LOSES	0.0804	(5.3)
AMS	−0.0011	(2.0)
Constant	3.6483	
R^2	0.3114	

significant (and rather large) effects of region, color, and schooling before service on AFQT, and the barely significant (and minor) effects of the parental status variables on it. The other interesting fact is that using occupational status rather than income as the dependent variable gives similar results: significance for the schooling variables and only marginal importance for parental status and AFQT.

4   Errors in the AFQT Variable and Other Extensions

We now consider the possibility that the AFQT is an erroneous measure of the "true ability" which is actually relevant for income determination. Since we have no direct knowledge of the errors in the AFQT, the discussion which follows is an essay: We assume the AFQT measures adult ability with random errors and specify a model for the explanation of earnings which takes these errors into account.[15]

The sources of random error in the AFQT are presumed to be grouping, unreliability, aggregation, and left-out components of ability. Grouping, the use of midpoints of score intervals rather than the individual scores themselves, creates random errors if the actual scores are distributed evenly within intervals. Reliability errors, though doubtless present, are probably

Table 11.6 Auxiliary regressions relating determinants of income

Dependent variable	COLOR	FO	FS	POC	POS	ROS	SB	SI	AFQT	AGE	AMS	R^2
					(Entries are *t*-ratios)							
SB	a	8	6	4	a	5						0.152
SB	-4	6	4	3	a	3			17			0.289
AFQT	5	5	5	2	a	6						0.139
AFQT	6	2	3	a	a	4	17					0.271
AMS	a	a	2	4	a	a	5	a	a	9	4	0.083
SI	a	3	4	a	a	3	-11	a	8	4	a	0.130
NM	a	-2	a	a	a	-3	a	a	a	-9	a	0.073
LCJ	a	a	a	a	a	2	3	3	a	18	4	0.208
RNS	3	a	a	a	a	46	a	3	a	a	a	0.625
RNW	a	a	a	-4	a	-5	3	2	a	a	3	0.051
SMSA	-3	a	a	12	6	-4	2	3	a	a	2	0.145
LOSES	a	2	3	5	a	12	12	10	3	5	a	0.290

[a]In the equation but estimated *t*-ratio less than 2.

minor because of the grouping procedure. Aggregation, in the sense of using a global index instead of its separate components, could create random differences between the ability index which maximally predicts income and the AFQT index. Left-out components of ability could also differ randomly from the AFQT.

The AFQT may also be subject to nonrandom errors, such as those contributed by test-wiseness and motivation. In this paper we make no adjustments for nonrandom errors in the AFQT.

Our revised model is presented in table 11.7 and in figure 11.2. The time subscripts 0, 1, 2 represent measurements taken before the start of formal schooling (approximately age 6), before entering military service (approximately age 18), and at the time of the survey (1964), respectively. Random disturbances appear only in equations with observable dependent variables. Basically we have an unobservable ability (or potential achievement, or human capital) variable (G) which is augmented by schooling, and which is estimable (subject to error) via test scores (T). We assume in this model that all of the influence of class (B) and heredity (H) are indirect, via the early ability variable (G_0).

We further assume that the contribution (γ) of a unit change in SI ($S_2 - S_1$) to ability is the same as that of a unit change in S_1 (SB), and that the schooling increment is uncorrelated with the error in observed test scores (t_1) and also with that part of heredity which is not already reflected in S_1 or correlated with B. The various random disturbance terms (e, t, w, and u) are assumed to be uncorrelated with each other and with the causally prior variables of the system. These assumptions are the important identifying restrictions in our model.

The present data are not sufficient to estimate this model in its entirety. We have no measures of G, T_0, or H. Yet, we can mesh our data with

Table 11.7   Schematic causal model of income determination[a]

(1)	$G_0 = a_1B + a_2H$
(2)	$T_0 = G_0 + t_0$
(3)	$S_1 = b_1B + b_2H + e$
(4)	$G_1 = G_0 + \gamma S_1$
(5)	$T_1 = G_1 + t_1$
(6)	$S_2 - S_1 = c_1S_1 + c_2B + w$
(7)	$G_2 = G_1 + \gamma(S_2 - S_1)$
(8)	$I_2 = \beta G_2 + u$

[a] Variables:

G	potential achievement or ability to earn income (unobservable).
B	background factors including social class of parents (Fa. Stat.) and location of adolescence (Reg. Bef.).
H	heredity or genotype (unobservable).
T	test score ($T_1 = $ AFQT).
S	schooling ($S_1 = $ SB, $S_2 = $ ST, $S_2 - S_1 = $ SI).
I	income (LINC).
e, t, w, u	random disturbances.

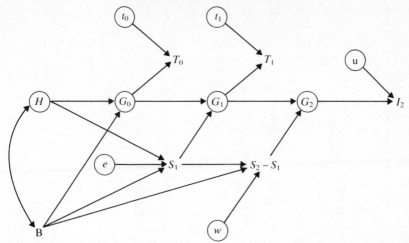

Figure 11.2 Revised causal model of income determination (circles denote unobservables).

this model in a way which may allow us to identify the effect of errors in AFQT and estimate the contribution of $S_2 - S_1$. Substituting eqs (4) and (1) into (5) gives

$$T_1 = \gamma S_1 + a_1 B + a_2 H + t_1 \qquad (9)$$

and substituting (7) and (5) into (8) results in

$$I_2 = \beta [\gamma (S_2 - S_1) + T_1 - t_1)] + u = \beta\gamma(S_2 - S_1) + \beta T_1 + (u - \beta t_1). \qquad (10)$$

In eq. (10) we have an errors-in-variables problem, or equivalently, a simultaneity problem, in that T_1 is correlated with the disturbance $u - \beta t_1$. To handle this problem, we can use the observable predetermined variables (S_1 and B) not appearing in eq. (10), in a two-stage procedure. In the first stage, we estimate (9), ignoring the unavailable H variable, and get a predicted value of T_1, denoted by $\hat{T}_1$ (AFQT Hat). This predicted value replaces T_1 in (10). In the second stage we regress I_2 (LINC) on $S_2 - S_1$ (SI) and $\hat{T}_1$ to estimate $\beta\gamma$ and β.

This procedure solves the problem of error in T_1, assuming that our model is correctly specified, but does little about the effect of the omitted variable H (except for its influence via S_1). Here we have to count on the presumed relative independence of the increment in schooling from H, net of their joint relationship with S_1 and the variables contained in B.

The model we actually use is even more complicated than the one outlined above because it includes as additional variables COLOR, Age, AMS, Reg. Now, and Curr. Exp. variables. All of these are assumed to have an independent effect on income (not only via G) in eq. (8) while COLOR is also assumed to enter in the previous equations for G and S [(1), (3), and (6)]. To carry all this along explicitly in table 11.7, figure 11.2, and eqs (1)–(10) would only have obscured the basic logic of our procedure.

Table 11.8 Two-stage regression for log income and log occupational SES

Reg. No.	COLOR	SI	AFQT Hat[a]	AFQT	Other variables in equation	R_2
			(Standard errors in parentheses)			
			Dependent variable is log income			
17	0.0351 (0.0494)	0.0504 (0.0069)	0.01051 (0.00078)		AGE, AMS	0.1876
18	0.0730 (0.0468)	0.0483 (0.0065)	0.00889 (0.00078)		AGE, AMS, Reg. Now, Curr. Exp.	0.2855
19	0.1982 (0.0458)	0.0331 (0.0067)		0.00298 (0.00038)	AGE, AMS, Reg. Now, Curr. Exp.	0.2526
			Dependent variable is log occ. SES			
20	-0.3979 (0.0815)	0.1320 (0.0114)	0.02554 (0.00129)		AGE, AMS	0.2636
21	-0.3517 (0.0809)	0.1277 (0.0113)	0.02626 (0.00134)		AGE, AMS, Reg. Now, Curr. Exp.	0.2880
22	0.0157 (0.0831)	0.0843 (0.0121)		0.00809 (0.00069)	AGE, AMS, Reg. Now, Curr. Exp.	0.1779
23	0.1014 (0.0787)	0.1151 (0.0117)		0.00253 (0.00073)	AGE, AMS, Reg. Now, Curr. Exp., Fa. Stat., Reg. Bef., SB	0.3034

Coefficient of

[a]The first-stage regression equation used to compute this variable is
AFQT Hat = $-19 + 17.85$ COLOR $+ 0.0735$ FO $+ 0.5505$ FS $+ 4.434$ SB $- 5.472$ ROS,
        (2.83)   (0.0309)   (0.1481)   (0.262)   (1.282)
with $R^2 = 0.271$.

The first part of table 11.8 reports our estimates of the revised model. Regressions 17 and 18 both use the constructed AFQT Hat variable and exclude the Fa. Stat., Reg. Bef., and SB variables assumed to affect income only via the unobservable G variable. The difference between these two regressions is that regression 18 includes the Reg. Now and Curr. Exp. variables. Regression 19 is comparable to 18 but uses the actual AFQT values instead of the estimated AFQT Hat. Comparing regressions 17 and 18 and 4, 11, and 13 in table 11.3, we observe that the coefficient of incremental schooling does not decrease when we switch to the AFQT Hat variable and eliminate the direct influence of personal background variables (except for COLOR, Age, and AMS) and preservice schooling.

Constraining the model as we have done here so that background factors and schooling before service work through the unobserved ability variable results in almost the same estimates for the coefficients of the remaining schooling variable as in the earlier, unconstrained regressions. Allowing for direct effects of measured AFQT, schooling before service, and social background improves the fit only marginally (regression 4 versus 17, or 11 versus 18). Thus, the approach taken here suggests that our initial estimate of the schooling effect on income is robust with respect to the presence of random measurement errors in AFQT. Moreover, the comparable levels of fit in the error model and the unconstrained regressions support the model outlined in table 11.7.

Considering next the AFQT Hat variable, note that its coefficient in regressions 17 and 18 is much larger and more highly significant than those for the original AFQT measure (table 11.3). The estimated $\hat{\beta}$ implies that an increase of a percentile in the "true ability" score adds about 1 percent to income, while the contribution of an additional year of schooling ($\hat{\gamma}$ in regression 18) is equivalent to a 5.4 percentile improvement in the true ability score. "Purging" AFQT of random errors thus increases its contribution to income, even though it does not modify the estimated contribution of education. Observe also that a bound can be set on the effect of ignoring the H variable in eqs. (9) and (10) derived from the error model. In particular, the gain in predicting income with the estimate of error-free AFQT more than offsets the loss due to lack of a measure of the direct influence of H. That is, the *ignored* systematic part of ability, the part of heredity that is uncorrelated with the variables defining AFQT Hat, has a smaller variance than the variance of error in observed AFQT, since the R^2 is higher for regression 18 than for 19.[16]

The only novel result in table 11.8 (regressions 17–19) pertains to the coefficient of the color variable in the presence of the AFQT Hat variable. It is insignificant now, indicating that all of the color effects were captured by AFQT Hat. Thus, we could have included the color variable in the definition of the B (background) set in table 11.7 and excluded it from the version of eq. (10) that we actually estimated. Taken at face value, this result implies that discrimination against blacks does not affect white-black differences in income once person-to-person differences in ability are adjusted

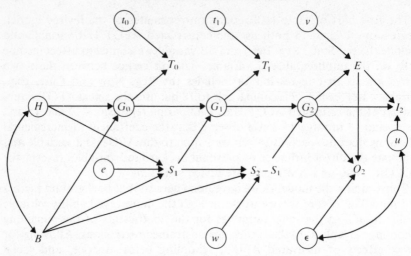

Figure 11.3 Extension of revised model of income determination to include occupational SES (O) and other current experience (E) variables (circles denote unobservables).

for random measurement error. This outcome could not have been forecast on the basis of any previous literature. Since the number of blacks in the sample is very small, the result cannot be taken for anything more than an invitation to further work.

The model of table 11.7 may be extended by adding equations connecting other indicators of social position, such as Occupational SES (O) and other current experience variables (Reg. Now, Curr. Exp.), to the unobserved ability variable. In figure 11.3 we sketch one possible extension. As shown, we now assume that the current experience variables (E) are causally dependent on ability, and that Occupational SES and Income are causally dependent on the current experience variables, as well as on ability. We leave unspecified the causal relationship between Occupational SES and Income.

In revising the structural equations of table 11.7, we add equations linking ability to Occupational SES and Income. Both E and the color variable (C) are included in the equations for Income and Occupational SES. We alter the initial equation for ability (G_0) to include C, and recognize the unobservability of heredity (H) by replacing it with a disturbance (h), defined to be that component of heredity which is uncorrelated with our measures of personal background, color, and incremental schooling. Thus, in the extended model, eqs (2), (4), (5), and (7) remain unchanged, and we replace eqs (1), (3), (6), and (8) with

$$G_0 = a_1 B + a_2 C + h \tag{11}$$

$$S_1 = b_1 B + b_2 C + e \tag{12}$$

$$S_2 - S_1 = c_1 S_1 + c_2 B + c_3 C + w \tag{13}$$

$$E = d_1 G_2 + v \tag{14}$$

$$I_2 = \beta_1 G_2 + \delta_1 C + \eta_1 E + u \tag{15}$$

$$O_2 = \beta_2 G_2 + \delta_2 C + \eta_2 E + \epsilon. \tag{16}$$

As in our first error model, this extended version also includes AGE and AMS, which are assumed to have an independent effect on Income and Occupational in eqs (15) and (16). For clarity we have omitted AGE, AMS, and C from figure 11.3 and AGE and AMS from the revised, augmented set of equations (2), (4), (5), (7), (11)–(16).

Lacking measures of G, T_0, and h, we cannot estimate the extended model in its entirety. Just as in our initial error model, however, we can use equations derived from the extended model to identify the effect of errors in AFQT and to estimate the contribution of $S_2 - S_1$, this time with respect to Occupational SES.

Substituting (11) into (4), and (7) into (15) and (16), we have

$$G_1 = \gamma S_1 + a_1 B_1 + a_2 C + h \tag{17}$$

$$I_2 = \beta_1 G_1 + \beta_1 \gamma (S_2 - S_1) + \delta_1 C + \eta_1 E + u \tag{18}$$

$$O_2 = \beta_2 G_1 + \beta_2 \gamma (S_2 - S_1) + \delta_2 C + \eta_2 E + \epsilon. \tag{19}$$

These together with

$$T_1 = G_1 + t_1 \tag{5}$$

form a four-equation system with one unobservable latent variable (G_1), an indicator of it (T_1), and two current dependent variables (I_2 and O_2), both affected by this same latent variable. We estimate such a system in two stages, first estimating $\hat{T}_1$ (AFQT Hat) from the reduced-form equation given by the substitution of (17) into (5), and then substituting the resulting estimates for G_1 in (18) and (19).

The lower panel of table 11.8 reports our estimates for the model extended to include (logarithmic) Occupational SES. The lists of regressors for regressions 20, 21, and 22 parallel those for regressions 17, 18, and 19 respectively. Regression 23 adds the background and SB variables to the list of regressors used in 22.

Before summarizing these results we check them for the extent the coefficient estimates satisfy the proportionality constraint of eqs (18) and (19), which is that $\beta_1 \gamma /_1 = \beta_2 \gamma / \beta_2$ (or $\beta_1 / \beta_2 = \beta_1 \gamma / \beta_2 \gamma$).[17] If our model is correct, then the coefficients of SI and AFQT Hat should be in the same ratio in regressions 17 and 20, or 18 and 21. These ratios, all of which are implied estimates of γ, turn out to be 4.8 and 5.1, and 5.4 and 4.9 respectively, values which seem close and which support the earlier interpretation of these results.

In general, the coefficients for SI and AFQT Hat in the Occupational SES regressions perform in a fashion similar to their behavior in the Income regressions. The coefficient of incremental schooling remains about the same when we switch to the AFQT Hat variable and eliminate the direct influence of personal background variables (except for COLOR, AGE, and AMS) and preservice schooling (compare regressions 21 and 23). The coefficient for "error-free" AFQT is markedly greater than the coefficient for AFQT (compare regressions 21 and 23). An increase of a percentile in the "true ability" score adds about 2.6 percent to Occupational SES while the contribution of an additional year of schooling ($\hat{\gamma}$ in regression 21) is equivalent to a 4.9 percentile improvement in the true ability score. And, allowing for direct effects of measured AFQT, schooling before service, and social background improves the fit only marginally (regression 21 versus 23). Thus, the results for Occupational SES suggest that ignoring (as in regression 23) the presence of random measurement errors in AFQT does not bias the estimate of the schooling effect (using SI) on Occupational SES.

5   Discussion and Summary

We have tried to compare our results to those of other similar studies, but without too much success. Most of them use total schooling instead of the incremental schooling variable on which we rely so heavily. Also, such studies tend to treat years of school as an error-free measure of educational attainment, a position that is hardly tenable in light of the extreme diversity of the educational system in the US.

Duncan's (1968) major study uses the same basic data set as we do, but defines the subsample of interest as white males age 25–34, includes both veterans and nonveterans, and introduces early intelligence and number of siblings. Instead of actual income he uses expected income. For his sample, when parental status, number of siblings, and early intelligence variables are introduced into a regression with expected income as the dependent variable, the coefficient of total schooling declines about 30 percent.[18] We cannot, however, be sure that this difference between Duncan's study and ours is due to the difference in populations sampled, because expected and actual income are imperfectly correlated (in our sample the correlation between the logarithms of these two variables is about 0.7), and because his results do not control for differences in labor force participation or the effects of different regions of origin, and do not allow for the correlation of the parental status variables with the left-out school quality variable.

Hansen, Weisbrod, and Scanlon (1970) analyze a sample of 17–25-year-old men rejected by the selective service system because of low AFQT scores, and conclude that schooling is a relatively unimportant income determinant. The education coefficient drops about 50 percent when the AFQT variable is introduced in the regression of income on age, color, size of family of origin, intactness of family of origin, and education. It drops even further when such current success variables as job training and marital status are added. Their sample is special in that it concentrates on the very young and on blacks (about half of their sample is nonwhite versus 9 percent in our subsample). It is well-known that schooling–income differentials are rather low at the beginning of the labor force experience, and there is little evidence for a strong schooling–income relationship among blacks (see Hanoch, 1967). Both of these facts could help to explain the differences in results. Moreover, the correlation between AFQT and the omitted school quality variable is likely to be higher for their population than for higher ability groups, so that including AFQT in the regression overstates the bias in the education coefficient due to neglecting ability. For these reasons, then, we are not ready to conclude that using a larger number of low-ability men than was available to us within our own sample would alter our estimate of the bias in the education coefficient due to omitting ability. All of these considerations do remind us again, though, that we cannot take our sample as representative of the entire labor force.

Several other studies, based primarily on samples of males with relatively high social status, have also found a relatively small bias in the schooling coefficient due to left-out ability variables: Ashenfelter and Mooney (1968), Rogers (1969), Taubman and Wales (1970), and Weisbrod and Karpoff (1968). This last study can also be interpreted to show a rather significant effect on income of the quality of college schooling.

Our findings support the economic and statistical significance of schooling in the explanation of observed differences in income. They also point out the relatively low independent contribution of measured ability. Holding age, father's status, region of origin, length of military service, and the AFQT score constant, an additional year of schooling would add about 4.6 percent to income in our sample. At the same time a 10 percentile improvement in the AFQT score would only add about 1 percent to income.

Using a "clean" schooling variable, incremental schooling, we concluded that the bias in its estimated coefficient due to the omitted ability dimension is not very large, on the order of 10 percent. The earlier (before military service) schooling coefficient falls more, but we interpret this to be the consequence of the interrelationship between test scores and father's status variables with the other important omitted variable – the quality of schooling. Unfortunately, given the restricted nature of our sample, both as to the selectivity inherent in being a veteran, and in being relatively young (under 35), these results cannot be taken as representative for all males. Nevertheless, this is one of the largest samples ever brought to bear on this problem and we would expect these results to survive extension to a more complete population.

Our results also throw doubt on the asserted role of genetic forces in the determination of income. If AFQT is a good measure of IQ and IQ is largely inherited, then the direct contribution of heredity to current income is minute. Nor is its indirect effect very large. Of course, the AFQT scores may be full of error and heredity may be very important, but then previous conclusions about the importance of heredity are also in doubt since many of them were drawn on the basis of similar data.

Appendix

The formulas used in the text are all repeated variations on the excluded-variable formula.[19] Let the true equation be

$$y = \beta_1 x_1 + \beta_2 x_2 + e,$$

where all the variables are measured around their means (and hence we ignore constant terms) and e is a random variable uncorrelated with x_1 and x_2.

Now, consider the least-squares coefficient of y on x_1 *alone:*

$$b_{y1} = \Sigma x_1 y / \Sigma x_1^2 = \Sigma\ x_1(\beta_1 x_1 + \beta_2 x_2 + e)/\Sigma\ x_1^2$$

$$= \beta_1 + \beta_2 \Sigma\ x_1 x_2 / \Sigma\ x_1^2 + \Sigma\ x_1 e / \Sigma\ x_1^2.$$

Since the expectation of the last term is zero, we can write

$$E(b_{y1}) = \beta_1 + \beta_2 b_{21}$$

where $b_{21} = \Sigma\ x_1 x_2 / \Sigma\ x_1^2$ is the (auxiliary) least-squares regression coefficient of the variable x_2 on the included variable x_1.

Moreover, if e were to refer to the computed least-squares residuals, $\Sigma\ x_1 e$ would equal zero by construction. Hence, the same formula holds also as an *identity* between computed least-squares coefficients of different order. That is,

$$b_{y1} = b_{y1.2} + b_{y2.1} b_{21}.$$

This same formula, with a suitable change in notation, applies also to higher-order coefficients:

$$b_{y1.2} = b_{y1.23} + b_{y3.12} \cdot b_{31.2}.$$

In what follows we shall assume that we are talking either about least-squares coefficients or about population parameters and we shall not carry expectation signs along. The discussion could be made somewhat more rigorous by inserting the plim (probability limit) notation at appropriate places.

The model we deal with can be written as

$$y = \beta_1 E + \beta_2 G + e = \beta_1 S + \beta_2 G + \beta_1 Q + e,$$

where $E = S + Q$ is education, S is quantity of schooling, Q is quality of schooling, G is a measure of ability (here assumed to be error-free), and Q is uncorrelated with S but is correlated with G. Then, estimating the equation with both G and Q excluded leads to

$$b_{yS} = \beta_1 + \beta_2 b_{GS} + \beta_1 b_{QS} = \beta_1 + \beta_2 b_{GS},$$

since $b_{QS} = 0$ by assumption. If G is included in the equation, then

$$b_{yS.T} = \beta_1 + \beta_1 b_{QS.G}.$$

Now, while b_{QS} is zero, $b_{QS.G}$ need not be zero. Given our assumptions we can write

$$b_{QS} = b_{QS.G} + b_{QG.S} b_{GS} = 0,$$

which implies that

$$b_{QS.G} = -b_{QG.S} b_{GS} < 0,$$

since both $b_{QG.S}$, the partial relationship of school quality to test scores, and b_{GS}, the relationship between test scores and levels of schooling, are expected to be positive. We also have

$$b_{QG} = b_{QG.S} + b_{QS.G} \cdot b_{SG}.$$

Substituting the formula for $b_{QG.S}$ into the formula for b_{QG} we get

$$b_{QG} = b_{QG.S} - b_{QG.S} b_{GS} \cdot b_{SG}.$$

Solving for $b_{QG.S}$, and recognizing that $b_{GS} b_{SG} = r_{SG}^2$, gives

$$b_{QG.S} = b_{QG}/(1 - r_{GS}^2),$$

which then gives

$$b_{QS.G} = - b_{QG} b_{GS}(1 - r_{GS}^2).$$

The algebra gets a bit more complicated when S is divided into two components, which for notational convenience will be called B (before) and A (after) here. The model now is

$$y = \beta_1 B + \beta_1 A + \beta_2 G + \beta_1 Q + e.$$

Then

$$b_{vB.AG} = \beta_1 + \beta_1 + \beta_1 b_{QB.AG} \quad \text{and} \quad b_{vA.BG} = \beta_1 + \beta_1 b_{QA.BG}.$$

Assume, as is approximately true in our sample, that A is uncorrelated with G. Since we have already assumed that Q is uncorrelated with both A and B, we have

$$b_{QB.A} = b_{QB.AG} + b_{QG.AB} \cdot b_{GB.A} = 0$$

$$b_{QA.B} = b_{QA.BG} + b_{QG.AB} \cdot b_{GA.B} = 0,$$

and hence

$$b_{QB.AG} = - b_{QG.AB} b_{GB.A}, \quad b_{QA.BG} = - b_{QG.AB} b_{GA.B}.$$

Thus, we can see immediately that the relative magnitude of the biases in the two schooling coefficients depends on the size of $b_{GB.A}$ relative to $b_{GA.B}$. Now because

$$b_{GA} = b_{GA.B} + b_{GB.A} b_{BA} = 0$$

by assumption, we have

$$b_{GA.B} = - b_{GB.A} b_{BA},$$

which we can substitute in

$$b_{GB} = b_{GB.A} + b_{GA.B} \cdot b_{AB}$$

to yield

$$b_{GB.A} = b_{GB}/(1 - b_{AB} b_{BA}) = b_{GB}/(1 - r_{AB}^2).$$

Now, if A (schooling after service) were entirely uncorrelated with B (schooling before service), then $b_{BA} = 0$ and its coefficient in the income-generating equation ($b_{yA.BG}$) would be unbiased:

$$b_{QA.BG} = -b_{QG.AB} \cdot b_{GA.B} = b_{QG.AB} \cdot b_{GB}b_{BA}/(1 - r^2_{AB}) = 0,$$

while the coefficient of schooling before service in the income-generating equation would be biased downward. In our sample, however, b_{BA} is actually negative and on the order of -0.3, implying that the coefficient of A is also biased downward, but only by about a third of the bias in the coefficient of B.

Acknowledgements

This work has been supported by NSF Grant no. GS 2762X. We are indebted to Mr. Paul Ryan for research assistance, and to E. Denison, O. D. Duncan, A. S. Goldberger, A. C. Kerckhoff, and K. O. Mason for comments on previous drafts. An earlier version of this paper appeared in *The Journal of Political Economy*, vol. 80, May/June 1972.

Notes

1 Concern with the accuracy of the education estimate due to the omission of ability may, of course, be readily extended to other factors associated with educational attainment and known to contribute to the determination of socioeconomic outcomes. Denison (1964), for instance, notes the salience of race, inherited wealth, family position, and diligence, and the list can easily be lengthened. In the present analysis we control for these factors to a considerable degree.

2 This is not too unreasonable an assumption, since there is a wide variation in quality of education at all levels of schooling. It is possible, however, that children going to better schools are also more likely to accumulate more years of schooling. If that is the case, we define Q to be that part of the "quality" distribution which is uncorrelated with "quantity." The rest follows in a similar manner.

3 These formulas hold as computational identities between least-squares coefficients. They can also be interpreted as expectations of computed least-squares coefficients from random samples from a population satisfying our assumptions.

4 See Klassen (1966) and Rivera (1965) for a description of the sample. These data have also been used by Duncan (1968), Mason (1968, 1970) and others.

5 Educational deferments have channeled substantial numbers of young men into entirely civilian careers, and a low score on the AFQT reduces the probability of being drafted. For a general discussion of this aspect of the Selective Service System see US President's Task Force on Manpower Conservation (1964). For an overview of Selective Service, see Davis and Dolbeare (1968).

6 Each of the education measures is based on eight categories of school years completed and is scored as follows: Less than eight years = 4, eight years = 8, 9 to 11 years but not high school graduate = 10, high school graduate = 12,

some college but less than two years = 13.5, two or more years of college but no degree = 15, BA = 16, graduate study beyond the BA = 18. As a matter of convenience we shall hereafter refer to SI as postservice schooling, ignoring the possibility that some of the increment may have occurred while the man was in service.

7 The percentile scores are the midpoints of each of the eight categories provided in the data. For a number of individuals in the sample, there were records of results for mental tests other than the AFQT. Prior to our acquisition of the data these scores were converted to AFQT equivalents following instructions provided by the Department of Defense. Despite the use of the AFQT to select individuals into the armed forces, all levels of performance on the AFQT are represented in our sample.

8 Our review of the literature on intelligence tests turned up nothing about the reliability of the AFQT or about correlations between it and civilian IQ tests. The only articles we have found which discuss the AFQT are those of Karpinos (1966, 1967), which are concerned not so much with the characteristics of the test as they are with the characteristics of those who fail it. We have seen fragmentary evidence about the AGCT, which is a predecessor of the AFQT. But extrapolation from experiences with the AGCT to the AFQT would merely be speculative. Nonetheless, we feel reasonably confident in using the AFQT as a measure of IQ, because of its face validity and because Duncan, Featherman and Duncan (1968, pp. 80–119) present evidence which may be interpreted as suggesting that several different mental ability tests, including the AFQT, have about the same relationships with socioeconomic variables.

9 If the AFQT is not virtually interchangeable with the standard civilian IQ tests, then Jensen (1969) could well be wrong in assuming that the heritability of the AFQT is the same as for the standard civilian tests. Griliches (1970, pp. 92–104) suggests that the heritability of the AFQT may be quite low, and pursues related issues.

10 These are of course only incomplete measures of the family's socioeconomic status, and are subject moreover to the possibility of recall error and mis-perception on the part of the respondents (sons), from whom this information was elicited. Blau & Duncan (1967, Appendices D and E) take up the issue of recall error for these two variables in their OCG sample. Conclusions drawn from their discussion should apply here, since the OCG sample is comparable to ours in the same age group. For evidence on this see Duncan (1968), who reports virtually identical correlations between father's education and occupation for the OCG and the CPS sample from which we draw.

11 The use of such a measure was suggested to us by Jacob Mincer.

12 There is scant reason (Mason, 1970) to believe that military service conveys an advantage in subsequent experience in the civilian labor force. Thus, we expect the AMS variable to have a negative coefficient in the income-generating equation.

13 The argument concerning the effects of the left-out variable of schooling quality is slightly more complicated than that outlined in the introduction because of the presence of *two* schooling variables. Considering only differences in the quality of schooling before military service and assuming that they are uncorrelated with both SB and SI, leads to the conclusion that

the introduction of the AFQT variable will bias the estimated SB coefficient downward (due to the assumed positive correlation of quality of schooling Q with AFQT and the observed positive correlation of AFQT with SB). The estimated coefficient of SI would remain unbiased provided that it really was uncorrelated with SB, AFQT, and the unobserved Q. The correlation of SI with AFQT is effectively zero but it does have a nonnegligible negative correlation with SB. This leads also to a downward but smaller bias in the coefficient of SI, the ratio of the two bias (in the coefficient of SI relative to the bias in the coefficient of SB) being equal to $b_{SB \cdot SI}$, which is about 0.3 in our data. See the Appendix for further details.

14 Since (apart from irrelevant constants) potential experience = age − SB − SI − AMS/12, in a regression that contains separately Age, SB, SI, and AMS, its coefficient is given by *either* the coefficient of Age or by 12 times the negative of the coefficient of AMS. The latter yields 0.0132 = (−12)(−0.0011) while the coefficient for Age is 0.0126. The two are thus quite consistent and support the interpretation that both calendar age and time spent in military service influence income via their effect on "experience." Another way of testing this is to concern the coefficient of Age to equal 12 times minus the coefficient of AMS. The computed F-statistics for such constrained versions of regressions 1 and 4 are 3.7 and 2.8 respectively, indicating that the data are consistent with the constraint at the conventional 5 percent significance level (the critical F is 3.8). For regression 4, the constrained version implies that a year of experience is worth a 2.3 percent increase in income on the average, and that holding "experience" (but not age) constant leads to estimated 7.3 and 7.8 percent increases in income per year of schooling, for pre- and postservice schooling respectively.

15 Ideally we would like to correct all of our variables for random errors. But although it is possible to adjust some others besides ability for random errors (Siegel and Hodge, 1968), we do not have enough information to adjust all of the variables. And, since our major interest is with changes in the education coefficient due to the inclusion of the ability measure, it is the errors in the latter that are most crucial to our analysis.

16 Let $G = S + H$, and H be defined so as to be uncorrelated with S, where S now stands for schooling and all other "environmental" effects. Then using the observed T as a variable implies leaving out from the regression $-\beta t$, the error of measurement in T. Using $\hat{T} = S$ implies the leaving out of βH. The latter causes a smaller reduction in the explained variance than the former.

17 Our procedure for estimating eqs (18) and (19) is not fully efficient since it does not take into account the proportionality constraint. Efficient methods for the estimation of systems such as this one have been developed by Zellner (1970), Hauser and Goldberger (1971), and Goldberger (1972). Since the system aspects of this model are peripheral to our main interest (the role of schooling in income determination), we did not use such procedures to improve the efficiency of our estimates further. In another round of estimation using more efficient methods we would not impose the proportionality constraint on the coefficients of the color variable, nor on the coefficients of the current experience variables. Equations (18) and (19) do not contain such constraints since we expect variables such as color, region (representing,

for example, regional cost of living differences), and marital status (representing differential supplies of effort in the market) to have rather different effects on earnings and on occupational status. This seems to be actually the case. For example, the estimated color coefficients in table 11.8 are quite different in the Income and Occupational SES regressions.

18 In addition to collating information from several samples, Duncan's study also uses correlations between the AFQT and other variables based on an extrapolation from the veterans subsample to the total sample. The use of these adjusted correlations would seem to be part of the reason for the discrepancy between our own results and those implied by Duncan's data. Although the assumptions which underlie the adjusted correlations appear reasonable, they do remain open to question.

19 These formulas are given, in a different context, by Griliches and Ringstad (1971, Appendix C). See Yule and Kendall (1950, pp. 284–5) for the notation used here.

References

Ashenfelter, Orley, and Mooney, Joseph D. "Graduate education, ability and earnings," *Rev. Econ. Statis.*, 50 (February 1968), 78–86.

Blau, Peter M. and Duncan, Otis Dudley. *The American Occupational Structure*. New York: Wiley, 1967.

Bowles, Samuel. "Schooling and inequality from generation to generation." *J. Pol. Econ.*, 1972, 80, S219–S251.

Davis, James W., Jr and Dolbeare, Kenneth M. *Little Groups of Neighbors: The Selective Service System*. Chicago: Markham, 1968.

Denison, Edward F. *The Sources of Economic Growth in the United States and the Alternatives before Us*. Supplementary Paper no. 13. New York: Committee Econ. Development, 1962.

——. "Measuring the contribution of education." In *The Residual Factor and Economic Growth*, pp. 13–55, 77–100. Paris: Org. for Econ. Cooperation and Development, 1964.

Duncan, Otis Dudley "A socioeconomic index for all occupations," and "Properties and characteristics of the socioeconomic index." In Albert J. Reiss, Jr. with Otis Dudley Duncan, Paul K. Hatt, and Cecil C. North (eds), *Occupation and Social Status*, chaps. 6 and 7, pp. 109–61. New York: Free Press, 1961.

——. "Path analysis: sociological examples," *American J. Sociology*, 72 (July 1966), 1–16.

——. "Ability and achievement," *Eugenics Q.*, 15 (March 1968), 1–11.

——. "Inheritance of poverty or inheritance of race?" In Daniel P. Moynihan (ed.), *On Understanding Poverty*, pp. 85–110. New York: Basic, 1969.

Duncan, O. D., Featherman, D. L., and Duncan, B. "Socioeconomic background and occupation achievement: extensions of a basic model." Final Report, Project no. 5–0074 (EO–191), Contract no. OE-5-85-072. US Office of Education. Ann Arbor, Michigan: University of Michigan, 1968.

Goldberger, A. S. "Maximum-likelihood estimation of regression models containing unobservable independent variables." International Econ. Rev, 1972; 13, 1–15.

Griliches, Zvi. "Notes on the role of education in production functions and growth accounting." In W. Lee Hansen (ed.) *Education, Income and Human Capital*, pp. 71–127. Studies in Income and Wealth, vol. 35. New York: Nat. Bur. Econ. Res., 1970.

—— and Ringstad, V. *Economies of Scale and the Form of the Production Function*. Amsterdam: North-Holland, 1971.

Hanoch, Giora. "An economic analysis of earnings and schooling," *J. Human Resources*, 2 (Summer 1967), 310–29.

Hansen, W. Lee, Weisbrod, Burton, A., and Scanlon, W. J. "Schooling and earnings of low achievers," *AER* 50 (June 1970), 409–18.

Hauser, Robert, M. and Goldberger, A. S. "The treatment of unobservable variables in path analysis" in H. L. Costner (ed.) *Sociological Methodology 1971*, San Francisco, California: Jossey-Bass, Chapter 4, 81–117.

Jensen, Arthur, R. "How much can we boost IQ and scholastic achievement?", *Harvard Educ. Rev.*, 39 (Winter 1969), 1–123.

Karpinos, Bernard, D. "The mental qualification of American youths for military service and its relationship to educational attainment." *Proc. of Social Statis. Sect. American Statis. Association, 1966*. Washington: ASA. 1966.

——. "Mental test failures." In Sol Tax (ed.), *The Draft*, pp. 35–53. Chicago: Univ Chicago Press, 1967.

Klassen, A. D. *Military Service in American Life since World War II: An Overview*. Report no. 117. Chicago: Nat. Opinion Res. Center. Univ. Chicago. 1966.

Mason, William. "Working paper on the socioeconomic effects of military service." Unpublished paper, Univ. Chicago, 1968.

——. "On the socioeconomic effects of military service." PhD dissertation. Univ. Chicago, 1970.

Rivera, Ramon, J. "Sampling procedures on the military manpower surveys." Chicago: Nat. Opinion Res. Center, Univer. Chicago, 1965.

Rogers, Daniel, C. "Private rates of return to education in the U.S.: a case study," *Yale Econ. Essays*, 9 (Spring 1969), 89–134.

Siegel, Paul, M. and Hodge, Robert, W. "A causal approach to the study of measurement error." In H. M. Blalock and A. B. Blalock (eds), *Methodology in Social Research*, pp. 28–59. New York: McGraw-Hill, 1968.

Taubman, Paul and Wales, Terence. "Net returns to education." In *Economics—a Half-Century of Research, 1920–1970; 50th Annual Report*, pp. 65–66. New York: Nat. Bur. Econ Res., 1970.

U.S. President's Task Force on Manpower Conservation. *One-third of a Nation: A Report on Young Men Found Unqualified for Military Service*. Washington: Government Printing Office, 1964.

Weisbrod, Burton, A., and Karpoff, Peter. "Monetary returns to college education, student ability, and college quality," *Rev. Econ. Statis.*, 50 (November 1968), 491–97.

Yule, G. Undny, and Kendall, M. G. *An Introduction to the Theory of Statistics*. 14th edn, rev. London: Griffin, 1950.

Zellner, Arnold. "Estimation of regression relationships containing unobservable independent variables," *Internat. Econ. Rev.*, 11 (October 1970), 441–54.

12
Capital–Skill Complementarity*

In a previous paper [3] I suggested the following three-input demand model for testing the possibility that "skill" or "education" is more complementary with physical capital than unskilled or "raw" labor:

$$\ln S/N = a_1 + (\eta_{sn} - \eta_{nn}) \ln W/Z$$
$$+ (\eta_{sk} - \eta_{nk}) \ln R/Z$$
$$\ln S/K = a_2 + (\eta_{sn} - \eta_{kn}) \ln W/Z$$
$$+ (\eta_{sk} - \eta_{kk}) \ln R/Z$$

where N is "raw" labor, K is capital, S is skill or schooling with the corresponding (rental) prices W, R, and Z. In getting to this set of equations we have assumed constant returns to scale and used the condition $\Sigma_i \eta_{ij} = 0$. Note also that $\eta_{ij} = \nu_j \sigma_{ij}$, where ν_j is the share of the jth factor in total costs and the σ_{ij}'s are the Allen-Uzawa partial elasticities of substitution ($\sigma_{ij} = \sigma_{ji}$ and $\sigma_{ii} < 0$). Thus the hypothesis can be restated as

$$0 < \sigma_{nk} > \sigma_{sk} \text{ and } \sigma_{kn} > \sigma_{sn},$$

and implies a negative coefficient for $\ln R/Z$ in the first equation and for $\ln W/Z$ in the second.

The basic difficulty in testing such a hypothesis is the lack of good price series for either K or S.[1] The strategy of this note is to use cross-sectional data for the United States and assume, which is not too implausible, that the price of "skill" or skilled labor (Z) has been largely equalized by the mobility of educated labor, and hence can be treated as approximately constant across states and industries. Of the two remaining prices, the quality of the data on wages of unskilled labor is much superior to that on the cost of physical capital. Hence, we shall concentrate on the coefficient of the "better" variable, in W, in the second equation and endeavor to show that the use of a "poor" variable for R cannot

*Reprinted from *Review of Economics and Statistics*, vol. LI, no. 4, November 1969, published by Harvard University.

account by itself for the observed negative coefficient (of ln W in the second equation).

We have two sets of data. The first is more extensive, using observation on two-digit manufacturing industries in individual states in 1954, but is restricted by the availability of only crude occupational distribution data for the construction of skill variables. The second body of data is based on the 1960 Census of Population data on the education of workers by industry and on the 1964 Annual Survey of Manufactures for data on capital per worker. These data are available for 60 three-digit manufacturing industries for the United States as a whole. There is no geographical breakdown, except for the differential average location of an industry, to provide us with significant variance in the real price of labor.

Table 12.1 Capital–skill complementarity regressions, US manufacturing industries (two-digit), 1954, by states, $N = 261$

				Coefficients of				
						Dummies		
Dependent variable			*ln*W	*ln*GRR	States	Industries	R^2	$\sigma^2 u$
I	(a)	*O/L*	1.	0.302	− 0.023			0.445	0.081
				(0.023)	(0.009)				
			2.	0.354	− 0.020	✓		0.534	0.077
				(0.028)	(0.008)				
			3.	0.054	− 0.009	✓	✓	0.920	0.033
				(0.022)	(0.005)				
	(b)	*S/US*		0.113	− 0.020	✓	✓	0.909	0.229
				(0.155)	(0.034)				
II	(a)	*O/B*	1.	− 1.055	0.966			0.763	0.384
				(0.111)	(0.041)				
			2.	− 0.615	0.715		✓	0.859	0.308
				(0.138)	(0.049)				
			3.	− 1.351	0.697	✓	✓	0.939	0.231
				(0.157)	(0.038)				
	(b)	*S/B*		− 0.615	0.715		✓	0.859	0.308
				(0.137)	(0.049)				

O/L – Logarithm of "Skill" per man (and female) based on *1950 Census of Population* data on occupations by industry, weighted by 1949 mean incomes by occupations and sex from *Census Technical Paper 17*, table 38. Basic form of index is similar to the one constructed from 1960 data and described in Griliches [1].

S/US – Logarithm of "Skilled" to "Unskilled" workers ratio in 1950. "Unskilled" defined as operatives, laborers, and service workers, from *1950 Census of Population*. "Skilled" – all others.

B – Bookvalue of Fixed Assets in 1954. Based on *Census of Manufacturers* book value data for 1957 and investment expenditures in 1954–57. Construction of variable described in Griliches [2].

S/B – Logarithm of "skilled workers" per unit of capital computed by multiplying the number of man-hours in 1954 by the fraction "skilled" in 1950 and dividing by *B*. Similarly $O/B = \ln O/L - \ln B/L$, where *L* are total man-hours in 1954.

W – Average wage rate per hour of production workers, 1954.

GRR – Gross rate of return = (Value added – Total Payrolls)/Bookvalue of fixed assets, 1954.

The results of fitting these two equations to the first set of data (two-digit Manufacturing Industries, by States, $N = 261$) are summarized in table 12.1. Two measures of "skill" are used: the first, O, is an all inclusive measure of "quality per man," using the observed occupational and sex distribution within an industry and state and national average earnings by occupation and sex to "predict" the earnings ("earning power") of the average worker in the particular industry and state (see previous paper [3], for additional discussion of such "quality" indexes). The second measure, S/US, is just a ratio of the number of skilled to unskilled workers, where "skilled" is defined as all workers who are not laborers, operatives, or service workers. The capital measure used, B, is an estimate of the gross book number value of fixed assets at the beginning of 1954; the average wage per hour of production workers, W, is assumed to be proportional to the relevant price of unskilled or "raw" labor; and the cost of capital is approximated by the "gross rate of return," GRR, which is computed as value added minus payrolls divided by the book-value of fixed assets. In addition, state and industry dummy variables are included in some of the regressions. The industry dummies are expected to adjust for general industry-wide errors in such *ex post* cost of capital measures.

To summarize briefly, the results presented in table 12.1 have all the expected signs. The coefficient of ln GRR in the first set of equations is always negative, though often not significantly so. Nevertheless, this is an interesting finding in light of the rather poor quality of this variable (which would tend to bias its coefficient toward zero) and the fact that there is no spurious relation between it and the dependent variable; they are computed from entirely separate sets of data. More interesting is the very "significant" negative sign of the coefficient of the "better" ln W variable in the second set of equations. If the model is accepted, it definitely implies a higher elasticity of substitution of "raw" labor for capial than for skilled labor. The positive coefficient for ln GRR is consistent with the model but could be in part spurious, since GRR is computed by dividing nonlabor payments by the capital measure used on the left-hand side of the same equation. The likely presence of errors in the capital measure would tend to bias the coefficient of ln GRR towards unity. Accepting the probability of such a bias we shall endeavor to show below that it cannot, by itself, explain the negative coefficient of ln W.

The results for the second set of data (60 United States Manufacturing Industries) are summarized in table 12.2. Here a similar all-inclusive quality index, Q, is constructed using the educational and sex distribution of workers in 1960 and mean United States earnings by these categories in 1959. The second measure of "schooling," H, measures it above a minimum unskilled level which is attributed to "raw" labor. Ignoring the fact that these indexes are first constructed separately for males and females and then aggregated, the relation between them is very simple:

$$Q_j = \Sigma_i \, w_i N_{ij} / N_j, \qquad H_j = \Sigma (w_i - w_o) N_{ij} / N_j$$

Table 12.2 Capital–education complementarity: 60 United States manufacturing industries, 1963 (1960)

Dependent variable		Coefficients of ln WUS	ln GRR	R^2	$\sigma^2 u$
I	$\ln Q/N$	0.913	-0.114	0.706	0.109
		(0.084)	(0.039)		
	$\ln H/N$	1.220	-0.050	0.729	0.131
		(0.100)	(0.046)		
II	$\ln Q/SK$	-1.136	0.595	0.352	0.430
		(0.330)	(0.152)		
	$\ln H/SK$	-0.830	0.659	0.314	0.446
		(0.343)	(0.158)		

The data on characteristics of workers are from the 1960 Census of Population, "Industrial Characteristics," PC(2)7F, table 21, and "Occupational Characteristics," PC(2)7A, table 28. The data on capital and rates of return are from the 1963 Census of Manufacturers and the 1964 Annual Survey of Manufactures, 1965 (AS)–6. The list of industries is the same as in table 21 of PC(2)-7F, starting with "Logging" and ending with "Leather products, except footwear."

GRR – gross rate of return = (Value Added – Total Payrolls – Total Rentals)/(0.51 Average Bookvalue of fixed assets + Average Inventories) where the bookvalue of fixed assets and the value of inventories at the end of 1962 and 1963 are averaged and 0.51 is an approximate translation of "gross" into "net" bookvalue based on data in the 1958 Census of Manufactures.

SK – capital services = 0.12 Average Gross Bookvalue + Total Rentals + 0.07 Average Inventories. This is consistent with a 7 percent rate of interest and about 15 percent net depreciation rate, which in turn is consistent with aggregate figures for 1957 given in the 1958 Census of Manufactures.

WUS – Median wage and salary income of male laborers who worked 50–52 weeks in 1959. From table 20 in PC(2)-7A.

Q – Education per man index, constructed from table 21 in PC(2)-7F for males *and* females using 1959 mean earning by education (for males) and 1959 median incomes of females who worked 50–52 weeks as weights. For additional discussion of such indexes see Griliches [2].

H – Human capital per person (up to a constant of proportionality) = $(Q_M - 2035)$ $(1 - FF) + (Q_F - 1026)FF$ where *FF* is the fraction of female workers and 2935 and 1026 are the weights of the lowest education class ("less than five years of schooling") for males and females respectively. This attributes to *H* all of the (average) earning power above the wage of the least educated class. To convert it into actual human capital units, it would need also to be divided by some discount factor which would capitalize these estimated flows.

H/SK and *Q/SK* are obtained by dividing the two education per man measures by capital services per total number of employees (in 1963, from Census of Manufactures).

and hence

$$N_j Q_j = N_j H_j + w_o\, N_j = N_j(H_j + w_o)$$

where the summation is over the various schooling classes, N_{ij} is the number of workers with i^{th} level of schooling in industry j, $N_j = \Sigma_i\, N_{ij}$ is the total number of workers in industry j, w_i are national earnings by schooling weights, and w_o is the average income of the lowest (less than five years) schooling class. Here the wage of unskilled labor, *WUS*, is based on independent data on the median earnings of full-time laborers by industry (from the 1960 Census of Population); the cost of capital, GRR, is computed as before (except that bookvalue is averaged for the beginning and end of the year and inventories are included in the definition of capital); and the measure of capital services, SK, is a more complicated concept, allowing for rentals of equipment and a role for inventories. (A more detailed definition of SK is given in the notes to table 12.2.)

Here again the estimated coefficients have the expected signs: negative but not "too strong" for the relatively poor GRR variable in the schooling per worker equations, and negative and significant for the better WUS variable in the schooling per capital equations.

Thus, in both sets of data, there is evidence for the hypothesis that "skill" or "schooling" is more complementary with capital than unskilled or unschooled labor. It remains to be shown that these results are not a spurious consequence of the use of relatively poor cost of capital measures. To show this we shall concentrate on the second equation. Let small letters stand for the logarithms of the variables and let "true" unobserved versions of these variables be identified by asterisks (we'll be interested in this distinction only for k and r), then we can write the second equation as

$$s - k^* = \alpha w + \beta r^* + u$$

let $k = k^* + v$ and $r^* = \pi - k^*$, then $r = \pi - k = r^* - v$, which in turn can be rewritten in terms of the observed variables as

$$(s - k) = \alpha w + \beta r - (1 - \beta)v + u.$$

Thus, running the regression in the observed variables is equivalent to leaving out a relevant variable θ with the coefficient $-(1 - \beta)$ that is negatively correlated with the included variable r (since $r = r^* - v$).[2]

It can be shown (see Appendix), if one makes all the standard assumptions and in addition assumes that v is a random error uncorrelated with w and r^*, that the estimated coefficient of GRR is biased towards one[3]

$$E(\beta' - \beta) = (1 - \beta)\lambda_v / (1 - r^2_{rw})$$

where

$$\lambda_v = \sigma^2_v / \sigma^2_r = \sigma^2_v / (\sigma^2_{r^*} + \sigma^2_v)$$

is the fraction of the observed variance in r due to "error" and r^2_{rw} is the square of the simple correlation coefficient *in the sample* between the observed r and w. On the other hand, the bias in the coefficient of w, the one we are interested in, can be written as

$$E(\alpha' - \alpha) = - \text{bias } \beta' \cdot b_{rw}$$
$$= -(1 - \beta) \lambda_v b_{rw} / (1 - r^2_{rw})$$

where b_{rw} is the regression coefficient, in the sample, of r on w. Since it is assumed that $\beta > 0$, and since all of our estimated β's are less than one, we expect the true β to be less than β'. Hence $1 > \text{bias } \beta' > 0$, and therefore the magnitude of the absolute bias in α' is bounded by the observable b_{rw}. To make a long story short, the observed b_{rw}'s are negative, -0.6 for the first sample and -0.3 for the second sample, and hence this type of bias would make the estimated (negative) α'''s appear to be less negative than they really are. Even if one accepted the possibility that all the error in r is due to the randomness of profits, rather than to errors in the measurements of K, which would imply a negative bias in β', this too could not explain away the observed negative α'''s, since again one would expect bias β' in this case too, to be less than one in absolute value. Thus, the observed negative α'''s cannot be explained as the consequence of using a bad cost of capital variable and therefore can be taken as an indication of greater capital-schooling (skill) complementarity.[4]

Appendix

Derivation of formulae on page 217:

Start with the general case of the left out variable v which is correlated with only one of the included variables (r). Then the "true" equation is

$$y = \alpha w + \beta r + \gamma v + u$$

while we estimate, by least squares,

$$y = \alpha w + b v + u'.$$

Then, given the usual assumptions

$$E\alpha = \alpha + \gamma b_{vw \cdot r}$$
$$Eb = \beta + \gamma b_{vr \cdot w}$$

where the $b_{vi \cdot j}$ si the regression coefficient of the ith variable in the "auxiliary" regression of the left out variable v on the included variables i and j. Moreover, we know that we have the following identities

$$b_{vw} = b_{vw \cdot r} + b_{vr \cdot w} b_{rw}$$
$$b_{vr} = b_{vr \cdot w} + b_{vw \cdot r} b_{wr}.$$

Now, by assumption, $b_{vw} = 0$. Hence

$$b_{vw \cdot r} = -b_{vr \cdot w} b_{rw}$$

which leads immediately to

$$\text{Bias } \alpha' = -\text{Bias } \beta' \cdot b_{rw}.$$

Also, substituting the previous expression for $b_{vw \cdot r}$ in the b_{vr} relation leads to

$$b_{vr \cdot w} = b_{vr}/(1 - r^2_{rw})$$

since $b_{rw} b_{wr} = r^2_{rw}$. Therefore

$$\text{Bias } \beta' = E(b - \beta) = \gamma b_{vr}/(1 - r^2_{vw}).$$

The formulae given in the text follow immediately upon noting that $\gamma = -(1 - \beta)$, and that $b_{vr} = -\lambda_v$, given the assumption that v is a random error uncorrelated with r^* and $r = r^* - v$.

Acknowledgement

I am indebted to the National Science Foundation for financial support of this and related work.

Notes

1 Even if one had the relevant data, the alternative of attacking the problem directly by estimating a general three-input production function, $Y = F(N, S, K)$, is not really promising since questions about the relative size of elasticides of substitution are questions about the curvature of the $F(\;\;)$ function which are usually very hard to answer on the basis of ordinary economic data. As far as the production function is concerned the σ's are second order parameters and hence almost impossible to estimate. They are, however, first order parameters in the derived demand equations.

2 We could have complicated the model some more by adding also an error to the "profit variable" $\pi = \pi^* + e$, which would have led to $r = r^* + e - v$. This would have only attenuated the effects we are looking for, since e and v would operate in opposite directions, and would have only strengthened the argument.

3 This, and the following statement require also the assumption $\lambda < 1 - r^2_{rw}$ which is likely to hold for our samples since the observed r^2_{rw} is quite small (about 0.04 in the first set of data).

4 The statement in the text does not distinguish adequately between two related but different hypotheses: $\sigma_{nk} > \sigma_{sk}$ and $\sigma_{nk} > \sigma_{ns}$. The evidence presented above supports the second one better than the first, though it does support both.

References

[1] Griliches, Z., "Production functions in manufacturing: some preliminary results," in *The Theory and Empirical Analysis of Production*, NBER Studies in Income and Wealth, 31, New York: Columbia University Press, 1967).

[2] ——, "Production functions in manufacturing: some additional results," *Southern Economic Journal*, XXXV (Oct. 1968).

[3] ——, "Notes on the role of education in production functions and growth accounting." To appear in *Education, Income and Human Capital*, L. Hansen (ed.), Studies in Income and Wealth, 35, NBER, 1970.

13
Postscript on Education and Economic Growth

The NORC-Veterans data-set, on which the Griliches-Mason study was based, was somewhat peculiar and unrepresentative both because of its limitation to army veterans and because we had to rely on AFQT as a measure of "ability." The AFQT was given rather late in a young man's career (ages 17–19) and hence was not really a measure of *native* ability. Moreover, given its slant as an army *qualification* test, it was not clear whether it represented adequately the range of abilities important for subsequent economic success. This led me to a series of studies based on the National Longitudinal Survey of Young Men, a random sample of the cohort of young men who were 14 to 24 years of age in 1966. The advantage of these data is in their representativeness, in the fact that these young men have been followed from 1966 through 1981, and in their inclusion of data on IQ-type scores collected from the high schools of the respondents. The drawback of these data is the youth of this cohort, but that has been diminishing over time, as more data are collected and as the cohort continues to age.

An analysis of the wage rates and earnings of the NLS Young Men, reported in Griliches (1976b and 1977) confirmed the earlier NORC-CPS Veterans-based results: the effect of leaving out IQ-type measures on the estimated schooling coefficients is minor, on the order of less than 0.01. Both ability and family background are important in determining the ultimate economic success of individuals but they work largely through their effect on schooling rather than independently.

These results, particularly the relatively small role attributed to "ability" (net of its effect on schooling) in determining earnings were not all that plausible and met with some resistance. Three areas of possible bias had to be examined before they could be given full credence: (1) the available test scores may be subject to large random errors at the individual level which would bias downward the estimated effect of true ability; (2) ability may be important but the particular abilities which are reflected in wage and income differentials may not be measured adequately or at all by cognitive tests such as AFQT or IQ; (3) there may be other yet unspecified sources of bias in our data and modes of analysis.

I first tackled the errors-in-variables problem. To get consistent estimates of the relationship of interest in the presence of random errors in one's "independent" variables one needs to bring in additional "instrumental" variables into the model, variables that are correlated with the systematic portion of the affected variables but not with the errors in it. These can be of three kinds: background variables such as mother's schooling which are assumed to affect true ability but not wages (holding schooling and actual ability constant), parallel replicated readings on the same type of variable on the same individuals, such as other test scores whose measurement errors are assumed to be independent of each other, and parallel measures (tests) on related individuals (siblings) which could be assumed to be correlated with the individual's true ability but not with the errors in its measurement.

The use of background variables as instruments has been described in the Griliches and Mason paper. Since the NLS data contained data on two test scores (IQ and KWW – a test of the Knowledge of the World of Work), a more elaborate instrumental variables procedure could be used, and was reported on in Griliches (1976b and 1977). The econometric problems associated with the use of such models are surveyed in Griliches (1974 and 1977). The actual results of these more elaborate estimation procedures were not all that different: the impact of true ability on earnings was estimated to be higher than before, but this change in the estimated ability coefficients led to very little change in the estimate of the schooling coefficient, or to any revision in the role attributed to ability in accounting for the observed differences in earnings across individuals. They did imply, however, very large errors in the observed test scores, amounting to about half or more of the observed variance in them. This raised the following problem: the standard psychometric literature indicates that simple random (test–retest) measurement errors do not account for more than 20 percent of the variance in such tests. We, however, got much higher numbers. Either these tests are significantly worse than psychologists have led us to believe or a large fraction of the true IQ variance may be irrelevant in determining wage differentials across individuals. Much of the differences in IQ may not matter in terms of their effect on the earnings potential of individuals *after* their effect on his achieved level of schooling has already been taken into account. Also the "ability" we want for our models may not be IQ. This view throws doubt on the assumption that "errors" in IQ are uncorrelated with everything else, especially schooling. They may be irrelevant for income but may still have an effect on schooling.

Such concerns, and the desire for more and better instrumental variables, led me to search for data on siblings. Given data on several individuals who share a common attribute one can devise a method for estimating the returns to schooling even if such attributes are not observed directly. The basic idea here is that the unobservable ability affects several variables and/or individuals at the same time. Thus there is, in a sense, replication; more than one observation on the effect of the same unobservable. This allows one to estimate the general properties of that variable and to adjust for it in estimating the net impact of schooling. The econometric details of such sibling (variance-components) models are given in Chamberlain and Griliches (1975 and 1977), Griliches (1974 and 1979a), and Bound, Griliches, and Hall (1986). The results of these studies are roughly as follows: if one identifies "ability" with the thing that is measured (albeit imperfectly) by test scores, and accepts the genetic model which postulates that family members are positively correlated on this, then the unobservable variable that fits these requirements seems to have little to do with earnings beyond its indirect effect via schooling. If one expands the model to allow for several unobserved factors or abilities, then the factor that is common to brothers and sisters and has an effect on earnings appears to have little to do with test scores. In either case the original estimates of the contribution of schooling are largely upheld.

During this period much work was also done by others on this range of problems. I will not attempt to review it all here, except to point out a few of the more important and closely related lines of research. (See Jencks et al., 1979 and Willis, 1986, for additional reviews, and Psacharopoulos, 1981, for an international survey of the estimates of the rate of return to education.) Jencks and his associates (Jencks and Brown, 1977) reanalyzed the NORC–CPS Veterans sample and also analyzed several data sets on brothers (NORC, Talent, and Kalamazoo); Hauser and his associates (Hauser and Daymont, 1977; Sewell and Hauser, 1975; Hauser and Sewell, 1986) have been following up the 1957 Wisconsin Cohort of high school seniors; Olneck (1977) has analyzed data on brothers from the town of Kalamazoo in Michigan; and Taubman and his associates (Behrman et al., 1980) have been analyzing data on twins, based on an NRC sample of veterans. All these studies differ in coverage, variable definitions, and modes of analysis. Where comparable their results, by and large, are consistent with mine. Studies that focus on the contribution of measured ability to the variance of earnings, net of schooling, find rather small effects and also relatively small absolute biases in the schooling coefficient. Whether the resulting bias is "large" percentage-wise depends on what else has been held constant in the model. Holding work experience constant, Hauser and Daymont found only a 13 percent upward bias in the estimated high school–college earnings differential for the period 1965–71, with part of this bias being due to the introduction of parental income in addition to the mental ability measures in the equation. Larger effects are obtained by Olneck in his analysis of Kalamazoo brothers, and by Behrman, Taubman, and Wales (1977) in their analysis of the NRC twins data. In both these cases the effects come not from the introduction of test scores *per se*, but from estimating the effect of schooling on earnings from the "within" samples, from differences in the experience of brothers or twins. The problem with this type of approach is that if there are also errors in the measurement of schooling, their *relative* importance will tend to be magnified when comparisons are made only between brothers or twins.

Moreover, adding additional variables such as test scores to the equation only exacerbates the problem (cf. Griliches, 1977, section VI, and Griliches, 1979a, for more detailed discussions of this point). Thus, for example, all of the rather large bias in the schooling coefficient estimated by Behrman, Taubman, and Wales from a comparison of earnings differences between monozygotic and dizygotic twins would disappear if one allowed for a 15 percent error component in the observed variance of schooling. That schooling could be subject to that much measurement error is suggested by the work of Bishop (1974) and Bielby, Hauser, and Featherman (1976). Their error of measurement estimates range from 10 to 20 percent of the observed variance in schooling. In short, twin and sibling studies have yet to tackle effectively the problem of measurement errors. Until that is done, it is difficult to interpret their results except to note again that while family members seem to share a common experience, there is enough variance in the data to belie the view that the family of origin is an "unriddable" millstone around the neck of the individual. Individuals do move and transcend their family origins, and do succeed or fail with only the weakest relation to the test scores attached to them early in their careers. While social science models are couched in deterministic language, actual individual experiences are much more open-ended. The vagaries of time, place, and accident overshadow the stylized order suggested by our models.

My work on the construction of labor quality indexes was continued by Jorgenson and his associates – see especially Chinloy (1980), Jorgenson (1984), and Jorgenson and Pachon (1983). Denison's most complete recent statement can be found in Denison (1974 and 1979). See Riley (1979) and Kroch and Sjoblom (1986) for attempts to test the screening hypothesis. What is missing in the literature is a detailed discussion of how to incorporate the educational system within an expanded framework of national income accounts and treat the production of human capital endogenously, allowing for productivity growth in this industry also. See Plant and Welch (1984) for a discussion of some of these issues, and Eisner [1985] for an attempt at some imputations.

To date, all of the new work and new data support the original conclusions. The contribution of the rising schooling level of the labor force to US economic growth during the 1940s, 1950s, and 1960s has been substantial, on the order of about one-third of the measured growth in total factor productivity. While this number is subject to various reservations, caveats, and uncertainties, it does not appear to be seriously and systematically biased upwards because of inadequate control for differences in ability.

Part III

R&D and Productivity

Part III

R&D and Production

14
Research Costs and Social Returns: Hybrid Corn and Related Innovations*

Introduction and Summary

Both private and public expenditures on "research and development" have grown very rapidly in the last decade. Quantitatively, however, we know very little about the results of these investments. We have some idea of how much we have spent but very little of what we got in return. We know almost nothing about the realized rate of return on these investments, though we feel intuitively that it must have been quite high. This paper presents a first step toward answering some of these questions. However, all that is attempted here is to estimate the realized social rate of return, as of 1955, on public and private funds invested in hybrid-corn research, one of the outstanding technological successes of the century. The calculated rate of return is an *estimate*, subject to a wide margin of error, but it should provide us with an order of magnitude for the "true" social rate of return on expenditures on hybrid-corn research. Actually, I believe that my estimate is biased downward, for, whenever I had to choose among alternative assumptions, I chose the assumption that led to the lowest estimate. This estimate will not tell us the global rate of return on research expenditures, but even a modest step in that direction may be of some use.

The following procedure is used to arrive at the estimate: First, private, and public research expenditures on hybrid corn, 1910–55, are estimated on the basis of a mail survey and other data. Then the annual gross social returns are estimated on the assumption that they are approximately equal to the value of the resulting increase in corn production plus a price-change adjustment. The additional cost of producing hybrid seed is subtracted from these gross returns to arrive at an annual flow of net social returns. Using first a 5 and then a 10 percent rate of interest, I bring all costs and returns

*Reprinted from *Journal of Political Economy*, vol. LXVI, no. 5, October 1958.

forward to 1955, when the books are closed on this development and a rate of return is computed. Research costs are expressed as a capital sum, and returns are converted into a perpetual flow. The estimated perpetual flow of returns is divided by the cumulated research expenditures to arrive at a rate of return that will equalize the present value of the flow of returns with the cumulated value of research expenditures. This procedure leads to the estimate that *at least* 700 percent per year was being earned, as of 1955, on the average dollar invested in hybrid-corn research.

Since this is not the only way in which a rate of return could be computed from these data, some alternative ways of defining and estimating the social rate of return are explored briefly. Comparisons are also made with estimates of returns in some other areas of technological change. Finally, I discuss the limitations of the procedure used and the implications of the results. In particular I shall emphasize that almost no normative conclusions can be drawn from these few estimates.

Research Expenditures

Inbred lines and hybrids have been developed by state agricultural experiment stations, the United States Department of Agriculture (USDA), and private seed companies. The distinction between the first two developing agencies is mainly in the source of funds. Except for funds spent on research and coordinating activities at Beltsville, Maryland, most of the USDA funds were spent on co-operative corn-breeding research at various experiment stations.

A mail inquiry to ascertain expenditures on hybrid-corn research was sent to all the agricultural experiment stations, and usable data were obtained from twenty of them. The twenty states represented by these replies include most of the important corn states in the country. Expenditures of non-responding stations were estimated by setting the expenditures of each of them equal to the expenditures of a "similar" station.[1]

Data on USDA expenditures on "corn-production research: agronomic phases" beginning with 1931 were obtained from the Agricultural Research Service and extrapolated back to 1910. They over-estimate the USDA contribution substantially, because they include various other aspects of corn research besides hybrid corn. Moreover, some of the USDA funds have already been counted in the expenditures of agricultural experiment stations.

The research expenditures of one of the major private seed companies for the years 1925–55 were extrapolated back to 1911 and divided by that firm's estimated share of the total market for hybrid seed corn to arrive at an estimate of the research expenditures of the "private" segment of the industry.[2]

The figures for 1955 may be used as an example of the numbers involved. I estimate that in 1955 the USDA spent about $300,000 on hybrid-corn research, the experiment stations about $650,000, and the private companies about $1,900,000.[3]

The historical research expenditure data, deflated by the Consumers Price Index (1955 = 100), are reproduced in column 1 of table 14.1. In view of all the assumptions made, these figures should be taken with several grains of salt, the dosage increasing as one goes back into the past. In particular, for the years 1910–25, the figures are little more than guesses.

For the purpose of estimating the rate of return on these expenditures, I assume that the public sector will continue to invest in hybrid-corn research at an annual rate of $1 million, and the private sector at an annual rate of $2 million. No incremental returns, however, will be ascribed to these expenditures. I assume them to be "maintenance" expenditures in face of a malevolent nature.

Cost of Additional Resources Devoted to Production of Hybrid Seed

I assume that the price of hybrid seed, approximately $11 per bushel in 1955, measures adequately the value of resources devoted to its production. If there were no hybrid corn, farmers would use mainly home-produced open-pollinated seed, which I value at $1.50 per bushel.[4] The quantity of hybrid seed used annually was estimated by multiplying the reported corn acreage planted with hybrid seed by the average seeding rate of corn. Multiplying the result by $9.50, the difference between the price of hybrid and non-hybrid seed, and subtracting $2 million research expenditures, I get $90 million as my estimate of the additional resources currently devoted each year to hybrid-seed production.[5] Using the average 1939–48 corn acreage (90 million), the 1951 seeding rate (7.5 pounds per acre), and the percentage planted with hybrid seed, I computed the additional cost of hybrid seed for the years 1933–55 and subtracted this from the subsequent estimate of gross returns to arrive at a net return figure.[6]

The Value of Hybrid Corn to Society

As everyone knows, hybrid corn increased corn yields. The figure most often quoted for this increase is 20 percent. Fo my purpose I assume that the superiority of hybrid over open-pollinated varieties is 15 percent, the lower figure in most estimated ranges.[7] The value of this increase to "society" will be measured by the loss in total corn production that would have resulted if there were no hybrid corn. This hypothetical loss will be valued at the estimated equilibrium price of corn plus a price-change adjustment, a procedure equivalent to computing the loss in "consumer surplus" that would occur if hybrid corn were to "disappear."

The amount of this loss will depend on our assumptions about the relevant demand-and-supply elasticities. As will be seen from the formulas presented below, these elasticities have only a second-order effect, and hence different reasonable assumptions about them will affect the results very little. I assumed that the price elasticity of the demand for corn is approximately -0.5.[8]

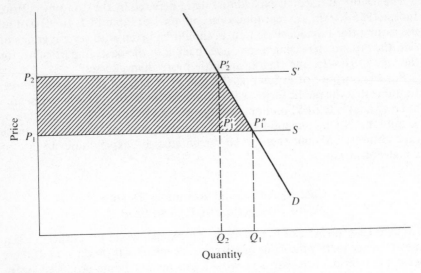

Figure 14.1.

Since we know much less about the supply elasticity of corn, I shall first explore the consequences of two different extreme assumptions about it.

Let us assume, first, that in the long run the supply of corn is infinitely elastic; that is, we face long-run constant costs. The "disappearance" of hybrid corn would shift the supply curve upward by the percentage reduction in the yield of corn. The "loss" to society, in this case, is the total area under the demand curve between the new and the old supply curves. In figure 14.1 this area is $P_1P_2P_2'P_1''$. This area can be interpreted as the increase in the total cost of producing the quantity Q_2 in the new situation, the rectangle $P_1P_2P_2'P_1'$, plus the loss in consumer surplus caused by the rise in price, the triangle $P_1'P_2'P_1''$. A linear approximation of this area is given by the formula

$$\text{Loss } 1 = kP_1Q_1(1 - \tfrac{1}{2}\,kn),$$

where k is the percentage change in yield (marginal cost and average cost), P_1 and Q_1 are, respectively, the previous equilibrium price of corn and quantity of corn produced, and n is the absolute value of the price elasticity of the demand for corn.

Alternatively, it could be assumed that the elasticity of the supply curve is zero. In this case the loss is measured by the area $Q_2P_2'P_1''Q_1$ in figure 14.2. Instead of assuming that the supply curve shifts upward, we now assume that it shifts k percent to the left. The rectangle $Q_2P_1P_1''Q_1$ measures the loss in corn production at the old price P_1. The triangle $P_1'P_2'P_1''$ can be viewed as the additional loss in consumer surplus or as an adjustment for the increase in price from P_1 to P_2. The total loss is now given by the formula[9]

$$\text{Loss } 2 = kP_1Q_1(1 + \tfrac{1}{2}\,k/n).$$

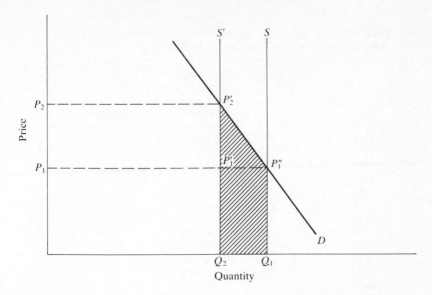

Figure 14.2.

It is easily seen that the second assumption leads to a higher estimate of the loss. It can be also shown that the two estimates bracket estimates implied by assuming other intermediate supply elasticities. The ratio of the loss under assumption 2 to the loss under assumption 1 is $(1 + \frac{1}{2} k/n)/(1 - \frac{1}{2} kn)$. In our case, this ratio is approximately 1.13.[10] The difference between these two extreme assumptions implies only a 7 percent difference in the final estimate of the total loss. Because this difference is so small and because I am striving for a lower-limit estimate, I have chosen the first assumption – that of an infinitely elastic long-run supply of corn.

To calculate the loss we must assume an equilibrium price of corn. I shall use $1.00 per bushel in 1955 dollars as a minimal estimate of the value of corn to society. The current price of corn about $1.25, is affected by the existence of price-support programs and probably overestimates the social value of corn.[11]

Because not all corn acreage was or is planted with hybrids, I multiply the percentage shift k by h, the percentage of all corn acres planted with hybrids (loss = $hkPQ[1 - \frac{1}{2} hkn]$). This procedure disregards the fact that the acres first planted to hybrids were higher-yielding acres than those planted later, and hence the procedure underestimates total returns.

In estimating past returns from hybrid corn I ignore annual fluctuations in prices and production, basing my computations on the average 1937–48 level of production of 2,900 million bushels[12] and a real price of corn of $1.00 per bushel in 1955 dollars. On the returns side, only the percentage planted with hybrid seed varies over time. To calculate the annual flow of future returns, I assume that the average 1943–52 level of production – approximately 3,000 million bushels annually – will continue and that the

Table 14.1 Hybrid corn: estimated research expenditures and net social returns, 1910–55 (millions of 1955 dollars)

Year	Total research expenditures (private and public)	Net social returns[a]
1910	0.008	
1911	0.011	
1912	0.010	
1913	0.016	
1914	0.022	
1915	0.032	
1916	0.039	
1917	0.039	
1918	0.039	
1919	0.044	
1920	0.052	
1921	0.068	
1922	0.092	
1923	0.105	
1924	0.124	
1925	0.139	
1926	0.149	
1927	0.185	
1928	0.210	
1929	0.285	
1930	0.325	
1931	0.395	
1932	0.495	
1933	0.584	0.3
1934	0.564	1.1
1935	0.593	2.9
1936	0.661	8.3
1937	0.664	21.2
1938	0.721	39.9
1939	0.846	60.3
1940	1.090	81.7
1941	1.100	105.3
1942	1.070	124.3
1943	1.390	140.4
1944	1.590	158.7
1945	1.600	172.6
1946	1.820	184.7
1947	1.660	194.3
1948	1.660	203.7
1949	1.840	209.8
1950	2.060	209.0
1951	2.110	218.7
1952	2.180	226.7

Table 14.1 *(continued)*

Year	Total research expenditures (private and public)	Net social returns[a]
1953	2.030	232.1
1954	2.270	234.2
1955	2.790	239.1
Annually after 1955	3.000	248.0

[a] Net of seed production cost but not net of research expenditures. Net social returns are zero before 1933.

percentage planted with hybrid seed will stabilize at 90. Both these assumptions are conservative and will result in an underestimate of returns.

Assuming that k, the relative shift in the supply curve, is 0.13 (15/115); that PQ is \$3,000 million; that n, the demand elasticity, is 0.5; and that h, the current and future fraction of all corn acres planted with hybrid seed, is 0.9, we can calculate the current and expected annual flow of gross social returns from hybrid corn as follows: $0.9 \times 0.13 \times \$3,000$ million $(1 - \frac{1}{2} \times 0.9 \times 0.13 \times 0.5) = 0.117 \times \$3,000$ million $(1 - 0.029) = 0.117 \times 0.971 \times \$3,000$ million $= \$341$ million. Subtracting the projected annual cost of hybrid-seed production and research – \$93 million – we get \$248 million as the current and expected annual flow of net social returns.

Similarly, returns for the past years, beginning with 1933, are calculated by multiplying 2,900 million 1955 dollars, the average total value of corn production, by the percentage of total corn acreage planted with hybrid seed in each year. Subtracting the estimated past costs of hybrid-seed production, we get the net social returns for the years 1933–55 shown in column 2 of table 14.1.[13]

Calculation of a Rate of Return

Table 14.1 presents estimates of costs and returns. There are several ways in which these figures could be summarized and a rate of return calculated. My procedure is as follows: consider the development closed as of 1955. Future expenditures will not increase returns, nor will there be an expansion of hybrid-corn acreage. Standing in 1955, I cumulate and bring forward to 1955 all past costs and returns at a reasonble *external* rate of interest. To explore the impact of two quite different rates of interest, I perform the calculations twice, using first a 5 percent and then a 10 percent rate of interest. Past research costs are cumulated and expressed as a capital sum. Past returns are cumulated to 1955, and a 5 or 10 percent rate of return on these cumulated returns is projected into the future. The estimated flow of future net returns

Table 14.2   Rate of return on hybrid-corn research expenditures as of 1955 (millions of dollars)

		$r = 0.05$	$r = 0.10$
(1)	Net cumulated past returns	4,405	6,542
(2)	Past returns expressed as an annual flow	220	654
(3)	Annual future gross returns	341	341
(4)	Annual additional cost of production and research	93	93
(5)	Total net annual returns, (2) + (3) − (4)	468	902
(6)	Cumulated past research expenditures	63	131
(7)	Rate of return $100 \times (5)/(6)$	743	689

is added to the flow from past returns to arrive at a perpetual flow of net social returns from hybrid corn. This flow, divided by the cumulated research expenditures, gives us our estimate of the realized perpetual rate of return.

Table 14.2 presents the calculations that lead to my estimate of approximately $7.00 as the annual return in perpetuity, as of 1955, for every dollar that has been invested in hybrid-corn research. Actually, even if we ignore all past returns completely the figure is still very high – approximately $4.00 annually (using the 5 percent interest rate) for every research dollar.

This way of calculating a "rate of return" is not really different from a benefit–cost ratio calculation. It may be useful to bring out explicitly the relationship between these two concepts. The preceding rate of return is defined as follows: $r = 100 \, (PR \times k + AFR)/RC$, where PR = cumulated past returns, k = the external rate of interest used to cumulate or discount returns, AFR = annual future returns, and RC = cumulated research costs.

A benefit–cost ratio from these same data would be $B/C = (PR + AFR/k)/RC$. Hence $r = 100 \, k \, (B/C)$, and we can translate our calculation into a benefit–cost ratio, and vice-versa. Using 5 and 10 percent as the external rates of interest, the benefit–cost ratios for hybrid-corn research expenditures are 150 and 70, respectively.[14] When we recall that most Bureau of Reclamation watershed projects have *ex post* benefit–cost ratios of 1 or less, this does imply a certain misallocation of public resources.[15]

These calculations use an external rate of interest to bring all sums forward to 1955. It is reasonable to assume that the marginal productivity of capital in alternative investments is between 5 and 10 percent and to use these rates as conversion factors for funds expended or earned at different dates. Alternatively, however, one could calculate an *internal* rate of return – that rate of interest which will equate the flow of costs and the flow of returns over time.[16] Such a rate has to be calculated using an iterative procedure, changing the rate used until the cumulated costs are equal to the discounted returns. The two procedures will give different answers when the time shape of costs differs markedly from the time shape of returns, as in our case. The internal rate of return on hybrid-corn research expenditures is between 35 and 40 percent.[17] My objection to this particular procedure is that it values a dollar spent in 1910 at $2,300 in 1933. This does not seem very sensible to me. I prefer to value the 1910 dollar at a reasonable rate of return on some alternative social investment. Also, this procedure gives tremendous weight to the early expenditures, which are subject to the largest error of measurement. Actually, however, the two estimates are not very far apart. The estimate using an *external* rate of interest says that a dollar invested in hybrid-corn research earned 10 cents annually until 1955 and $7.00 annually thereafter. The *internal* rate estimate says that the dollar earned 40 cents annually throughout the whole period. If the average delay between investment and fruition is about 10 years, then the two figures represent different ways of saying the same thing.[18]

Limitations

The estimate of 700 percent is probably too low. At almost every point at which there was a choice of assumptions to be made, I have purposely chosen those that would result in a lower estimate. This is an attempt to arrive at an estimated lower limit of the social rate of return from hybrid corn.

Both the public and the private research expenditures are probably overstated substantially. In fact, the expenditure estimates supplied by some experiment stations are obviously too high. Of the USDA expenditures, perhaps less than half were devoted to hybrids. I did leave out all expenses incurred before 1910, but these in total could not have been more than a few thousand dollars. This should remind us, however, that the estimated rate of return is mainly a rate of return on applied rather than basic research. The basic idea of hybrid corn was developed between 1905 and 1920, with the help of very little money. The rate of return on this basic research, if it could be calculated, would be much higher. However, the idea had to be translated to commercial reality, and separate adaptable hybrids had to be developed for different areas. These are the activities reflected in my estimate of research cost.

The returns, on the other hand, are understated. The assumed price of corn of $1.00 per bushel in 1955 dollars and the assumption of a 15 percent superiority of hybrids are both conservative and probably result in an underestimate of the real returns to society. I have also assumed that all past research has already borne all its fruit and that all future research on hybrids will result in no benefit whatsoever. Nor has credit been given for the impact of hybrid corn on other fields: the research on hybrid poultry and hybrid sorghum which it stimulated or the reduction of farmer resistance to new technology as a result of the spectacular success of hybrid corn.

Hence, as far as costs and returns from hybrid corn *per se* are concerned, the estimate is too low. One troublesome problem, however, remains to haunt us. Does it really make sense to calculate the rate of return on a successful "oil well"? What is the point of calculating the rate of return on one of the outstanding technological successes of the century? Obviously, it will be high. What we would like to have is an estimate that would also include the cost of all the "dry holes" that were drilled before hybrid corn was struck.

The estimate does include the cost of all the "dry holes" in hybrid-corn research itself. Hybrid corn was not a unique invention – it was an invention of a method of inventing. Many different combinations were tried before the right ones were found. One major seed company annually tests approximately fifteen hundred different combinations of inbred lines. Of these, at most three or four prove to be successful. The cost of the unsuccessful experiments is included in my estimate. What is excluded is investment in various other areas of agricultural research which has not borne fruit.

The problem of dry holes, however, can be reduced *ad absurdum*. What is the relevant segment for which an aggregate rate of return is to be computed? Is it really reasonable to ascribe the cost of unsuccessful gold exploration to the oil industry? If one takes this kind of reasoning too seriously, there is only one rate that has any meaning – the rate of return on research for the economy as a whole.

Nevertheless, the rate of return on a successful innovation may be of some interest. In particular, it may be useful, ex ante, to break down the probable rate of return into two components: the rate of return if the development turns out to be a success and the probability that it will be a success. The approach outlined here is a way of estimating the first component. An estimate of the probability of success, however, must be made on the basis of data other than those presented in this paper.

Returns in Some Other Areas

T. W. Schultz has provided us with estimates of costs and returns of research for United States agriculture as a whole.[19] His data can be used to estimate the rate of return on agricultural research as a whole. Schultz gives an upper- and a lower-limit estimate of how much more input it would have taken to produce the 1950 output with 1940 techniques and inputs. His upper estimate is that it would have taken 18.5 percent more input; his lower estimate is 3.7 percent. Let us use Schultz's figures to estimate the perpetual gross annual returns, beginning with 1951, from the agricultural research that bore fruit between 1940 and 1950. Using the lower estimate, there would be a loss of 3.7 percent in output if the new technology were to disappear. Taking the total annual value of farm output as $30 billion ($32 billion cash receipts from farm marketings in 1951, minus approximately 10 percent to allow for the impact of the support programs), and assuming a price elasticity of demand for agricultural products of -0.25 and an infinite supply elasticity, we get

$$k(1 - \tfrac{1}{2}\,kn) \times \$30\,\text{billion} = 0.037 \times (1 - 0.5 \times 0.037 \times 0.25) \times$$
$$\$30\,\text{billion} = 0.037 \times 0.995 \times \$30\,\text{billion} = \$1,110\,\text{million} .$$

Using the upper limit estimate of 18.5 percent saving in inputs, we get

$$0.185(1 - 0.5 \times 0.185 \times 0.25) \times \$30 \text{ billion} = \$5,430 \text{ million}$$

as an upper-limit estimate of the gross annual social return from the technical change that occurred between 1940 and 1950.

Schultz also provides an estimate of total public expenditures on agricultural research for the years 1937–51. I assume that all these expenditures were used to produce the increase in output in 1951. This leaves out the contributions developed from funds spent before 1937, but, on the other hand, it disregards the possible returns after 1951 from the 1937–51 expenditures. On balance, we will probably overestimate the funds spent on the 1940–50 improvement in technology.

I will assume that total private agricultural research expenditures were of about the same magnitude as the public expenditures. This is approximately half the corresponding ratio for hybrid corn but is probably an overestimate for agriculture as a whole.[20]

Multiplying the 1937–51 public expenditures by 2, deflating them by the Consumers Price Index, and cumulating them at the rate of 5 percent, I get the figure of 3,180 million 1951 dollars as my estimate of total cumulated research expenditures in agriculture. Comparing this with the two estimated limits of the annual social returns of \$1,110 and \$5,430 million, I get a lower limit of 35 and an upper limit of 171 percent as estimates of the annual rate of return per dollar spent on agricultural research. These are substantially lower than the estimated returns from hybrid corn but are comparable to estimates made by Ewell for the economy as a whole (100–200 percent per year per dollar spent on "research and development") and to figures quoted by major industrial companies on their returns on research.[21]

Of course, these estimates are quite consistent with the estimated returns on hybrid corn, if the probability of success in research on innovations like hybrid corn is on the order of one-tenth or one-twentieth. Nevertheless, all these figures indicate that, in spite of the large growth in research expenditures during this century, the social returns to this activity are still very high.

Hybrid Sorghum

The approach previously outlined can be used to estimate the probable rate of return on a new development: for example, hybrid sorghum. The development of hybrid sorghum began seriously only after World War II but has gained momentum rapidly since. Hybrid sorghum is now being introduced commercially. Very little was planted in 1956, but substantial amounts were planted in 1957. The experimental data to date suggest that the superiority of hybrid sorghum over previous seed may be even greater than that of hybrid corn.[22]

The 1956 value of the grain sorghum crop was approximately \$232 million. Assuming a 15 percent superiority of hybrids, a demand elasticity of −1.0, and an infintely elastic supply curve, the estimated gross social returns from sorghum hybrids would be about $37 million annually.[23] Assuming that the extra cost of hybrid seed will be about $7 million annually, the net annual returns would be about $30 million.[24] I have no official data on research expenditures on hybrid sorghum. The head of one of the major seed companies has estimated that to date all public and private expenditures on hybrid sorghum total approximately $1 million and that current expenditures are at an annual rate of $300,000. Doubling his estimate of past expenditures and projecting into the future an annual research expenditure rate of $500,000, I get $10 or $13 million, depending on the rate of interest used, as my "estimate" of cumulated hybrid-sorghum research expenditures in 1967. I choose 1967 as the reckoning date on the assumption that it will take ten years for hybrid sorghum to capture most of the sorghum seed market. While the projected rate of development is faster than that of hybrid corn in the United States as a whole, it is equivalent to the rate of acceptance of hybrid corn in Iowa. It is reasonable to assume that hybrid sorghum will spread much faster than hybrid corn did. Sorghum production is more localized than corn production, almost all sorghum is grown for commercial purposes, and hybrid sorghum will probably encounter much less resistance than hybrid corn encountered. I also assume that the use of hybrid sorghum in the United States will follow the same time path that the use of hybrid corn followed in Iowa.[25] This assumption allows me to estimate the social returns during the "transition period," 1957–67.

Table 14.3 outlines the calculation of the estimated rate of return on research expenditures on hybrid sorghum, which is approximately 400 percent per annum. While this is somewhat lower than the estimated rate of return on expenditures on hybrid-corn research, it is still very high indeed.

Some Implications

One might have expected, on a priori grounds, that the rate of return on expenditures on hybrid-sorghum research would be higher than the return on hybrid-corn research. The cost of hybrid sorghum has been and will be lower than the cost of hybrid corn both because of the cumulated experience

Table 14.3 Estimated costs and returns of hybrid-sorghum research as of 1967 (millions of dollars)

	$r = 0.05$	$r = 0.10$
Cumulated social net returns, 1957–66	155	171
Value in 1967 of returns beyond 1967	590	295
Total value of net returns in 1967	745	466
Cumulated research expenditures	9.4	13
Benefit–cost ratio	79	36
"Rate of return" in percent per annum	395	360

in hybrid-corn breeding and because sorghum-growing is much more localized than corn-growing. Therefore, adaptable hybrids will have to be developed for a much smaller portion of the United States. Nevertheless, the estimated returns from hybrid sorghum are lower than those from hybrid corn. Why? To some extent this lower rate may be a result of overestimating research costs for hybrid sorghum, but the principal explanation is that the total value of the sorghum crop is substantially smaller than that of corn; sorghum is a relatively unimportant crop. It has recently been suggested that we should redirect our research efforts away from the major commodities that are in "surplus" and away from commodities with low elasticities of demand, where technical improvements result in reduced total returns to farmers in the long run. However, if we assume, in the absence of any other information, that technological change operates somewhat like a percentage increase in yield, and that the cost of achieving a given percentage boost in yield is the same for different crops or at least independent of their price elasticities and relative "importance," then the highest *social* returns per research dollar are to be found in the important, low-elasticity commodities. For it can be easily shown that the absolute social gain from a given percentage increase in yield will vary proportionately with the total value of the crop and that the impact of different demand-and-supply elasticities is of a second order of magnitude. The latter affect only the "triangle," or the magnitude of the price-change adjustment. Among all the different factors, the total value of the crop is by far the most important determinant of the absolute social gain from a given percentage increase in yield.

No matter how we calculate them, there is little doubt that the overall social returns on publicly supported technological research have been very high. It is not clear, however, whether or not this fact has any normative implications. I am afraid it has very few. More knowledge than is now at hand is required for prescription.

It is clear that we have not succeeded in equalizing the returns on different kinds of public investments. The returns from technological research have been much higher than the returns from reclamation and watershed projects. But should we have had more technological research? Surely, we have yet to reach the optimal level of expenditures on research, but this can only be a hunch. Should the public support agricultural research? My analysis illustrates and quantifies one of the major arguments for *public* investments in this area – the divergence between the social and private rates of return. Almost none of the calculated social returns from hybrid corn were appropriated by the hybrid-seed industry or by corn producers. They were passed on to consumers in the form of lower prices and higher output. Entry into the hybrid-seed industry was easy, and in the long run no "abnormal" profits were made there. By valuing the extra cost of seed production at the market price, I have counted as a cost whatever profit was made in this area by private producers, and the resulting estimate of returns consists almost entirely of social rather than private returns. These social returns were diffused widely among consumers of corn and corn products. Given the difficulty of patenting most of the valuable ideas in this area, the short life of a patent, and the general precariousness of a monopoly position in the long run, the incentive for private investment was very much smaller than that implied by the social rate of return.

While a divergence between social and private rates of return is a necessary reason for public intervention, it is not, by itself, a sufficient reason. We must ask not only whether social returns are higher than private – this is also true of many private investments – but also whether the private rate of return is too low, relative to returns on alternative private investments, to induce the *right* amount of investment at the *right* time. The social returns from nylon were probably many times higher than DuPont profits, but the latter were high enough to induce the development of nylon without a public subsidy, although, perhaps, not soon enough. To establish a case for public investment one must show that, in an area where social returns are high, private returns, because of the nature of the invention or of the relevant institutions, are not high enough relative to other private alternatives. This was undoubtedly true of hybrid corn, and it is probably true of many other areas of agricultural research and basic research in general. But it is not universally true. Hence a high social rate of return is not an unequivocal signal for public investment.

In this paper I have estimated that the rate of return on public investments in one of the most *successful* ventures of the past has been very high. This may give support to our intuitive feeling that the returns to such ventures in general have been quite high and to our feeling that "research is a good thing." But that does not mean that we should spend any amount of money on anything called "research." The moral is that, though very difficult, some sort of cost-and-returns calculation is possible and should be made. Conceptually, the decisions made by an administrator of research funds are among the most difficult economic decisions to make and to evaluate, but basically they are not very different from any other type of entrepreneurial decision.

Acknowledgements

This paper is an outgrowth of a larger study of the economics of hybrid corn. See my article, "Hybrid Corn: An Exploration in the Economics of Technological Change," *Econometrica*, October 1957 [chapter 2 in this volume]. I am indebted to A. C. Harberger, Martin J. Bailey, Lester G. Telser, and T. W. Schultz for valuable comments and to the National Science Foundation and the Social Science Research Council for financial support.

Notes

1 The pairing was made on the basis of geographic proximity and general information about the industry. For example, Indiana expenditures were

assumed to equal the reported Illinois expenditures; Oklahoma's to equal Kentucky's, and so forth. These pairings probably overestimate the total expenditures on hybrid-corn research.

2 This will again overestimate expenditures, because "public" hybrids make up 25–30 percent of the total market, and research expenditures on these have already been counted once.

3 In 1951 M. T. Jenkins, of the USDA, estimated the total annual expenditures on hybrid-corn research as follows: USDA, $220,000; states, $600,000; and private industry, $1,100,000. My own independent estimate for 1951 is: USDA, $190,000; states, $550,000; private industry, $1,300,000. The two totals are $1,920,000 and $2,040,000 respectively. The agreement is very close, considering how arbitrary some of my assumptions are (see M. T. Jenkins, "Corn breeding research – whither bound," *Proceedings of the Sixth Annual Hybrid Corn Industry – Research Conference* (Chicago: American Seed Trade Association, November 1951), pp. 42–5).

4 This is somewhat higher than the market than the market price of corn because of the better quality of seed corn and the labor that would go into its selection. Since open-pollinated seed is now quoted at about $3.00 to $4.00 a bushel, this assumption also contributes to an overall overestimate of cost.

5 This result is reached as follows: 80 million acres × 90 percent in hybrids × (8.6/64) average seeding rate × $9.50 = $92 million. Subtracting $2 million research expenditures, we get $90 million (source: *Agricultural Statistics, 1955*).

6 Throughout the period these computations use the average corn acreage planted in 1939–48; they disregard annual fluctuations in total corn acreage and seeding rates. For any year before 1956, the extra cost of seed equals the percentage planted with hybrid seed times $98 million (90 million × [7.5/64] × $9.50 – $2 million (research expenditures) = $98 million).

7 For example: "Plant breeders conservatively estimate increase in yields of 15 to 20 percent from using hybrid seed under field conditions. They expect about the same relative increases in both low – and high – yielding areas" (USDA, *Technology on the Farm* (Washington, 1940), p. 7).

8 This figure is based on a USDA demand analysis. See R. J. Foote, J. W. Klein, and M. Clough, *The Demand and Price Structure for Corn and Total Feed Concentrates* (Technical Bull. 1061, Washington: USDA, October 1952).

9 Corrected from the original at the suggestion of T. D. Wallace of North Carolina State University and R. C. Lindberg of Purdue University. They also raise the question whether the difference between these losses is due merely to different assumptions about the supply elasticity. Would not the area defined as $P_1P_2P_2'P_1''$ be a more relevant measure than the area $Q_2P_2'P_1''Q_1$ in figure 14.2 which I actually used? The ratio of loss 2 so defined to loss 1 would be the order of the reciprocal of the elasticity of demand, and thus larger than 1 for all of the relatively inelastic demand situations considered in the paper.

It is true that, in defining these losses, I mentioned consumer surplus, and that as defined loss 2 does not take all of the consumer loss into account. To that extent the original text is in error. But the actual definition of loss 2 as used in the paper is the more sensible of the two. It is simply the gain (or loss) in output due to hybrid corn valued at average (pre- and post-hybrid) prices. The objection to the alternative suggestion can be illustrated by considering the case of an infinitely elastic demand function. The suggested definition would indicate no social gain from hybrid corn, which is clearly wrong, whereas the definition I used would still value the increase in output at the constant price.

In general, refinements in valuing the social gain are probably not worth the confusion they may create. As far as the substantive issues are concerned, the above reservations only reinforce the conclusion that, if anything, the estimated social gains are on the low side.

10 Assuming $k = 0.13$, i.e., 15/115, and $n = 0.5$, the ratio is $(1 + 0.5 \times 0.13 \times 2)/(1 - 0.5 \times 0.13 \times 0.5) = 1.13$.

11 An approximate formula for determining the price of corn in the absence of price supports is given by Marc Nerlove in "Estimates of the Elasticities of Supply of Selected Agricultural Commodities," *Journal of Farm Economics*, XXXVIII (May 1956), 497; $dp/p_0 = (dq/q)/(n + e)$, where p_0 is the equilibrium price, n and e are the demand-and-supply elasticities, and dq is the quantity placed under loan. In recent years about 7 percent of the annual corn crop, on the average, has been placed under loan with the Commodity Credit Corporation. The assumptions $n = 0.5$, $e = 0.2$, imply that the current price is about 10 percent above the equilibrium price. The current price is about $1.25 per bushel, which implies an "equilibrium" price of corn of about $1.13. But this estimate does not take into account the impact on corn prices of the elimination of price supports on all other agricultural commodities. Taking this into account, the equilibrium price would be closer to $1.00 per bushel. In any case, it is unlikely to be lower than $1.00.

12 This assumption was made to simplify the calculations. In the first part of the period, production was below this figure, and, since I use a relatively high rate of interest, this will result in an overestimate of returns. But the percentage planted with hybrids was also low then, while it was much higher during the period when production exceeded its average, and this will result in an underestimate of returns. On balance, the second effect should outweigh the first by a fair margin.

13 These are calculated from the following formula: $h \times [0.13 \times \$2,900$ million $(1 - \frac{1}{2} \times 0.9 \times 0.13 \times 0.5) - 90$ million $\times (7.5/64) \times \$9.50 - \$2$ million $] = h \times \$268$ million. This procedure is an approximation, since h should also have entered into the second part of the first term of the formula, the "triangle." I neglect this. Because h is less than 1 and because the part is always subtracted, this procedure again underestimates total returns.

14 If before cumulating we were to subtract the research costs from net returns annually (that is, have in our denominator only research expenditures before 1934, the year when net returns began to exceed research costs), the benefit-cost ratios would be substantially higher (about 700 and 200, respectively), and so would also the rate of return as defined in the text.

15 For an evaluation of public investments in watershed projects see E. F. Renshaw, *Toward Responsible Government* (Chicago: Idyia Press, 1957).

16 I am indebted to Martin J. Bailey for suggesting this alternative way of calculating the rate of return.

17 That is, 40 percent is too high and 35 is too low. The iterative procedure was not carried further.

18 An annual flow of 40 cents discounted at 40 percent has a present value of $1.00. An annual flow of $7.00, discounted at 40 percent, will also have a present value of $1.00 if there is a lag of approximately 10 years between the date of investment and the date at which the perpetual flow of returns begins.

19 *The Economic Organization of Agriculture* (New York: McGraw-Hill Book Co., 1953), pp. 114–22.

20 Mighell has estimated that recent annual expenditures by industry for research on agricultural products and on machinery and materials used in agriculture were in excess of $140 million. At the same time the USDA and state agricultural experiment stations spent about $118 million annually on research. However, only one-third of the industry research was in aid of farm production, mainly in machinery and chemicals: the rest was used in product research, while four-fifths of public expenditures were for research in aid of farm production – see R. Mighell, *American Agriculture* (New York: John Wiley & Sons, 1955), p. 130.

21 R. H. Ewell, "Role of research in economic growth," *Chemical and Engineering News*, XXXIII (1955), 298–304.

22 *Sorghum Hybrids* (USDA, ARS, Special Report 22–26 Washington, May, 1956).

23 This figure is derived as follows: $0.15 (1 + \frac{1}{2} \times 0.15 \times 1.0) \times \$232$ million = $0.16 \times \$232 = \$37$ million. The figure would have been approximately twice as large if I had used the value of the 1957 crop – $493 million – as my base.

24 The additional cost of hybrid seed, seven million dollars, is estimated from the following data: 10 million acres sown to sorghum; 10 dollars difference between the average prices per hundredweight of hybrid and open-pollinated grain sorghum seed (*Agricultural Prices*, April 1958, p. 42); and an average seeding rate of 7 pounds per acre.

25 The percentage of all corn planted with hybrid seed in Iowa followed the following time path by years: 0.02, 0.06, 0.14, 0.31, 0.52, 0.73, 0.90, 0.99, and 1.00.

15
Research Expenditures and Growth Accounting*

1 The Existing Literature

Many aspects of the economics of research have been discussed ably in a number of recent books and articles and will not be resurveyed here.[1] I shall concentrate instead on a relatively limited topic, the possible contribution of public and private research expenditures to the growth in the "residual" as conventionally measured. Sections 1–3 of this paper review and summarize earlier work on returns to research, explain the logic behind them, and present some additional estimates of the impact of research expenditures in United States manufacturing industries on subsequent growth in their total factor productivity. Sections 4–6 discuss how research might be treated consistently in a set of real product and input accounts and explore what traces, if any, such expenditures leave in the conventional United States productivity accounts. Section 7 closes the paper with a discussion of some, only slightly related, policy implications.

Investment in research, both public and private, has been thought to be one of the major sources of growth in output per man in this century. That it has been a good investment, in the sense that it yielded a positive rate of return which has been as good and often better than the rate of return on other private and public investments, is reasonably clear. Whether it can account for a major part of the observed growth in productivity is another matter. That will be considered in the latter half of this paper.

The evidence for the statement that R&D investments yielded a high social rate of return is scattered; much of it is second-hand; but it is still quite strong. It is of three kinds: individual invention returns calculations, industry studies, and aggregate residual attribution calculations. The individual invention calculations have been done by among others, Griliches (1958) for hybrid corn and hybrid sorghum, by Peterson (1967) for poultry breeding research, by

* From B. R. Williams (ed.), *Science and Technology in Economic Growth*, MacMillan Press, 1973.

Ardito-Barletta (1971) for agricultural research in Mexico, by Eastman (1967) for military transport aircraft, and by Weisbrod (1971) for polio vaccines. In such studies the total social returns from a particular invention are estimated and compared with *all* the research costs in the particular area of research, not just the costs of the successful part of the research. The internal rates of return implied by these estimates are quite high (10 to 50 percent per annum), even though they are usually based on conservative assumptions, and even at their lower end (such as the 12 percent estimate for polio vaccines by Weisbrod) are at a level worth considering investing more if the opportunity to do so were to arise again.[2]

A major objection to such studies is that they are not "representative," having concentrated on the calculation of rates of return only for "successful" inventions or fields. This objection can be met by econometric studies of productivity growth in specific industries. Here all of the productivity growth is related to all of the research costs and an attempt is made to estimate by correlational techniques the part of productivity growth that is attributable to the past investments in research.[3] The list of detailed studies is longer here: the most prominent being Mansfield's (1965) study of the private rate of return to research for chemical and petroleum firms, Minasian's (1962 and 1969) study of private return to chemical and drug research, Griliches' (1964) and Evenson's (1968) studies of the social rate of return to public investments in research in agriculture, and studies of the social rate of return to research investments in manufacturing industries by Terleckyj (1960, 1967, and Kendrick 1961), Brown and Conrad (1967), and Mansfield (1965).[4] While each of these studies is subject to a variety of separate reservations, together they all point to a reasonably consistent relationship between productivity growth and research expenditures and to relatively high (30–50 percent) rates of return on average to both public and private investments in research.

Finally, a number of studies having measured as well as they could the contribution of other sources to economic growth, attribute the rest (the "residual") to research and compute the implied rate of return. Such calculations have been performed in the past by Schultz (1953) and Griliches (1958) for total research in agriculture, and by Fellner (1970) for the total economy. Again, the implied rates of return have been very high.

Perhaps it is not surprising that different studies yield rather high estimates for the rate of return to research in the private sector. After all, research is an investment just as any other and should yield a positive return.[5] Given the great uncertainties associated with it (including the uncertainty about the degree of appropriability of its results) it is plausible that the *ex post* returns should be rather high on the average. They include a non-negligible risk premium. What is more surprising is that public investments in research, which are not constrained or guided by the profit motive, and where there is wide scope for potential bureaucratic bungling, do seem to yield not only positive but actually rather high rates of return. This, while not unexpected, is heartening.

There are two other noteworthy points to be drawn from the above mentioned studies: first, there is a significant lag between the time that the investments in research are made and the time that the products of such research activity begin to affect the average productivity of an industry or economy. For public and primarily basic research the average lag appears to be of the order of 5–8 years. For the bulk of industrial research (applied and development) the lag is much shorter, of the order of 2–3 years, but still significant. Thus we are unlikely to observe the effects of the current drought in research support very soon, but it may come back to plague us in the late 1970s. Second, research investments both depreciate and become obsolete. They depreciate in the sense that much of knowledge would be forgotten and rendered useless without continued efforts at exercising, retrieving and transmitting it. This is what much of higher education is about. Also, some of the newer findings displace, make obsolete, large parts of previously acquired knowledge. Thus, their net contribution is smaller than would appear at first sight. In short, a non-negligible rate of investment in research may be required just to keep us where we are and to prevent us from slipping back.

2   The Model in Common Use

Common to most analyses of the contribution of research to productivity growth is a model that can be summarized along the following lines

$$Q = TF(C,L) \tag{1}$$

$$T = G(K,O) \tag{2}$$

$$K = \Sigma w_i R_{t-i} \tag{3}$$

where Q is output, C and L are measures of capital and labor input respectively, T is the current level of (average) technological accomplishment (total factor productivity), K is a measure of the accumulated and still productive (social or private) research capital ("knowledge"), O represents other forces affecting productivity, R_t measures the real gross investment in research in period t, and w_i's connect the levels of past research to the current state of knowledge.

An important problem arises as soon as we write down such a system, a problem that will stay with us throughout this paper. Ideally, we would like to distinguish between capital and labor used to produce current "output" and capital and labor used in research (the production of future knowledge and the maintenance of the current stock). In fact we are usually unable to observe these different input components and are forced to use totals for C and L in our investigations. This leads to a mis-specification of (1). Moreover, if components of L and C are weighted in proportion to their current private returns, the resulting estimates of the contribution of K (or R) represent, errors in timing apart, the excess of social returns over private. The model, as written, is strictly applicable only where all (or almost all) of the research activity is performed outside the accounting boundaries of the sector in question; as for example in the case of agriculture where most of the research is public and where the inputs used in this research are not included in the definition of agricultural capital or the agricultural labor force.

For estimation purposes the F and G functions are usually specialized to the Cobb-Douglas form and O is approximated by an exponential trend. The whole model then simplifies to

$$Q_t = A_e^{\lambda t} K_t^\alpha C_t^\beta L_t^{1-\beta} \tag{4}$$

where A is constant, λ is the rate of disembodied "external" technical change, and constant returns to scale have been assumed with respect to the conventional inputs (C and L). Equations like this have been estimated by Griliches (1964) from several agricultural cross-sections and by Evenson (1968) and Minasian (1969) from combinations of time-series and cross-section data for agricultural regions and chemical firms respectively. Alternatively, if one differentiates the above expression with respect to time and assumes that conventional inputs are paid their marginal products, one can rewrite it in terms of total factor productivity

$$f = q - \hat{\beta}c - (1 - \hat{\beta})1 = \lambda + \alpha k \tag{5}$$

where f is the rate of growth of total factor productivity, lower case letters represent relative rates of growth of their respective upper case counterparts $[x = \dot{X}/X = (dX/dt)/X]$, and $\hat{\beta}$ is the estimated factor share of capital input.[6] (5) is a constrained version of (4). Versions of it were run by Evenson (1968) for agriculture and by Mansfield (1965) for manufacturing industries, among others. In either form the esimates of α have tended to cluster around 0.05 for public research investments in agriculture (Evenson and Griliches) and around 0.1 for private research investments in selected manufacturing industries (Mansfield, Minasian, and Terleckyi).[7]

Up to now I have been deliberately vague as to the operational construction of the various variables. The difficulties here are myriad. Perhaps the two most important problems are the measurement of output (Q) in a research-intensive industry (where quality changes may be rampant), and the construction of the unobservable research capital measure (K). Postponing the first for later consideration, we note that $K_t = \Sigma w_i R_{t-i}$ can be thought of as a measure of the distributed lag effect of past research investments on productivity. There are at least three forces at work here: the lag between investment in research and the actual invention of a new technique, the lag between invention and complete diffusion of the new technique, and the disappearance of this technique from the currently utilized stock of knowledge due to changes in external circumstances and the appearance of superior techniques (depreciation and obsolescence).[8] These lags have been largely ignored by most of the investigators. The most common assumption has been one of no or little lag and no depreciation. Thus, Griliches and Minasian have defined $K_t = \Sigma R_{t-i}$ with the summation running over the available range of data, while Mansfield assumed that since R has been growing at a rather rapid rate, so has also K (i.e. $K/K \approx R/R$). Evenson (1968) has been the only one to investigate this question econometrically, finding that an "inverted-V" distributed lag form fitted his data best, with the peak influence coming with a lag of 5–8 years and the total effect dying out in about 10–16 years. There is some scattered evidence based largely on questionnaire studies (see Wagner, 1968), that in industry, where the bulk of research expenditures are spent on development and applied topics, the lag is much shorter and so also is the expected length of life of the product of this research.[9]

Several additional points can be made concerning the measurement of K: (1) Ideally we should be measuring the output of the research industry directly, not its inputs. Unfortunately, attempts to use direct measures such as patents as a proxy for research output have been largely unsuccessful (see, e.g., Nordhaus, 1969b).[10] For inventive activity the relationship between input and output is likely to be stochastic and unstable. We shall come back to this below. (2) The time series on R have to be deflated somehow; a hard but not impossible task. (3) The level of productivity in an industry and the productivity of research expenditures in it depends also on research activities of other related industries and on the research success of other countries.[11] These cross-effects are, of course, somewhat less important at more aggregative levels but may be of the essence at the firm level.[12]

Because of the difficulties in constructing an unambiguous measure of K, many studies have opted for an alternative version of equation (5), utilizing the fact that:

$$\alpha = \frac{dQ}{dK} \cdot \frac{K}{Q}$$

and

$$\alpha k = \frac{dQ}{dK} \cdot \frac{K}{Q} \cdot \frac{\dot{K}}{K} = \frac{dQ}{dK} \cdot \frac{\dot{K}}{Q'}$$

allowing one to rewrite (5) as

$$f = \lambda + \alpha k = \lambda + \rho I_R/Q \tag{5'}$$

where ρ is the rate of return to research expenditures (the marginal product of K) while I_R/Q is the net investment in research as ratio to total output.[13] This is the kind of framework that underlies the calculation of the rate of return presented in Fellner (1970), for example. In practice, to make some connection between gross and net investment in research one needs information about the "depreciation" of research which if available would have allowed us to construct a measure of K in the first place.

Form (5') is particularly suitable for back-of-the-envelope calculations. It requires only a "few" guesses as to (a) the fraction of R&D expenditures that represent net investment, and (b) the rate of return (ρ) to these expenditures. For example, between 1963 and 1967 the ratio of total R&D expenditures to total GNP in the United States hovered around 3 percent. We shall see below that only about half of it is likely to have an effect on national output as currently measured. If we assume that only half of the remainder represents net investment, the rest being devoted to keeping up with where we are,[14] and that the gross rate of return was 30 percent, we get an estimated contribution to growth of $0.3 \times 0.75 \approx 0.22$ percent.

3 Rough Estimate of the Contribution of Research to Growth

Equation (5′) can be estimated, albeit crudely, utilizing data from the National Science Foundation publication *Industrial R&D Funds* (NSF 64–25), which reports (in table C–1) the 1958 research intensity of companies for a number of rather finely defined manufacturing industries. For 85 two-, three-, and four-digit manufacturing industries these data could be matched with the *Annual Survey and Census of Manufactures* data and a "Residual" technical change measure could be computed. It was defined as follows:

$$T(\text{residual technical change}) = \tfrac{1}{5}[(\ln VA_{63} - \ln VA_{58} - \ln DP) - ALSV (\ln N_{63} - \ln N_{58}) - (1 - ALSV) (\ln GBV_{63} - \ln GBV_{58})]$$

where VA is valued added, DP is the 1963 price index ($1958 = 1.00$ from 1963 *Census of Manufactures*, vol. 4, table 5), N is total employment, GBV is gross book value of fixed assets, and $ALSV$ is the average share (for 1958 and 1963) of payroll to value added. Another productivity related measure is given by the "partial price change."

$$PP = \tfrac{1}{5} (\ln DP - ALSV = \ln DW)$$

where PP is "partial price change," and $\ln DW$ is the change in the average wage rate ($\ln DW = (\ln \text{Payroll}_{63} - \ln \text{Payroll}_{58}) - (\ln \text{Employment}_{63} - \ln \text{Employment}_{58})$).

Additional variables of interest are:

R = R&D expenditures as a fraction of net sales.
ROV = $R \times$ average sales to value added ratio =
  R&D as fraction of value added.
D_5 = Dummy variable for $R > 0.15$.

All the observations were weighted by value added in 1958. Consider first the result of estimating an approximation to (5′):

$$T = 0.0043 + 0.397\ ROV - 0.107\ D_5;\ R^2 = 0.574,\ SE = 0.0215$$
$$(0.0033)\ (0.038)\qquad (0.013) \tag{6}$$

Note the highly significant coefficient of ROV and the large "discount" (D_5) for $R > 0.15$. The latter corresponds to two industries (SIC 19 and 372), Ordnance and Aircraft and parts, where government financed R&D constitutes about 86 percent of the total and the results of which are unlikely to show up in the residual as measured. Since the average ROV for this group is 0.305, the discount implied by D_5 coefficient is consistent with the above ratio:

$$0.107/(0.397 \times 0.305) = 0.88 \approx 0.86.$$

The measure of T used above is based on a rather dubious estimate of the rate of growth of tangible capital. An alternative approximation to T is available, however, from the price side by the duality of the price and quantity growth accounts:

$$T = [ALSV \cdot \ln DW + (1 - ALSV) \ln D\Pi - \ln DP] \tag{7}$$

where $\ln D\Pi$ is an index of the rate of change in the service price of tangible capital. Assuming that the latter is the same for all these manufacturing industries (which is not unreasonable), implies that:

$$PP = \ln DP - ALSV \cdot \ln DW = -T - ALSV \cdot \ln D\Pi + \ln D\Pi \tag{8}$$

Treating $\ln D\Pi$ as a constant, we can use PP as the dependent variable in an equation paralleling (6) but containing also $ALSV$ (the share of labor in value added) as an additional variable. Doing this we get:

$$PP = 0.0207 - 0.319 \ ROV + 0.084 \ D_5 - 0.043 \ ALSV; \ R^2 = 0.635 \tag{9}$$
$$(0.0085) \ (0.027) \qquad (0.010) \qquad (0.016) \ SE = 0.0155$$

The coefficients of ROV and D in this equation should be similar to those of (6) but of opposite sign, while the coefficient $ALSV$ provides an estimate of the unobserved rate of change in the service price of capital. The estimates are in fact close and largely independent of each other. Moreover, those based on equation (9), do not depend on the use of questionable capital data.

They are also quite consistent with the assumption that $\rho \approx 0.3$ made earlier.[15] Subtracting 86 percent of the Aircraft and Missiles R&D from the total, we have an average (weighted) ROV ratio of 0.04 in total manufacturing in 1958. This implies that *in manufacturing*, research investments may have been contributing as much as a full percentage point or more $(0.3 \times 0.04 = 0.012)$ to the measured growth in output. Note, however, that in this period manufacturing accounted for only about a quarter of total GNP.

4   The Difficulties of Measurement

From the point of view of "real" national income or growth accounting, the suggestion that R&D should be treated as another investment founders on the difficulty of measuring the corresponding real output stream. There are a number of problems here and I can do no better than just catalog them:

1 The bulk of the R&D "product" is sold to the public sector. It consists of research on defense and space exploration for which no adequate market valuation exists. By convention these "outputs" are measured by costs, resulting in zero contribution to measured productivity, by definition.[16]
2 Much of the product of private R&D investments is in the form of new commodities or in improvement in the qualities of old commodities.[17]

Whether or not this product shows up in current dollar GNP depends on the extent of the short and long run monopoly position of the firms engaging in research and on the fraction of the social gain (consumer surplus) appropriated by these firms. Whether or not this product shows up in "real" GNP (in constant prices) depends on what happens to the price indexes with which the current net output of the particular industry is deflated. If the price indexes were to recognize these improvements in "quality" fully, the resulting real GNP measures would reflect the social product of this research. But this is unlikely.[18] If the price indexes fail to reflect these quality improvements, only the private product will show up in the real accounts, to the extent that firms succeeded in appropriating it via higher prices for the newer higher quality products.

3 If some of these unmeasured improvements are attached to products that are used in turn as inputs in the production of other private products, their contribution will show up in the productivity measures of the industries that purchased them. Thus, for example, even though the contribution of research towards improving the performance of farm tractors may not show up in the output–input account of the tractor industry, it will have an effect on measured productivity in agriculture.

4 There is the technical accounting problem connected with the fact that most private R&D is treated as a current expense (an intermediate good) and does not appear explicitly in the value of output. Moreover, we have no explicit income stream (return) to associate it with. Thus, any accounting scheme based on factor shares is very difficult to implement, even disregarding the fact that social returns to R&D would in any case not be reflected in these factor shares even if we could compute them. The latter fact is the basic reason for the various attempts recounted above to estimate the social contributions of research econometrically.

5 The basic difficulty with treating research as an investment is that it is largely an "internal" investment, without an explicit intermediate market for its product. When we talk of "tangible" investment, meaning equipment and construction, these can be valued by what was spent on purchasing them. That is, we have a separate measure of the output of the machinery industry, independent of our measure of the resources (labor and capital) used to produce these machines. Having separate measures on two sides of the accounts is a prerequisite for the measurement of productivity of the investment of productivity of the investment goods industry.[19] But research, like advertising and other costs of change, is largely internal to the firm and does not show up in its output accounts.[20] More importantly, it may not even show up, except with a very long and random lag, in the income account of a firm. Consider, for example, the parallel case of a firm that engages in drilling for oil. It will incur some costs in doing so. The success of this activity should be measured in new wells brought in and in new reserves discovered. Obviously, they are unlikely to show up in the receipts of the firm in the current or even following year. Eventually, there will be a return, though its relationship

to these costs may be by then quite obscure. On the other hand, the finding of oil should be reflected in the "capital gains" column, in the appreciation of the market value of this firm. Thus, potentially there may be a way of measuring the private product of research, but it cannot be derived from the current income account.[21]

Having listed all these caveats we are now ready to look at what might be the consequences of the way we treat current research expenditures in the national income accounts. First, let us dispose of a few easy cases:

Research performed in the public sector is measured by input and does not directly contribute to measured productivity, since productivity growth in the public sector is zero by national accounting conventions (and lack of real data on the "real" value of the output of this sector). To the extent that it affects the productivity of resources in the private sector, as in the case of agricultural research or the possible "spillovers" from the space programme, it is an "externality" that might be caught by the conventional total factor productivity measures.

Research performed by the private sector but sold to the public sector, i.e. contract research on defense and space topics, again does not contribute directly to measured productivity since this part of private output is deflated by cost indexes, implying zero productivity growth in this endeavor. Only the "externalities" of this research could show up in the conventional accounts.

The more interesting case is that of private research investments, spent internally on the future improvement in the productivity and profitability of the firm. To reduce the problem to manageable proportions we shall first consider a special and perhaps trivial case: All research is private, there are no externalities, and the aggregate production function is one of constant returns to scale in C, L and K. Subject to these assumptions, we shall investigate the consequence of two "errors":

1 The expensing of R&D instead of the more correct treatment of it as a type of capital, and

2 The application of the same rate of return to this investment as to the other investments while, in fact, the private rate of return to it may be higher than that for tangible capital.

For this special case a potentially correct growth accounting scheme would start from the output-input value identity:

$$P_I I + p_g G + p_K I_K = q_L L + q_c C + q_K K \qquad (10)$$

where $p_I I$, $p_g G$, $q_L L$, and $q_c C$ are the conventional measures of investment, consumption ("goods"), labor, and capital returns respectively, all in current prices. The additions to the usual accounts are the $p_K I_K$ term, on the left-hand side, representing current gross investment in research ("knowledge") or other intangible capital times term and the $q_K K$ term on the right-hand side, standing for the current returns to the previously accumulated knowledge "capital." The notation used distinguishes between prices of current outputs, the p's, and the service prices (rents) of the various inputs and capital stocks, the q's.[22] Given this definition, the conventional product measure can be rewritten as:

$$P_I I + p_g G = q_L L + q_c C + q_K K - p_K I_K \qquad (10')$$

Now, it is immediately obvious that the conventional productivity measures will ignore the contribution (positive and negative) of the last two terms of this expression. In addition, there will be another source of bias to the extent that the contribution of tangible capital (its service price, or the rate of return) is estimated residually, as is usually the case. The wrong estimate of the service price of tangible capital, q^*, will be given by

$$q_c^* = \frac{q_c C + q_K K - p_R R}{C} \qquad (11)$$

where $p_R R$ are research and development expenditures which have been "expensed" from the profit accounts and treated as if they were a cost of producing current output. Note that we are making the distinction between $p_K I_K$, which is the current marketable product of this activity, and $p_R R$ which are the current identified costs of it. Except in stable equilibrium with perfect foresight and correct accounting schemes, the two need not be equal, a point stressed earlier. For further convenience in manipulating these expressions I will introduce the identity:

$$p_K I_K = p_R R + p_R U \qquad (12)$$

where $p_R U$ is the current excess of growth in knowledge over current costs, evaluated at the factor prices, p_R, of current research costs. With this distinction in mind, we will put it aside for a while and return to the derivation of the bias in the conventional total factor productivity measures. Given (11), the bias in the measurement of the service price of tangible capital is given by

$$q - q_c^* = \frac{p_R R - q_K K}{C} \qquad (13)$$

The total (absolute) bias in the measurement of total factor productivity is given by

$$\dot{T}^* - \dot{T} = q_K \dot{K} - (q_c^* - q_c)\dot{C} - q_K \dot{I}_K \qquad (14)$$

where $X = dX/dt$. Thus, measured growth in total factor productivity T^* exceeds the "true" measure T by the contribution of the growth in the "stock of knowledge," but falls short of it by the overestimate in the contribution of the growth in tangible capital and by the omission of the growth in the investment into this type of knowledge from the conventional output measures. Substituting (13) into (14) and expressing the various variables as rates of growth, we have

$$\dot{T}^* - \dot{T} = q_K K \frac{\dot{K}}{K} - (q_K K - p_R R)\frac{\dot{C}}{C} - p_K I_K \frac{\dot{I}_K}{I_K} \qquad (15)$$

Collecting terms, and using (12) we get

$$\dot{T}^* - \dot{T} = q_K K \left(\frac{\dot{K}}{K} - \frac{\dot{C}}{C} \right) + p_R R \left(\frac{\dot{C}}{C} - \frac{\dot{I}_K}{I_K} \right) - p_R U \frac{\dot{I}_K}{I_K} \tag{16}$$

If we were willing to assume that $K/K \simeq I_K/II_K$, i.e. that the rates of growth in the investment in knowledge and its stock are approximately equal, which may not be too bad an assumption for a fast-growing item in its early history, and that there is no discrepancy between current research inputs and outputs ($p_R U = 0$), then (12) would simplify to

$$\dot{T}^* - T = (q_K K - p_R R) \left(\frac{\dot{K}}{K} - \frac{\dot{C}}{C} \right) \tag{17}$$

Noting in addition that we can write $R = \dot{K} + \delta K$, where δ is the rate of depreciation of the knowledge stock, and assuming that the service price of this capital is given by $q_K = p_R(\rho + \delta)$, where ρ is the rate of return to this stock, (17) simplifies further to

$$\dot{T}^* - T = p_R K \left(\rho - \frac{\dot{K}}{K} \right) \left(\frac{\dot{K}}{K} - \frac{\dot{C}}{C} \right) \tag{18}$$

That is, as long as the rate of return to this stock exceeds its rate of growth, and as long as its rate of growth exceeds the rate of growth of tangible capital, the conventional total factor productivity measure will be biased upward.[23]

We can use this formula and some of the estimates for the United States domestic private economy from the Kendrick and Wagner studies mentioned earlier to get an order of magnitude estimate for this type of bias. Their estimates imply that private K (R&D capital) grew at about 7 percent per year during the 1948–66 period, or at about 75 percent of the rate of growth of total R&D capital. The estimated stock of business research capital was about $35 billion (in 1958 prices), and the estimated current research investment of industries' own money was about $7 billion (in current prices). Additional numbers of use (from Kendrick) are the total private product in 1966 of $520 billion (in current prices), implicit total private product deflator 110 (1958 = 100), and rate of growth of tangible net capital stock of 3.5 percent for the 1948–66 period.[24]

Assuming $\rho = 0.3$, i.e. that the private rate of return to R&D capital is 30 percent, we get (dividing through by the total product):

$$\frac{\dot{T}^*}{T} - \frac{\dot{T}}{T} = \frac{p_R K}{p Q} \left(\rho - \frac{\dot{K}}{K} \right) \left(\frac{\dot{K}}{K} - \frac{\dot{C}}{C} \right)$$

$$= \frac{35}{520/1.1} (0.3 - 0.07) (0.07 - 0.035)$$

$$= 0.074 \cdot 0.23 \cdot 0.035 = 0.0006$$

That is, assuming a rather high private rate of return to R&D we can account for only 0.0006 in our estimates of the "residual" by the failure to capitalize past private R&D expenditures. This is a minuscule fraction of any of the estimates of the rate of growth in total factor productivity.

This does not mean, of course, that the contribution of these investments to

$$\frac{q_K K}{pQ} \cdot \frac{\dot{K}}{K}$$

the actual output growth is nil. *That* is measured by which given the same numbers and assuming that $\rho + \delta = 0.4$ is 0.0021. This while small, is not negligible. It is not to be found in the "residual," however, because of offsetting errors.

The above computation was made assuming that R&D investment grew at about the same rate as its associated stock, and that there is no discrepancy between the current R&D costs and output. If we go back to equation (16) we can note that the type of calculation we just performed will overestimate the bias in periods when research investment is growing very fast ($\dot{I}_K/I_K > \dot{K}/K$) and underestimate it in periods, like the current one, when its growth is slow.[25]

We can try to extend these calculations to a more recent period. Using approximately the same methods as Kendrick yields an estimate of the private R&D stock of about \$50 (in 1958 prices) at the beginning of 1971, as against \$35 in 1966.[26] This implies an average $\dot{K}/K$ of 0.06 per annum between 1966 and 1971. At the same time, "real" private investment in R&D was growing only at about 0.04 percent (or less) in 1970. If we assume that net tangible capital was also growing at about 4 percent, the second term in (12) cancels, and still assuming that the third term is zero ($p_R U = 0$), we get

$$\frac{\dot{T}^*}{T} - \frac{\dot{T}}{T} = \frac{1.3 \ (0.3 + 0.1) \ 50}{864} (0.06 - 0.04) = 0.03 \cdot 0.02 = 0.0006,$$

which is the same as the previous estimate.

To raise this estimate by relaxing the assumption that $p_R U = 0$ we would have to assume that it is negative, i.e. that the observed research costs exceed the actual current investment in knowledge. This is unlikely in a period of decelerating growth in real research expenditures. One would assume that new returns will be still forthcoming to some of the past investments. In general, one would expect U to be negative in periods of rapid growth in R, given the lags between research investments and their actual product.

The findings in this section can be objected to because they do not allow for economies of scale, the externalities of private R&D, and the contribution of government R&D to private productivity. We shall examine these in turn in the next section.

5 The Role of Increasing Returns to Scale

Before we proceed to an evaluation of the potential externalities we should discuss briefly the role of increasing returns to scale to private R&D investments. The discussion in the previous section concentrates on *private*

Table 15.1 Rough distribution of research and development expenditures in the United States, 1970, by source, performer, and possible purpose (US$ millions)

Performer, source, and type	Total	Defense AEC, and space	Other expenditures with probable effect on measured private productivity
Federal intramural			
Basic	620	320	60[a]
Applied	1,300	850	240[b]
Development	1,740	1,610	50[c]
Total	3,660	2,780	350
Industry			
Federally financed			
Basic	400	400	
Applied	830	730	30[d]
Development	6,900	6,590	220[e]
Company funds			
Basic	570	80[f]	480
Applied	2,480	250[g]	2,230
Development	7,700	900[h]	6,800
n.e.c.		180[i]	− 180
Total	18,880	9,130	9,520
Universities, colleges, non-profit institutions, and federal research centers			
Federal funds			
Basic	960	480	240[j]
Applied	960	310	180[k]
Development	950	830	40[l]
Industry funds	160		160
Other funds	1,290		650[m]
Total	4,320	1,620	1,270
Total	26,860	13,530	11,140

Notes: Totals are from NSF 70–46 (1970). Breakdown of federally financed expenditures based on "obligations" for 1970 from NSF 70–38 (1970). Breakdown of industrial expenditures based on 1968 data from NSF 70–29 (1970).

[a] 20 percent of all other.
[b] All of USDA and interior, 20 percent of other.
[c] All of transportation, 20 percent of other.
[d] All of interior, 20 percent of other.
[e] All of transportation, 20 percent of other.
[f] Half of aircraft and missiles and electric equipment and communications.
[g] Half of aircraft and missiles and communication equipment.
[h] Half of aircraft and missiles, 1/10 of chemicals, machinery and electric equipment.
[i] 1/10 of motor vehicles and instruments.
[j] 50 percent of rest.
[k] All of USDA and interior financed, 20 percent of rest.
[l] All of transportation, 20 percent of rest.
[m] 50 percent of total.

R&D investments and assumes explicitly (along with the usual national income accounting conventions) constant returns to scale with respect to labor, tangible and intangible private capital. The production function we wrote down in (4) assumes, however, constant return to scale of conventional inputs and a multiplicative effect for the stock of knowledge (K), private or public. One way out of this inconsistency is to assume that private firms have enough monopoly power to "exploit" all factors proportionately. That is, observed market shares underestimate the true factor elasticities by a factor $1/(1+\alpha)$ where α is the elasticity of output with respect to private K. This would lead to a rewriting of equation (15) as

$$\frac{\dot{T}^*}{T} - \frac{\dot{T}}{T} = \alpha \frac{\dot{X}}{X} + (1+\alpha) \left[\frac{p_R R - q_K K}{pQ} \frac{\dot{C}}{C} + \frac{q_K K}{pQ} \frac{\dot{K}}{K} \right] - \frac{p_K I_K}{pQ} \frac{\dot{I}_K}{I_K} \quad (15')$$

where $\dot{X}/X$ is the rate of growth in the conventional total input index. The second part of this formula is essentially the same as before (the difference is well within the range of error of these numbers), but the first term, $\alpha\dot{X}/X$ is new. Given the numbers used earlier, private $q_K K/pQ$ and hence, α were both about 0.02 in 1966, while $\dot{X}/X$ is estimated (by Kendrick, for 1958–66) at 0.023. This would add only about 0.00046 to the earlier estimates. Since the effect of this is the same as that of any other source of economies of scale, I will not pursue it further here.

It is very hard to assess the claim of externalities for publicly supported research investments adequately. Two contradictory facts stand out: first are the obviously large gains from public research in agriculture, reported on in section 1. Second is the fact that over half of the total R&D spending has been on defense, space, and related objectives. Whatever the social benefits of defense and going to the moon, they do not show up in the usual productivity accounts. The direct effect of these expenditures on measured productivity is nil, while the evidence for indirect effects is not particularly impressive. There is another major category of public research whose social returns are much less disputable, medical research, but they too are unlikely to show up in the productivity accounts as currently defined.

Table 15.1 represents an attempt to gauge the possible magnitude of research expenditures that could (potentially) increase measured productivity. It brings together official statistics from several slightly inconsistent sources and tries to allocate them between defense and space-oriented research, and research that could conceivably show up in real product per man as currently measured. The allocation is somewhat arbitrary and is based on "guess-work," particularly for the figures in the last column of table 15.1. But they should provide some impression of possible orders of magnitude and a basis for discussion. In 1970, of the $27 billion or so identified as "research and development" expenditures, half could be attributed to defense and space activities. Of the remainder, about $11 billion could be thought as having a potential impact on aggregate productivity. These $11 billion consisted of $9.5 billion spent in industrial research, about a third of a billion in intramural

Table 15.2 The distribution of research and development expenditures in the United States in 1970 by their potential direct and indirect effect on private productivity (percentages)

	Total	Productivity related
Industrial R&D		
Private funds	40	35
Public funds	30	1
Public R&D		
Federal intramural	14	1
Universities, research centers, and related institutions	16	4
Indirect (externalities)		12
Total	100	53

From Table 1. Indirect = 20 percent of remainder.

federal research, and $1.3 billion in research in universities, institutes, and similar organizations.

At the same time, private for-profit research in industry (as distinct from private non-profit research in institutions outside government) also equaled about $11 billion. This balancing out of about $1.5 billion in private expenditures attributed to defense (since defense is a major purchaser of the outputs of many firms, it pays to invest private resources in this field) with $1.5 billion in public expenditures with potential impact on private productivity is entirely fortuitous but instructive.

We have still not allowed for any "spillovers" from defense and space research. An upper limit estimate might be one that took 20 percent of these expenditures to have the same effect and the same rate of return as private expenditures devoted directly towards affecting private productivity. Table 15.2 summarizes the rough allocation of various R&D investments as to their potential impact on measured productivity in the private sector, including an allowance for externalities generated by that part of R&D (such as the space effort) not deemed to have a direct effect on productivity. About half of total R&D expenditures is estimated to be "productive" in this sense.

Before we proceed to incorporate these estimates into a revised accounting scheme we have to decide how much of the estimated returns to private R&D are social rather than private. In the previous sections of this paper we used $\rho = 0.3$ interchangeably, both for private and social returns calculations. The econometric data which support such a number are obscure, however, as to the exact allocation of these returns. If the price indexes used to deflate output were perfect, almost all of the return estimated from total factor productivity or "residual" studies would reflect social returns. Since these indexes are far from perfect, some significant portion of the estimated returns is in fact private. To proceed with some order of magnitude calculations I shall make the following arbitrary but perhaps not unreasonable assumptions:

1 Private R&D expenditures earn a "normal" gross rate of return of about 20 percent (10 for depreciation and 10 net).
2 They yield an externality in the private sector of 20 percent. That is, I am assuming that of the earlier estimate $\rho = 0.3$, one-third is actually part of the private returns.[27]
3 Public R&D investments (direct and indirect) yield a gross rate of return of 30 per cent.

Applying these assumptions and the allocations in table 15.2 (which are based on 1970 data) to the earlier estimates of cumulated stocks of R&D investments in 1966 and noting that the share of private R&D in the total was lower in 1966 and earlier years (32 percent versus the 40 percent in 1970), we have:

K_{dp1966} = \$31 billion directly productive stock of private R&D investments in 1958 prices.

K_{dg1966} = \$7 billion directly productive stock of public R&D investments.

K_{ge1966} = \$13 billion of indirectly productive stock of private and public R&D.

The appropriate accounting formula is now

$$\frac{\dot{T}^*}{T} - \frac{\dot{T}}{T} = q \frac{(K_p + K_{pe})}{pQ} \frac{\dot{K}_p}{K_p} + qe \frac{(K_{dg} + K_{ge})}{pQ} \frac{\dot{K}_g}{K_g} - \frac{P_R R}{pQ} \frac{\dot{R}}{R} - \frac{P_R R}{pQ} \frac{\dot{C}}{C}$$

where the first two terms represent the contribution of private and public R&D to the growth of output in the private sector while the last two terms adjust for two errors in the conventional measures of total factor productivity: the exclusion of private investments in R&D from the output measure and the over-weighting of the contribution of tangible capital.

Setting $qK_p + qK_{pe} = p \times 0.4$ and $qK_g = p \times 0.3$, and taking as before $qK_p + qK_{pe} = 0.07$, $K_g/K_g = 0.1$, $p_R R = \$7$ billion, $pQ = \$520$ billion, $p = 1.1$, $Q = \$473$ billion, $\dot{C}/C = 0.035$, and $\dot{R}/R = 0.08$, we get

$$\frac{\dot{T}^*}{T} - \frac{\dot{T}}{T} = \frac{0.4 \times 31}{473} \times 0.07 \frac{+ 0.3 \times (7 + 13)}{473} \times 0.1$$

$$- \frac{7}{473} (0.08 + 0.035) = 0.0050 - 0.0016 = 0.0034$$

The total contribution of R&D to growth is now estimated at 0.005 (or half a percent per year) of which 0.00185 is attributed to the private returns of private R&D, 0.00185 to the externalities of private R&D, 0.00046 to the contribution of "directly productive" public R&D, and 0.00084 to the "externalities" of all other research. The net effect of all this on measured total factor productivity is somewhat less, 0.0034.

Noting that private returns and the errors in their treatment in the measurement of total factor productivity almost cancel out (0.00185 − 0.00155 = 0.0003), we can concentrate on the externalities alone. The above calculations estimate their annual contribution at about 0.0023 for "directly productive" investments (public and private) and attribute a potential 0.0008 to the "spillovers" from other research.

Assuming the cancellation of the private returns and the associated errors in the measurement of total factor productivity, we can use formula (9) directly for an evaluation of current contribution of R&D to growth. Total R&D with potential externalities (including spillovers) in the private sector was about $14 billion in 1970 (see tables 15.1 and 15.2). Assuming that about half of this gross investment was for replacement, and noting that total private domestic GNP was $855 billion, gives a net investment ratio of 0.0082. Our earlier assumptions, and the allocations in table 15.2, imply a weighted social rate of return of 0.234, and hence a 0.0019 contribution of R&D to the current rate of growth of private output. This is about two-thirds of the comparable number for 1966 and reflects the rather significant slowdown in the rate of growth of R&D expenditures. Given the lags involved, this reduction may not show up, however, until the mid-1970s.

6   Reasons for Thinking the Above Estimates to be Too High

The estimate of the contribution of R&D investments to the rate of growth in measured total factor productivity is largely the product of three fractions: (1) the assumed excess of the social rate of return over the private rate of return; (2) the fraction of actual R&D expenditures which can be expected to have an impact on total factor productivity as currently measured; and (3) the fraction of these expenditures that represents net investment rather than replacement. In the previous sections we have provided some arguments for setting these numbers at 0.23, 0.53, and 0.5 respectively. These, when multiplied together and applied to the current (1970) ratio of total R&D expenditures to private domestic GNP of 0.03, lead to the estimated effect on the residual of 0.0018.[28] The effect of these expenditures on the actual rate of growth of output is about double of that but that second half (the private returns portion) is almost canceled out by offsetting errors in the measurement of output and the contribution of tangible capital.

These attributions are not negligible. The earlier estimates as of 1966 amount to between one-fifth and one-quarter of the "residual" rate of growth as it has been recently estimated.[29] In my opinion, this is close to an upper-bound estimate, but the reader should feel free to insert his own "guestimates" into the formula outlined above.[30]

The reason why I believe this to be an upper-bound estimate rests on the application of a rather high social rate of return, derived from scattered evidence on actual returns in a few, largely successful, areas of research, to all research expenditures. Also I have been lenient, I think, in drawing the boundaries as to the type of research which can have an effect on total factor productivity as measured. The major omission is the possible contribution of the scientific education process as a whole above and beyond what is counted under the label of "research." But that is not as large as all that.[31] Moreover, if one expands the boundaries of the relevant concept of R&D, one should probably adjust the estimated rates of return downward accordingly since, if the productivity measures were correct, they would already contain the returns to all such R&D. Thus, for example, when we estimate the rate of return to manufacturing R&D in section 3 at about 0.3, we did not include in the definition of R&D the externalities assumed to flow from the other sectors. If these had been distributed proportionately to the industries' own R&D expenditures, the rate of return to all relevant R&D would be significantly (about one-quarter) lower.

Another reason why the above estimates could be too high is our implicit assumption that the relevant stock of knowledge has grown in proportion to the growth in research expenditures. But the rapid rise in the latter could have entailed significant diminishing returns, the actual stock of knowledge not growing anywhere near the same rate. For all these reasons, the estimates presented above are probably too high perhaps by as much as 50 percent.

7 Some Policy Implications

A major objection to drawing policy conclusions from such findings is that they are all based on "average" rates of return, while decisions have to be made on the margin. This objection contains an important half-truth: the next research project in the stack may turn out to be unsuccessful *ex post* and this, perhaps, could even have been clear *ex ante*. But this is only a half-truth. Any one research project may be wasteful without the same being true for other potential research endeavors in this and other fields.

Perhaps an analogy will help here. An economist can examine the accounts and experience of the wildcat oil prospecting and drilling industry and conclude that on the whole investments in this industry appear to have been profitable, and that the people running it seem to know what they are doing. He could not guarantee, however, that the next well drilled would not turn out to be dry, and much less tell them where to drill it.

At the more abstract level there is another answer to this objection. There does not seem to be any evidence that the average rate of return to research is declining, and hence there is no presumption that the average and marginal rates of return are very far apart.

Most economists, if queried, would assert that there is underinvestment in research by private firms because much of its product is not capturable (appropriable) by the private firm (see Arrow, 1962, and Nelson, 1959). This is the major argument for the patent system and for government support of this activity. There are only a few contrary voices. In a recent paper Hirschleifer (1971) points out that to the extent that the new knowledge may provide a firm (or an individual) with a competitive advantage, there may be a private incentive to invest in it even in excess of its social return. His point is well taken, but seems to be irrelevant for most types of research supported by the government. The results of such research are unlikely to be of a form directly translatable into a market advantage. In any case, government-supported research is rarely directly competitive with private research in the same area. Usually, if it is a flourishing private research area, there will be little public support for research in the same field.

Not all research contributes to growth or to the capacity of the economy to produce goods and service and to consume them wisely. Some of it, as pointed out by Johnson (1965), is a kind of public consumption like mountain climbing. We investigate certain secrets of Nature because they are there, not just and not even primarily because their solution may prove "useful." That some of research is of this type does not mean that it shouldn't be supported; only that it has to compete for the scarce public dollar with other public consumption activities such as the architecture of public buildings and the support of fine arts.

Both the empirical and the theoretical literature provide some arguments why public investment in science and research may be a good thing. They also provide some evidence, in the form of higher average rates of return in this sector relative to other sectors of the economy, for the argument that more investment might be even better, but almost no information on where particular investments should be made, and no assurances that any particular set of investments will in fact pay off.

Acknowledgements

The work embodied in this paper has been supported by a grant from the National Science Foundation. I am indebted to P. Ryan for research assistance, and to Y. Ben-Porath, R. B. Freeman, F. M. Fisher, R. J. Gordon, H. G. Johnson, D. W. Jorgenson, and S. Kuznets for comments on an earlier version. The usual caveats apply.

Notes

1 See Machlup (1962), Mansfield (1967), Nelson (1962) and Nordhaus (1969a), among others.
2 Private rates of return of comparable magnitude are derivable from data on returns to specific refining inventions collected by Enos (1962).
3 That conventional measures of productivity growth may not be very good measures of the output of research is something that we shall return to below.
4 There are a number of other studies, somewhat less rigorous in spirit, relating differences in rates of growth of output among industries to differences in the intensity of their investment in research; e.g. see Freeman (1971) and Leonard (1971).
5 That private inventive activity is subject to economic influences and calculations, just like other aspects of private investment, is documented in detail in Schmookler's (1966) important book.

6 To the extent that research inputs are included among the conventional input measures, they have already been imputed the average private rate of return.

7 Similar results were obtained recently for agriculture by Huber (1970), using 1959 census data by the type and size of farm.

8 See Evenson (1968) for a more detailed discussion of some of these issues.

9 Actually much of this "depreciation" is obsolescence, induced and hence not independent of the rate of research in the rest of the industry or economy. This point is emphasized by Fellner (1970). A model in which depreciation is a function of the rate of research could be constructed, but it would take us too far afield.

10 Somewhat more positive results have been recently achieved using counts of scientific agricultural publications in a yet unpublished study of agricultural yields by Evenson and Kislev at Yale. See also Baily (1970), where the output of research in the pharmaceutical industry is measured by the number of new drugs introduced.

11 These issues were discussed by Brown and Conrad (1967) and under the title of "pervasiveness" by Latimer and Paarlberg (1965) and Evenson (1969).

12 In spite of all these reservations, some estimates have been attempted. In particular, in a forthcoming NBER study Kendrick presents an estimate of the net stock of "research and development capital" for the United States for the year 1929, 1948, 1966. He estimates it to have grown at about 9 percent per annum between 1948 and 1966 and to have reached the level of $100 billion (in 1958 prices) in 1966. Wagner (1968) gives additional detail on the derivation of these figures. Essentially it was assumed that "basic research" does not depreciate, that "applied research and development" had a life of about 10–15 years, and the totals were deflated by a cost index based largely on salaries of scientists.

13 Whether this is the social rate of return or something less than that depends on whether the research inputs are already included in the conventional input measures or not.

14 This is a reasonable assumption for series whose growth rate is approximately equal to the depreciation rate of the stock.

15 The estimated coefficient may be too low because we used total investment in R&D rather than just net investment as is called for by formula (5'). To the extent that depreciation (or stock) levels are positively correlated across industries with the gross investment levels, the coefficient of ROV would be a downward-biased estimate of ρ.

16 This is true for the bulk of defense and space research which is contract research and whose end-products are blueprints, formulae, or prototypes. It is also largely true for the research components of the "hardware" sold to government because of the lack of appropriate price indexes to deflate these expenditures by.

17 This is an important dimension of the problem. It is estimated that over three-quarters of the applied research and development expenditures in the United States are devoted to "product innovations" as against "process innovations" (see Wagner, 1968). How this may affect the measured returns to research has been discussed by Gustavson (1962), Griliches (1962a) and Millward (1964).

18 For a recent review of the status of the quality problem in price indexes see Griliches (1971).

19 This is of course the problem with the conventional estimates of real capital formation by construction, where we do not have an independent measure of output or an output price index.

20 This point is made in a more general context by Treadway (1969).

21 It does suggest, though, a line of research to be pursued further elsewhere.

22 See Jorgenson and Griliches (1967) for a more detailed exposition of this type of accounting algebra.

23 Similar expressions can be found in Gordon (1968), which deals with the same question but in a slightly different context.

24 We are going to assume throughout that $\dot{P}/P \simeq \dot{P}_R/P_R$, i.e. the price index of research investment is equal to the implicit private product deflator. This probably underestimates P_R/P_R and overestimates the actual increase in research investment in constant prices.

25 This assumption is about right, for the figures at hand for the year 1966.

26 The basic data used are from NSF 70–46 (1970), deflated by the implicit GNP deflator and cumulated on the assumption of a 10 percent declining balance depreciation for applied research and development expenditures.

27 Implicitly this assumes that these investments yield less externalities than public investments which are directly oriented towards the production of externalities.

28 This is somewhat higher, but not greatly so, than Denison's estimate of a decade ago (Denison, 1962, pp. 239–46). I have been more liberal in ascribing to a larger fraction of R&D the potential to effect uncovered to top factor productivity.

29 See Jorgenson and Griliches (1972) and the literature cited here.

30 There is an additional "causal" contribution of R&D investments not discussed in this paper. Improvements in productivity make additional capital investments profitable and induce thereby additional capital accumulation with a consequent effect on the rate of growth of output. But according to the usual growth accounting conventions, this contribution is ascribed to tangible capital rather than R&D, and rightly so in my opinion. For a contrary view, see Gordon (1968).

31 In 1964, total spending at universities on instruction in sciences and engineering, exclusive of explicit R&D expenditures, was only $2.4 billion.

References

Ardito-Barletta, N., "Costs and social benefits of agricultural research in Mexico." Unpublished PhD dissertation, Department of Economics, University of Chicago (1971).

Arrow, K. J., "Comment," in *The Rate and Direction of Inventive Activity* (National Bureau of Economic Research). (Princeton: Princeton University Press, 1962), pp. 353–8.

Baily, Martin N., "Research and development costs and profits: the U.S. Pharmaceutical Industry." MIT, unpublished manuscript (1970).

Brown, M. and Conrad, A. H., "The influence of research and education on CES production relations," in *The Theory and Empirical Analysis of Production*, ed. M. Brown (NBER: Studies on Income and Wealth, 1967), XXXI, 341–72.

Denison, Edward F., *The Sources of Economic Growth in the United States and the Alternatives Before Us* (Supplementary Paper no. 13) (New York: Committee for Economic Development, 1962).

Eastman, S. E., "The influence of variables affecting the worth of expenditures on research or exploratory development: an empirical case study of the C-141A aircraft program." Unpublished Institute for Defense Analyses memorandum (1967).

Enos, J. L., "Invention and innovation in the petroleum refining industry," *The Rate and Direction of Inventive Activity*, ed. R. Nelson (NBER, 1962), pp. 299–322.

Evenson, R., "The contribution of agricultural research and extension to agricultural production." Unpublished PhD thesis, University of Chicago (1968).

——, "Economic Aspects of the Organisation of Agricultural Research," presented at the Symposium on Resource Allocation in Agricultural Research, University of Minnesota, Minneapolis. Unpublished mimeograph (February 1969).

Fellner, N., "Trends in the activity generating technological progress," *American Economic Review* (March 1970).

Freeman, Richard B., "Engineers and scientists in the industrial economy." Unpublished manuscript submitted to the National Science Foundation (1971).

Gordon, R. J., "The disappearance of productivity change." Unpublished ED Report no. 105, Project for Quantitative Research in Economic Development, Harvard University (1968).

Griliches, Zvi, "Research costs and social returns: hybrid corn and related innovations," *Journal of Political Economy*, 66 (October 1958).

——, "Comment," in *The Rate and Direction of Inventive Activity*, NBER (1962), pp. 346–51.

——, "Discussion," *American Economic Review*, Proceedings Issue (May 1962), pp. 186–7.

——, "Research expenditures, education, and the aggregate agricultural production function," *American Economic Review*, LIV (December 1964).

—— (ed.) *Price Indexes and Quality Change* (Cambridge: Harvard University Press, 1971).

Gustavson, W. E., "Research and development, new products and productivity change," *American Economic Review*, Proceedings Issue (May 1962).

Hirschliefer, J., "The private and social value of information and the reward to inventive activity," *American Economic Review*, LXI (4, 1971), 561–74.

Huber, Paul H., "Disguised unemployment in U.S. agriculture, 1959." Unpublished PhD dissertation, Yale University (1970).

Johnson, H. G., "Federal support of basic research: some economic issues," in *Basic Research and National Goals* (Washington: Government Printing Office, 1965).

Jorgenson, D. W. and Griliches, Z., "The explanation of productivity change," *Review of Economic Studies*, XXXIV (1967), 249–83.

——, "Issues in Growth Accounting; A Reply to Edward F. Denison," *Survey of Current Business*, May 1972, 65–94.

Kendrick, J. W., *Productivity Trends in the United States* (Princeton: Princeton University Press, 1961).

——, forthcoming NBER volume on *Postwar Capital Formation*.

Latimer, R. and Paarlberg, D., "Geographic distribution of research costs and benefits," *Journal of Farm Economics*, 47 (1965).

Leonard, W. N., "Research and development in industrial growth," *Journal of Political Economy*, 79 (2, 1971), 232–56.

Machlup, F., *The Production and Distribution of Knowledge in the United States* (Princeton: Princeton University Press, 1962).

Mansfield, Edwin, "Rates of return from industrial research and development," *American Economic Review*, LV, pp. 310–322.

——, *Econometric Studies of Industrial Research and Technological Innovation* (New York: W. W. Norton and Co., 1967).

Millward, R., "Research and development, new products and productivity" (Manchester unpublished, 1964).

Minasian, Jora R., "The economics of research and development," in *The Rate and Direction of Inventive Activity* (NBER) (Princeton: Princeton University Press, 1962).

——, "Research and development, production functions, and rates of return," *American Economic Review*, LIX (Proceedings Issue, 2, 1969), 80–5.

National Science Foundation, *Research and Development in Industry, 1968*, NSF 70–29 (Washington: Government Printing Office, 1970a).

——, *Federal Funds for Research, Development, and Other Scientific Activities, Fiscal years 1969, 1970, and 1971*, NSF 70–38 (Washington: Government Printing Office, 1970b).

——, *National Patterns of R&D Resources, 1953–71*, NSF 70–44 (Washington: Government Printing Office, 1970c).

——, *Methodology of Statistics on Research and Development*, NSF 59–36 (Washington: Government Printing Office, 1959).

Nelson, Richard R., "The simple economics of basic scientific research – a theoretical analysis," *Journal of Political Economy*, LXVII (June 1959).

——, (ed.), *The Rate and Direction of Inventive Activity* (Princeton for NBER, 1962).

Nordhaus, W. D., *Invention, Growth, and Welfare* (Cambridge: MIT Press, 1969a).

——, "An economic theory of technological change," *American Economic Review*, LIX (Proceedings Issue, 2, 1969b), 18–28.

Peterson, Willis L., "Returns to poultry research in the United States," *Journal of Farm Economics*, August, 1967.

Schmookler, J., *Invention and Economic Growth* (Cambridge: Harvard University Press, 1966).

Schultz, T. W., *The Economic Organization of Agriculture* (New York: McGraw-Hill, 1953).

Terleckyj, N. E., "Sources of productivity advance," unpublished PhD, dissertation, Columbia University, 1960.

——, *Research and Development: Its Growth and Composition* (New York: National Industrial Conference Board, 1963).

——, "Comment." In M. Brown (ed.), *The Theory and Empirical Analysis of Production* (NBER: Studies on Income and Wealth), 1967, vol. 31, pp. 372–9.

Treadway, A. B., "What is output?" In V. R. Fuchs (ed.), *Production and Productivity in the Service Industries* (NBER: Studies in Income and Wealth, 1969), vol. 34, 53–83.

US Department of Commerce, *Concepts and Methods of National Income Statistics*. National Technical Information Service, Washington, 1976.

Wagner, L. U., "Problems in estimating research and development investment and stock." In ASA, *1968 Proceedings of the Business and Economic Statistics Section*, Washington, 1968, pp. 189–97.

Weisbrod, Burton A., "Costs and benefits of medical research: a case study of poliomyelitis," *Journal of Political Economy*, 79 (3, 1971), 527–44.

16
Postscript on R&D Studies

There are a number of articles that survey the subsequent literature. Besides the literature already discussed in "Research Expenditures and Growth Accounting" additional references can be found in Griliches (1979b) and my introduction to the NBER volume (Griliches, 1984). Subsequent estimates of returns to agricultural research are reviewed by Evenson et al. (1979), Evenson (1982), and Weaver (1985). Major original contributions to the analysis of R&D processes and the returns to them outside of agriculture were made by Mansfield and his co-workers (see Mansfield et al., 1977b and Mansfield, 1984 for examples and additional references) and by Scherer (1984a and b). Other contributions of note in this area were made by Evenson and Kislev (1975), Grabowski and Mueller (1978), Grabowski and Vernon (1982), Link (1981), Nelson, Peck and Kalachek (1975), Terleckyj (1974), and others. A recent review of a subset of these issues can be found in Reiss (1985).

An attempt to extend the national income and productivity accounts by treating R&D as a form of social investment has been made by Kendrick (1976) and Eisner (1985). The number of studies which relate some measure of productivity growth to some measure of R&D intensity or other measure of R&D "capital" is too large to summarize adequately here. Interesting examples of current work can be found in Bernstein and Nadiri (1986), Gordon, Schankerman, and Spady (1986), Griliches and Mairesse (1983), Jaffe (1986), and Mohnen, Nadiri, and Prucha (1986), among others.

Part IV

Explanations of Productivity Growth

17
The Sources of Measured Productivity Growth: United States Agriculture, 1940–1960*

1 Introduction

Public interest in economic growth as an objective policy has stimulated a substantial number of studies into the causes of economic growth and possible ways of affecting it. All of these studies use the concept of an "aggregate production function" either explicitly or implicitly. This function can be represented in a general way by

$$Y_t = g(X_t, u_t, T_t),$$

where Y_t is an index of physical output (or value added) in an industry, X_t is a set of "measurable" inputs, usually labor and capital indexes, u_t is a random, or short-term, cyclical variable such as weather in agriculture or unemployment in manufacturing, T_t is the "level of technology," a postulated unobserved "latent" variable usually to be inferred from the data in a residual fashion, and $g(\ )$ is the function describing the connection between these variables. It is also often assumed that this function is homogeneous of degree 1 in X (constant returns to scale), and that the industry operates in perfectly competitive product and factor markets and is in equilibrium (or at least was in equilibrium in the weight-base period). This last set of assumptions allows one to approximate the relevant coefficients of the production function by relative factor shares, evading thereby a direct estimation of it.

Several recent studies of the historical record of United States growth have used such a framework together with conventional measures of labor and capital and found that very little of the growth in output can be accounted for by changes in these inputs, most of the growth being explained by the residual factor T – "technical change."[1] These findings have in turn

*Reprinted from *Journal of Political Economy*, vol. LXXI, no. 4, August 1963.

encouraged further attempts to refine and improve the measurement of what appears to be the most important source of economic growth.[2]

This formulation of the problem and the direction in which the resulting research has evolved are, in my opinion, not very helpful to the understanding of growth. The whole concept of a production "function" is not very useful if it is not a *stable* function, if there are very large unexplained shifts in it. Moreover, it does not further our understanding of growth to label the unexplained residual changes in output as "technical change." Nor does it help much to measure these changes accurately if we do not know what they are.

The purpose of this paper is to suggest an alternative, potentially more fruitful, approach to the problem and to illustrate it by reference to the growth of productivity in agriculture in the United States. According to this approach, changes in output are attributable to changes in the quantities and *qualities* of inputs, and to economies of scale, rather than to "technical change," the production function itself remaining constant (at least over substantial stretches of time). Conventionally derived residual measures of productivity growth are viewed not as measures of technical change but rather as the result of errors in the measurement procedure. These "errors" are due to several sources: (1) the list of variables affecting output may be misspecified, excluding some relevant factors from the calculations; (2) changes in included variables may be mismeasured, particularly if changes in their quality are disregarded; and (3) wrong weights may be used in estimating the contribution of changes in individual inputs to the growth in output. "Correcting" such errors will, I believe, lead to a substantial reduction in what has been conventionally measured as growth in "total factor productivity," reducing thereby the proportion of growth that had to be previously attributed to this essentially "unexplained" category. Such an approach does not, of course, remove technical change from the explanation of growth; it aspires, rather, to transform what is currently a catch-all residual variable into movements along a more general production function and into identifiable changes in the qualities of inputs.

This approach can be illustrated by bringing together the results of a number of studies of production relations and input quality change in United States agriculture.[3] These studies indicate that the main sources of conventionally measured productivity increases in United States agriculture during the 1940–60 period appear to have been:

1 Improvements in the quality of labor as a consequence of a rise in educational levels.
2 Improvements in the quality of machinery services that had been disguised by biases in the standard price indexes used to deflate capital equipment expenditures.
3 Underestimation of the contribution of capital and overestimation of the contribution of labor to output growth by the conventional factor-share based weights.
4 Economies of scale.[4]

A separate estimate of each of these sources is derived from various pieces of data.[5] The main point of this study, however, is not to define technical change away but rather to explain it.[6]

The plan of the rest of this article is as follows: section 2 discusses several questions that could be investigated in an econometric production function study but that are usually assumed away in the standard productivity or technical change measurement study. Section 3 reports the results of a cross-sectional production function study that was designed to investigate some of these questions. It is found, in particular, that the estimated production function differs substantially from what one would infer on the basis of the factor shares approach, both in the relative weight that it assigns to different inputs and in its finding of economies of scale. Moreover, evidence is found to support the view that education is a relevant quality dimension of the labor variable. Section 4 brings together results from a number of previous studies on input change, indicating a substantially larger growth in inputs over time when they are measured in more relevant units and with more regard for quality change. Section 5 brings together the results of the previous two sections and shows that in this particular case this approach leads to an almost complete accounting of the sources of output growth in United States agriculture during 1940–60, leaving no "unexplained" residual to be identified with unidentified "technical changes."

2 Why Estimate Production Functions?

Econometric production function studies can be used to investigate the appropriate algebraic form for the assumed aggregate production function, the number and type of variables that should be included in the list of inputs and the appropriate way of measuring them, and what numerical values should be assigned to the coefficients to be attached to each of these variables. Even though these questions are basic to any attempt to allocate the observed growth in output to its various "causes," they are usually assumed away rather than investigated in most of the studies measuring productivity and technical change.

The choice of a particular algebraic form for the production function is associated with the question of ease of substitution between different inputs. It is a question concerning the curvature of the isoquants. The elasticity of substitution is usually assumed to be either infinite, as illustrated by Kendrick's arithmetic total input indexes,[7] or unity, as is the case when the Cobb–Douglas function or a geometric input index with constant or shifting weights is used.[8] It has been suggested recently that the elasticity of substitution should also be estimated rather than assumed beforehand. This suggestion leads to the use of an exponentially weighted harmonic average of input indexes[9] in estimating the "residual" to be attributed to technical change. Which form of the production function is the most appropriate one is an empirical question and could be settled by testing different forms on the same set of input-output data. Unfortunately, the appropriate "curvature" is probably too fine a question to be settled on the aggregate level. Our data are just not good enough to discriminate among these various alternatives. Fortunately, however, this does not matter much so far as growth accounting is concerned. Using the conventional input measures but different index number formulas leaves us with pretty much the same large "residual." The curvature question is, of course, of much greater importance for the theory of functional income distribution. But that is another matter.

The main candidates for addition to the conventional list of inputs are research and development capital, education of the labor force, and "external" inputs such as research and extension activities of the government and other firms, and other non-market priced services such as the provision of transportation and communication facilities.[10] Ideally, we should investigate the relevance and actual numerical importance of such variables before we introduce them into our growth accounts. Most of the previous work on education, including my own, simply imputes part of the observed productivity increase to changes in education, using cross-sectional income-by-education tabulations as the source of its weighting scheme.[11] Many difficult questions are raised by the use of income-by-education data, and one cannot rule out the possibility that the observed associations may be in large part spurious. Moreover, until detailed tabulations of the 1960 Census become available, the only available income-by-education data are for the United States as a whole. One may not doubt the proposition that education is an important source of growth in the economy as a whole, and still not be convinced that it is very important in agriculture. By introducing the education of the labor force as a separate variable in an econometric production function study, it becomes possible to *estimate* rather than to *assume* its coefficient.

To measure that part of output change that results from a change in the level of a particular input, we must weight it by its respective production function coefficient. Assuming linear, or linear in the logarithms, production functions, constant returns to scale, and competitive equilibrium (at least in the weight-base period), these coefficients can be approximated by input market prices or their relative shares in total costs. But if, as has been alleged to be true for agriculture, a sector is in continuous disequilibrium, a weighting scheme based on factor shares will be incorrect for productivity comparisons. Many agricultural economists have argued for years that the marginal product of labor in agriculture is substantially below the going wage rate for hired labor and that the marginal product of capital is substantially above the conventional bank or mortgage rates.[12] They have been supported by the historically observed large outflow of labor from, and inflow of capital into, agriculture. A statistically estimated production function provides an alternative and conceptually more appropriate system of weights for compiling inputs into a "total" input index. For a sector such as agriculture, where different inputs have had very different time trends, the conventional productivity estimates are quite sensitive even to a small shift in weights.

All the conventional productivity indexes assume constant returns to scale. So do also many of the estimated production functions. On the other hand, most of the cost curves and much of the programming and budgeting literature imply the existence of substantial economies of scale, both in agriculture and in industry. This whole area, however, is quite controversial. From my point of view, it is not very important whether the economies of scale go on indefinitely or the cost curve finally turns up. The interesting question is whether there were and are *some* additional economies to be had at existing scale levels. Unfortunately, the fitting of standard Cobb–Douglas type of production functions, such as I will report on below, is not very well suited to answering this question, since it assumes that the production function is homogeneous in all inputs. It may provide an answer whether the function is homogeneous of degree more or less than 1, and this may be interesting and valuable by itself, but it may miss many aspects of what we usually think of as sources of economies of scale, such as indivisibilities and disproportionalities by its assumption of homogeneity. To study the subject of economies of scale adequately will require the use of a production function that is not homogeneous over at least some range of the inputs.

3 The Results of One Cross-sectional Study

I have recently investigated some of the problems discussed in the previous section by estimating an aggregate agricultural production function based on 1949 data for 68 regions of the United States. This study is described in some detail elsewhere, including the presentation of alternative estimates and of detailed reservations.[13] For the purposes of this paper only its main results are reproduced in table 17.1.[14] The estimates summarized in this table indicate that (1) education as measured is a statistically significant variable with a coefficient that is not very different from the coefficient of the man-years-worked variable. Thus, it turns out that it would not have been very wrong to "inflate" the labor variable by the computed "quality" (education) per man index before estimating the production function. This makes it much easier to apply such a framework to time series, since one can adjust the labor variable beforehand for quality change rather than carry education along as a separate variable.[15] (2) The estimated coefficients differ significantly from what has been assumed to be true on the basis of factor-shares data and equilibrium assumptions. Table 17.2 compares the coefficients as estimated in this study with the official estimates based on 1947–9 factor-shares data. The production function estimate of the labor coefficient is relatively smaller and the machinery coefficient is relatively larger than the official factor-shares-based estimates of these same coefficients. (3) There is evidence of substantial economies of scale.

These findings, particularly the last two, if accepted, will account for a substantial fraction of the conventionally measured productivity increases. Before we accept them, however, we should look for other pieces of evidence

Table 17.1 Aggregate agricultural production function, United States, 1949, 68 regions[a]

				Coefficients					
Regression	X_1 (livestock expense)	X_2 (other current expense)	X_3 (machinery)	X_4 (land)	X_5 (buildings)	(man-years)	X_6 E (education)	R^2	Sum of coefficients (excluding E)
(U17)	0.169 (0.023)	0.121 (0.032)	0.359 (0.048)	0.170 (0.033)	0.094 (0.044)	0.449 (0.072)		0.977	1.362
(R6)	0.140 (0.025)	0.111 (0.031)	0.325 (0.049)	0.167 (0.032)	0.085 (0.042)	0.524 (0.076)	0.431 (0.181)	0.979	1.352

[a] All variables are logarithms of original values unless otherwise specified; units = averages per commercial farm. Numbers in parentheses are the calculated standard errors of the respective coefficients.

X_1, log of (purchases of livestock and feed and interest on livestock investment).

X_2, log of (purchases of seed and plants, fertilizer and lime, and cost of irrigation, water purchased).

X_3, log of (purchases of gasoline and other petroleum fuel, repairs of tractors and other machinery, machine hire, and depreciation and interest on machinery investment).

X_4, log of interest on value of land.

X_5, log of (building depreciaton and interest).

X_6, log of average full-time equivalent number of workers per commercial farm.

E, logarithm of the average education of the rural farm population weighted by total US income by education class weights; not per commercial farm but per man.

Dependent variable – log of value of farm production per commercial farm.

Table 17.2 Alternative input weighting schemes, United States agriculture

| | Inputs | | | | | |
Source of weights	Labor (1)	Real Estate (2)	Power and machinery (3)	Feed, seed, and livestock (4)	Fertilizer and lime (5)	Other (6)
USDA, 1947–9 adjusted[a]	0.40	0.13	0.14	0.18	0.03	0.12
Cross-sectional aggregate production function, 1949[b]	0.33	0.19	0.26	0.13[c]	0.02[d]	0.07[d]

[a] From R. A. Loomis, "Production inputs of US agriculture," *The Farm Cost Situation*, May 1960, p. 3. Adjusted by substituting the estimated total value of purchased feed, seed, and livestock in 1947–9, instead of the "value added by the non-farm sector" concept, and recomputing the weights accordingly.

[b] From table 17.1, eq. (U 17), computed by summing the relevant coefficients and dividing through by this sum.

[c] The measure used in the cross-sectional regression does not include "seed" here but in the "Other" category. Luckily, this is a minor item.

[d] The cross-sectional measure included fertilizer and lime in the "Other" category. The coefficient of "other current expense" in the regressions is distributed between the two categories proportionately to the USDA weights.

that would either confirm or contradict these findings. The somewhat smaller coefficient for labor found in the production function study relative to the one estimated officially on the basis of factor shares implies that the marginal product of labor was less than was assumed in the estimation of factor shares.[16] This is most likely due either to an overestimate of the amount of work actually performed by family labor or to what is almost the same thing, an overvaluing of this (family) work by the application of the hired wage rate, a rate that is heavily affected by the price necessary to attract sufficient labor into the industry at peakload (harvest) time. Its hard to "prove" that the marginal product of labor in agriculture was below the estimated farm wage rate, but a variety of evidence points in this direction. Most of the cross-sectional agricultural production-function studies based on individual farm data seem to arrive at estimates of the marginal product of labor that are substantially below the comparable hired wage rate in the locality. Out of 43 production-function estimates compiled by J. G. Elterlich,[17] estimated from data for the years 1950–3 and for the states of Alabama, Illinois, Indiana, Iowa, Kentucky, Michigan, and Montana, 19 estimated the marginal product of labor (at the geometric mean of the sample) at less than $70 a month, seven of the estimates were between $70 and $100, eight between $100 and $150, and nine were over $150 a month.[18] The average monthly hired wage rate during these years was around $137 to $151, and higher than that in Illinois, Iowa, and Michigan, where most of these studies were made. This finding is also consistent with the very large and continuing migration out of agriculture. In 1950, the year after our study, out-migration was equal to 5.2 percent of the farm population and has averaged about 3.5 percent per year since.

The higher estimate of the coefficient of power and machinery inputs is also supported by the finding of relatively high marginal returns to machinery investment in many farm production functions, programming, and budgeting studies. From 1950 to 1961 the ratio of machinery inputs to labor inputs increased by 74 percent.[19] In the same period farm wages rose only by 9 percent relative to machinery prices,[20] implying a substantial decline in the "share" of labor in agriculture.[21] This, of course, could be explained by an elasticity of substitution that was larger than unity, but it would have to be larger than eight, and evidence presented elsewhere implies a much smaller elasticity of substitution.[22] One could, of course, assert that technical change during this period has been non-neutral and especially labor-saving, but there is little independent evidence for such a hypothesis. It is my belief that these facts can be best rationalized by the recognition that during most of the 1947-9 period agriculture was not in equilibrium, and hence that the "base-period" factors shares did not reflect well, if at all, the underlying production conditions.

Perhaps the most controversial finding is the one of economies of scale. Our equations indicate that a 10 percent proportional increase of all inputs in agriculture leads to a 13 percent or more increase in output. This finding is subject, however, to several possible sources of bias.[23] In particular, the use of values rather than quantities for some of our variables could lead to biased estimates of the coefficients and their sum. But this, if anything, should have biased our estimates downward.[24] More serious is the objection that the form of the equation does not allow for the most interesting source of economies of scale – indivisibilities and the non-homogeneity of the production function. An experiment with a somewhat more complicated form of the production function did not lead, however, to superior results.

The finding of substantial economies of scale is consistent with much scattered information about United States agriculture. In the previously cited survey of micro-farm production function estimates – of 43 surveyed production-function estimates, 36 had coefficients whose sum exceeded 1.0, and in 23 equations it exceeded 1.1. Similarly, much of the farm budgeting (including linear programming) literature implies that farmers could earn a higher rate of return on their owned capital and labor if their farms were larger. Perhaps the most striking confirmation of the hypothesis of economies of scale is to be found in what has happened to the distribution of farms by size of output since 1950.[25] Table 17.3 presents the distribution of commercial farms by size of sales (in 1954 prices) in 1950 and 1959, and illustrates clearly the rapid growth in the relative number of larger farms and the decline of the smallest size class. This by itself, however, need not indicate the presence of economies of scale, because a growth in total productivity could increase the output of each farm without really changing its "scale" (as measured by inputs) but still move some farms into the next size class

Table 17.3 Percentage distribution of commercial farms, by size classes, United States, 1950 and 1959

		Percentage distribution of commercial farms by economic class		
Economic class	Value of sales per farm (in 1954 dollars) (1)	1950[a] (2)	1959 (actual)[b] (3)	1959 ("predicted")[c] (4)
I	25,000 and over	3.0	8.9	6.2
II	10,000 – 24,999	11.0	21.5	12.8
III	5,000– 9,999	20.8	25.0	25.6
IV	2,500– 4,999	25.4	23.6	22.0
V and VI	250– 2,499	39.8	21.0	33.4

[a] From Jackson V. McElveen, *Family Farms in a Changing Economy* (US Department of Agriculture Information Bulletin no. 171 (March 1957)).

[b] From Karl A. Fox, "Commercial agriculture: perspectives and prospects," *Farming, Farmers, and Markets for Farm Goods* ("CED Supplementary Paper" no. 15 (New York: Committee for Economic Development, 1962)), table 15.

[c] Computed on the assumption that the USDA's 21 percent estimated increase in total productivity since 1950 would have increased each unit's output in the 1950 distribution by 21 percent. The resulting shift from one class to the next is estimated on the assumption of a uniform distribution within each class interval. Thus, for example, a 21 percent increase of output for farms in classes V and VI would have moved all farms that were above $2,066 ($2,500/1.21) into the above $2,500 class. Using more detailed information from the 1950 *Census of Agriculture* (vol. II, chap. ix), we know that 14.7 percent of farms were in the $1,500–2,499 class in 1950. Assuming a uniform distribution over this (smaller) class implies that about 6.4 percent of all farms were to be found in the $2,066–2,499 interval, which are then shifted to the row above. Similar calculations were performed for the other classes. Even with the use of more detailed class data, the assumption of uniformity biases these "predicted" shifts upwards, since the actual distribution is very much skewed to the right. The mean of the $10,000–24,999 class is, for example, around $14,500 instead of the assumed $17,500.

(as measured by sales). Column (4) in table 17.3 makes an outside estimate of the amount of shifting in the size distribution that could be due to a 21 percent "neutral" increase in total input productivity[26] on the assumption that all farms benefitted from from this increase equally and that farms are distributed uniformly within each size class. Both of these assumptions would lead to an overestimate of the "technical change effect," but even as computed, column (4) is about midway between columns (2) and (3) of table 17.3, underestimating substantially the relative decline of small farms and the relative increase of the larger ones. Thus there must have been a "real" shift toward larger-scale farms since 1950, such as would have been predicted by the estimated production function.

The most important test of an estimated production function is not how well it fits the data it was derived from but rather whether and how well it can "predict" and interpret subsequent behavior. The computed production function together with the assumption of some degree of economizing behavior "predicts" a decline in labor inputs, a rise in machinery inputs,

Table 17.4 Various input and output indexes, United States agriculture, 1940 and 1960 (1947–9 = 100)

	Official[a]		Adjusted	
	1940 (1)	1960 (2)	1940 (3)	1960 (4)
Inputs:				
Labor	122	62	115	67[b]
Real estate	98	106	98	106
Power and machinery	58	142	54	152[c]
			66	181[d]
Feed, seed, livestock	63	149	71	157[e]
Fertilizer and lime	48	192	48	207[f]
Other	93	138	89	144[g]
Output	82	129	81	132[h]
No. of commercial farms[i]	120	71		

[a] US Department of Agriculture, *Changes in Farm Production and Efficiency*.

[b] For a detailed description of this and subsequent adjustments see my "Measuring inputs in agriculture. . . ," op. cit. The official labor-input index was multiplied by an index of formal schooling per man in agriculture. See table 17.5 for the derivation of this index.

[c] Adjusted for estimate bias in the deflators. Past investment expenditures were inflated upward from 1947 on the basis of the discrepancy between the USDA indexes and other indexes covering the same items. Automobile purchases were adjusted upward by the discrepancy between the CPI and USDA automobile price indexes all the other price indexes were adjusted by the discrepancy between comparable components of the WPI and USDA indexes, except that for tractors Fettig's (op. cit.) index of new tractor prices was used from 1950 on. See table 17.6 for details. The resulting gross investment series were then substituted in the USDA formula for computing the capital stock series. This adjustment affects only about half of this index. The operations and maintenance components were left unadjusted.

[d] Based on a gross (15-year moving sum of past investment) concept of capital. Computed from the previously described adjusted investment series. Again, the operation and maintenance component was unaffected by this adjustment.

[e] Total value of purchased feed, seed, and livestock in 1947–9 prices from *Farm Income Situation* (deflated separately by the prices paid indexes for feed, seed, and livestock from *Agricultural Prices*), substituted for the "value added by the non-farm sector" concept.

[f] The fertilizer and lime index was increased by 8 percent from 1947–9 to 1960 to reflect improvements in quality (mainly the shift to nitrogen) not caught by the unweighted plant nutrient measure (see my "Measuring inputs in agriculture. . . ," op. cit., for details of this adjustment).

[g] The same 8 percent increase was applied also to the "other inputs" category on the assumption that quality improvements in this category have been *at least* of the same order of magnitudes as those not caught by an already quality-oriented measure of fertilizer inputs. This adjustment was distributed equally between the pre- and post-1947–9 periods.

[h] Adjusted by adding to it the purchased feed, seed, and livestock index weighted by the *difference* in 1947–9 between all purchased feed, seed, and livestock and the "value added by the non-farm sector" concept.

[i] See notes to table 17.3 for sources. Interpolated for 1940, 1947–9, and 1960 by changes in the number of all farms (from *Farm Income Situation*, July 1962, Table 8H).

and an expansion in the share of larger scale farms. All of these "predictions" were borne out by subsequent events. Moreover, the existence of economies of scale would imply a rise in the relative price of the most specialized input to agriculture, the one having the lowest supply elasticity. Land fits this description best and its price rose 83 percent from 1947–9 to date, while during the same period, the index of prices paid by farmers for all (other) production items including labor rose only 26 percent.[27]

4 Measuring Input Change

This section brings together the results of several studies of input quality change and other aspects of input measurement. These will be summarized in the form of a series of adjustments performed on the official input series for United States agriculture. The rationale and evidence for these various adjustments is discussed at greater length in the original papers.[28] The magnitude of these adjustments is illustrated in table 17.4, which presents the official input and output indexes for United States agriculture for 1940 and 1960 (1947–9 = 100) and the adjusted series for the same dates. The following major adjustments were made.[29]

1 Labor-force Quality

Our production results indicate that we can proceed directly to multiply the man-years figures by our index of education per man (since the coefficients of these two variables are almost the same). This index was computed by weighting each school-year-completed class by the average 1950 income of all United States males (25 years and over) in this class. The resulting measure of education is almost proportional to mean school years completed, except for a non-zero weight for "no education" and a somewhat higher relative weight for college education. Its derivation is given in table 17.5.

The adjustment itself amounts to about 5 percent between 1940 and 1947–9 and 8 percent from 1947–9 to 1960. The total increase in the "quality" of the agriculture labor force between 1940 and 1960 was about 14.4 percent. Additional adjustments were calculated for changes in the age, sex, and race distribution of the agricultural labor force based on 1950 income data by these characteristics. The resulting adjustments were quite small and hence were not included in the final analysis.

2 Bias in the Deflators

Since most machinery input estimates are based on cumulated deflated expenditure series, if the deflators are poor, so will also be the resulting "constant price" capital estimates. All the deflators used for these purposes in agriculture are based on USDA-collected price statistics and are components or a recombination of components of the Prices

Table 17.5   Index of educational level of rural farm males

School years completed	Rural farm population: males, 25 years old and over, by school years completed			1950 Mean income of males (25 years old and over), by school years completed, all US
	1940 (1)	1950 (2)	1960 (3)	(4)
None	5.2%	3.6%	2.7%	$1,378
Grade school 1–4	18.3	15.8	11.1	1,699
Grade school 5–7	24.7	23.5	20.2	2,164
Grade school 8	29.7	26.9	26.8	2,676
High school 1–3	10.8	12.8	14.4	3,096
High school 4	6.2	10.7	17.9	3,784
College 1–3	2.5	3.1	4.4	4,449
College 4 or more	1.2	1.8	2.6	6,318
Not reported	1.4	1.8	—	2,471[a]
"Mean weighted income"	$2,502	$2,644	$2,860	
Index, 1950 = 100	94.6	100	108.2	

[a] Computed from *1950 Census of Population, Education*, Series PE, no. 5B, p. 108, using Houthakker's midpoints (H. S. Houthakker, "Education and Income," *Review of Economics and Statistics*, vol. XLI, no. 1 [1959]).

Source: col. (1), *1940 Census of Population*, vol. IV col. (2), *1950 Census of Population*, vol. II, part 1 col. (3), *1960 Census of Population*, PC(S1)-20 (June 4, 1962), table 76; col. (4), computed by averaging for the appropriate age groups the mean incomes given by Houthakker op. cit., table 1. The weights for these averages were taken from *1950 Census of Population, Education*, series PE, no. 5B, pp. 42–43.

Paid Index. While most machinery price indexes do poorly as far as quality change is concerned, the Prices Paid by Farmers Index has been especially affected by the official USDA insistence that it wants an index of "unit values" rather than of prices and the consequent practices of relatively loose commodity specification, pricing the quality or brand "most commonly bought by farmers," and pricing items with all "the customarily bought attachments." As the result of these practices, the USDA indexes have drifted upward over time relative to other similar price indexes (table 17.6). The adjustments performed to counteract some of these biases consisted of computing for each of the machinery categories the drift of the USDA price index relative to a more tightly specified price index (the CPI new-automobile price index in the case of automobiles, and the comparable WPI components for motor trucks and other farm machinery). These relative indexes were averaged using 1947–9 gross investment in the respective categories as weights, and the resulting index was used to inflate the USDA estimates of farm gross investment in motor vehicles and farm machinery from 1947 to date. But even these more tightly specified price indexes do not take satisfactory account of the changes that have occurred in more complicated pieces of machinery such as automobiles or tractors.[30] In the case of tractors, estimates were available of the margin "price" of horsepower

Table 17.6   USDA prices paid by farmers indexes for machinery and motor vehicles relative to other comparable indexes, 1947 to 1960 (1947–9 = 100)

Year	Automobiles (USDA/CPI)	Motor trucks (USDA/WPI)	Tractors[a] (USDA/WPI and Fettig)	Other farm machinery (USDA/WPI)	Average[b] (1947 = 100)
1947	97.7	99.1	96.9	97.4	100
1948	101.4	101.3	98.6	99.1	102
1949	100.7	99.9	103.9	103.5	105
1950	100.8	100.5	103.8	102.7	105
1951	101.4	98.9	104.1	101.1	104
1952	103.3	101.4	103.1	104.3	106
1953	103.9	102.1	103.0	104.8	106
1954	105.5	104.0	102.9	104.7	107
1955	114.4	101.8	107.0	103.7	108
1956	114.9	100.2	106.0	103.1	107
1957	116.8	104.1	111.0	104.1	110
1958	116.1	103.7	107.6	104.0	109
1959	120.1	105.7	113.3	104.8	111
1960	118.3	108.4	114.2	105.2	112

[a] Fettig's index linked to the WPI in 1950; 1951 and 1952 (not computed by Fettig) interpolated by the WPI.
[b] Weighted average, with weights based on the 1947–9 rate of gross investment in these different categories (0.102 for automobiles, 0.160 for motor trucks, 0.264 for tractors, and 0.474 for other farm machinery).
Source: USDA unpublished component indexes used to deflate gross farm investment. They are components or a recombination of components of the USDA prices paid by farmers index. The CPI and WPI figures are taken from various BLS releases (in particular Bulletin nos. 1295 and 1296). Fettig's index is taken from his unpublished dissertation (op. cit.). From among the various indexes computed by him, I am using the chain-linked one, with weights based on his linear single-year regressions.

and the differential value of diesel versus gasoline engines.[31] These estimates were used to construct an alternative tractor price index that was then substituted for the USDA deflators. The resulting adjusted gross investment series was then used to recompute the capital stock series using the same procedures as in the official series.[32] These adjustments amounted to about 12 percent between 1947–9 and 1960 in the gross investment series, and to somewhat less for the resulting capital series. None of these adjustments, except for the tractor price index, are as far reaching as I have previously advocated in my work on automobile price indexes.[33] Even the tractor adjustment is relatively small (10 percent since 1950), because the USDA priced tractors within relatively narrow horsepower classes, preventing its index from drifting too far.

3   Differences in the Concept of Capital Services

The official USDA machinery input series (excluding maintenance and operation) is proportional to its estimate of the value of the stock of machinery on farms in constant prices. It is based on a declining balance depreciation formula using relatively high rates of depreciation (about 16.5 percent per year on the average). It can be shown that these rates are too high, even as approximations to market rates of depreciation. But what is even more important is that for productivity comparisons we want an estimate of the *current flow of services* rather than of the market value of all present and future services. The relevant measure would be approximated, in a perfect market, by the rental price per machine-hour times machine-hours used. The rental price would be a "constant" price, which would not only adjust for fluctuations in the general price level but also for relative price changes between new and used equipment (obsolescence) induced by the availability of superior machines. This rental would change over time only if the physical flow of services were to deteriorate due to wear and tear (and age) or change its character of use due to the appearance of new and different machines. It would not change just because the appearance of new machines makes the use of the old ones less profitable and leads to capital losses by the owners of the old machines. Nor would it change if new machines are made more durable at a higher price but still provide the same *annual* service flow.

The current official measures assume that the flow of annual services from a machine falls by more than half in the first four years of its life, and that very little of it is left (about 16 percent) after ten years. What little evidence we have on the mortality and use of machines with age indicates that this is not the case. An alternative, somewhat extreme, assumption is that the flow of services does not decline with age.[34] This would imply the use of "gross" instead of "net" measures of capital in productivity comparisons.[35] I have accordingly substituted a 15-year moving sum of past gross investment for the USDA measure. The new measure assumes that services are constant during the first 15 years of life of a machine and fall to zero thereafter.[36] The two measures move closely together over long periods of time, but they can have very different time trends in the intermediate run. In our particular case, this adjustment leads to a substantially lower estimate of the growth of the stock of machinery on farms between 1940 and 1947–9, and a much higher estimate of the subsequent (to 1960) growth in this stock.[37]

4   Other Quality Changes

The main adjustment here is to the fertilizer and lime index. The USDA measure does not use tons as its measure, converting them into units of effective ingredients (plant nutrients). It does, however, simply add up the three main ingredients (nitrogen, phosphoric acid, and potash), ignoring differences in their relative value. Since the consumption of the most expensive and important ingredient – nitrogen – has grown more over time, the resulting official fertilizer input series underestimate the actual growth in "effective" fertilizer consumption. The adjustment for this bias consists of substituting a weighted plant nutrient measure for the unweighted one used by the USDA, the weights having been derived from a cross-sectional study of prices of different fertilizer mixtures in 1955. The adjusted index rises by about 8 percent more from 1947–9 to 1960. A similar adjustment, distributed equally between the pre- and post-1947–9 period, is applied to the "other inputs" category on the assumption that quality improvements in this category have been *at least* of the same order of magnitude as those not caught by an already quality-oriented measure of fertilizer inputs.

5 Summary

The adjusted input series together with the estimated production function weights are brought together in table 17.7 to account for the growth in aggregate agricultural output between 1940 and 1960, and to compute the "residual," unexplained increase in productivity. For comparison purposes, several intermediate measures are computed, using the official input series and weighting schemes. Also, since the estimated production function is of the Cobb–Douglas form, implying the use of geometric rather than arithmetic input indexes, the final estimates are presented on a geometric index base. As can be seen from this table, there is little difference between the original official indexes and those adjusted for the different concept of output. Nor is there much difference between the arithmetic and geometric indexes for comparable combinations. Substituting "corrected" input series into the official weighting scheme reduces the estimated productivity increase by about one third.[38] Using the estimated production function weights, without allowing, however, for scale effects, leads to another one-fourth to one-sixth in reduction in the original productivity growth estimates. If one allows, in addition, for economies of scale at the cross-sectionally estimated rate, they account for all (and somewhat more) of what is left.[39]

The production-function study results together with the estimates of input quality change account for all (if not more than all) of the observed increases in agricultural productivity.[40] They imply that (very) roughly about one-third of the observed productivity increases are due to improvements in the quality of inputs (among which the rise in education per worker plays an important part), about a quarter or so is due to a move toward the elimination of relative disequilibria due to the overpricing of labor (in particular of family labor) and the underpricing of capital services by the conventional market measures, and that the rest is due to the expansion that occurred in the scale of the average farm enterprise.[41] This "complete" accounting for the observed productivity increases does not mean that there were no meaningful increases in agricultural productivity over this period. It means, rather, that we may have succeeded in providing an explanation for what were previously unexplained increases in farm output.

While the particular sources of measured productivity growth will be different in other sectors of the economy, the methodology presented in this paper is quite general and could be applied to quite different industries or sectors. Such breakdowns of measured productivity growth into its source

Table 17.7 US agriculture: various total input and total input productivity indexes, 1940 and 1960 (1947–9 = 100)

	Total input indexes		Total productivity indexes		
	1940	*1960*	*1940*	*1960*	*1960 (1940 = 100)*
Arithmetic indexes:					
1. Official	97	102	85	126	148
2. Official adjusted to cross-sectional output concept	95	109	85	121	142
3. "Corrected" input series (for quality change and deflator bias) and official (adjusted) weights	91	114	89	116	130
4. "Corrected" input series (for quality change *and* capital stock concept); adjusted official weights	93	118	89	112	129
5 "Corrected" inputs (as in 4); production function weights (U17), excluding scale effect (sum = 1.0)	90	124	90	106	118
Geometric indexes:					
6. Official adjusted for output concept	91	100	89	131	148
7. "Corrected" inputs (as in 4), adjusted official weights	90	112	90	118	132
8. "Corrected" inputs, production function weights, excluding scale effect	88	114	93	116	125
9. "Corrected" inputs, production function weights, including scale effect (sum = 1.36)			104	98	94

Sources: 1. From US Department of Agriculture, *Changes on Farm Production and Efficiency*.
2. Weights: row 1 of table 17.2; input series: cols (1) and (2) of table 17.4, except for "Feed, seed, and livestock" and "Output," where cols (3) and (4) were used.
3. Weights: same as in 2; input series: cols (3) and (4) of table 17.4; first row for power and machinery.
4. Same as 3, except that the second set of estimates was used for power and machinery.
5. Weights: row 2 of table 17.2; input series – same as 4. Geometric indexes of productivity computed as antilog of

$$[y - \Sigma a_i x_i],$$

where y is the log of output, x's are the logarithms of the input series and the a_i's are the respective weights.
6. Same sources as 2.
7. Same sources as 4.
8. Same sources as 5.
9. Productivity indexes equal antilog of

$$[y - \Sigma a_i x_i - k (\Sigma a_i x_i - n)],$$

where k is the excess of the sum of the coefficients in the original production function over unity (in this case $k = 0.36$) and n is the logarithm of the index of number of farms. The a_i's used here still sum to unity. The above expression is equivalent to antilog of

$$[(y - n) - \Sigma a_i (1 + k)(x_i - n)].$$

components should prove helpful in forecasting future rates of output growth and in any future attempts to manipulate these rates.

Since this approach goes after particular numbers it is somewhat ad hoc and does not lend itself well to an incorporation into a standard "growth model." It does suggest, however, that the concept of "embodiment" of technical change should be extended to all inputs rather than just to capital[42] or just to labor,[43] that the possibility of disequilibrium and its consequences should play a much larger role in such models, and that some of the more interesting questions raised by such models may eventually resolve themselves into questions concerning the determinants of the demands for and supply of different "qualities" and the feedback mechanism, if any, operating through the changing pattern of "quality" prices.

Acknowledgements

This paper is part of a larger econometric study of sources of productivity growth. I am indebted to both the National Science Foundation and the Ford Foundation for their generous support of this work.

Notes

1 Among others see Robert M. Solow, "Technical change and the aggregate production function," *Review of Economics and Statistics*, XXXIX (1957), 312–20, and Solomon Fabricant, *Basic Facts on Productivity Change* (National Bureau of Economic Research "Occasional Papers," no. 63 (New York: National Bureau of Economic Research, 1959)).

2 All these studies also assume "neutrality," that is, that changes in T do not affect the form of the relationship between Y and X. Some such assumption is necessary because of the latency of T_t.

3 See in particular my "Measuring inputs in agriculture: a critical survey," *Journal of Farm Economics* (Proceedings issue), XLII (December 1960); and "Estimates of the aggregate agricultural production function from cross-sectional data," XLV (May 1963) ibid.

4 The imputation of part of the observed growth in output to the last two sources arises out of the denial of the conventional assumption of equilibrium.

5 In this, my approach is very similar to that taken by Edward F. Denison (*The Sources of Economic Growth in the United States and the Alternative Before Us*. "CED Supplementary Paper" no. 13 (New York: Committee for Economic Development, 1962), except that I concentrate on providing more detailed and, I hope, better documented estimates for a smaller number of factors. My approach is also related to Robert M. Solow's discussion ("Technical progress, capital formation, and economic growth," *American Economic Review*, LII(2) (1962), 76–86, of "embodiment" of technical change in new capital, except that my concept of input quality change allows for embodiment in other inputs besides capital, and the estimation procedure does not require that the rate of embodiment or improvement in quality be constant over the period in question.

6 One could, of course, first compute the conventional measures of technical change and then proceed to explain them on the basis of this same set of factors. I find the direct approach more useful than the two-stage one, although, in principle, the answers should be the same.

7 John W. Kendrick, *Productivity Trends in the United States* (Princeton, NJ: Princeton University Press, 1961).

8 As in Solow, "Technical change and the aggregate production function," op. cit.

9 See K. J. Arrow, M. B. Chenery, B. S. Minhas, and R. M. Solow, "Capital-labor substitution and economic efficiency," *Review of Economics and Statistics*, XLIII(3) (1961).

10 It does not really matter whether we treat some of these as variables to be added to the production function or as variables that modify the "quality" of already included inputs. Thus, one could talk equally well about the services of educational capital or about education as an aspect of labor-force quality. Which is more convenient will depend on the form the data come in and on the interactions with the other variables that we are willing to assume (or assume away).

11 See, for example, Denison op. cit., and my "Measuring inputs in agriculture . . . ," op. cit.

12 See, for example, Theodore W. Schultz, "How efficient is American agriculture?" *Journal of Farm Economics*, XXIX(3) (1947), and the literature cited there.

13 See my "Estimates of the aggregate agricultural production function . . . ," op. cit.

14 These estimates are based on the fitting of a Cobb–Douglas type of equation. Several alternative forms of the production function were also tried but did not lead to any appreciable improvement in the results. A function allowing for cross-sectional differences in the mix of output (crops versus livestock) did fit the data somewhat better. But since the aggregate output mix has not changed much over time, these results are not reported here. For details see the above-cited papers.

15 Both the fit and the coefficients of the equation remain practically unchanged if we substitute labor times education for the two separate variables.

17 Whether this difference is significant can be tested by computing a predicted output for each observation in the sample using the official factor-shares as coefficients of the production function. The R^2 of the factor-shares equation is 0.89 against 0.98 for the estimated production function. The differences in the coefficients between ($U17$) and the factor-shares equation account for about 81 percent of the residual variance left over from the latter equation, indicating that they are "highly significant."

17 Joachim G. Elterlich, private communication (Michigan State University, East Lansing, June 12, 1961).

18 See also E. O. Heady and J. S. Dillon, *Agricultural Production Functions* (Ames: Iowa State University Press, 1961), table 17.1

19 See US Department of Agriculture, *Changes in Farm Production and Efficiency* (Statistical Bulletin no. 233 (rev. July 1961)), table 22.

20 From the US Department of Agriculture, *The Balance Sheet of Agriculture 1961* (Agricultural Information Bulletin no. 247), and *Agricultural Prices, 1961 Annual Summary*, PR 1-3(62). Strictly speaking, we want here not a price index of machinery but a price index of machinery services. This would involve us in a multiplication of these indexes by interest and depreciation rates. Since interest rates rose on balance over this period and the expected life of machinery is unlikely to have lengthened much (given the observed rate of obsolescence), such an adjustment would have resulted in a relatively higher price index of machinery services and hence an even lower increase in the relative price of labor. All this, of course, does not allow for any quality changes in machinery or for that matter, in labor.

21 For additional evidence on the decline of the "share" of labor in agriculture see Vernon W. Ruttan and Thomas T. Stout, "Regional differences in factor shares in American agriculture: 1925-1957," *Journal of Farm Economics*, XLII(1) (1960).

22 See my "Estimates of the aggregate agricultural production function . . . ," op. cit.

23 For a discussion of statistical sources of bias see my "Specification bias in estimates of production functions," *Journal of Farm Economics*, XXXIX (1957), and Irving Hoch, "Simultaneous equation bias in the context of the Cobb–Douglas production function," *Econometrica*, XXIV(4) (1958).

24 The direction of the bias will depend, among other things, on whether the demand elasticity for the particular input is larger or smaller than unity. In the case of the Cobb–Douglas function all these elasticities should be larger than unity, biasing thereby the coefficients of values downward and hence also their sum. This is supported by the finding of a substantially lower coefficient of labor if labor "expense" is used instead of man-years in the regression.

25 For a discussion of this test see George J. Stigler, "The economies of scale," *Journal of Law and Economics*, (1953), 54-71.

26 As estimated by the US Department of Agriculture for the period 1950-9.

27 Here again we would like to have the price of land services rather than the price of farm real estate. Scattered data (of dubious quality) on the ratio of gross cash rent of value (for entire farms rented wholly for cash) for the North Central states indicate that this ratio has not declined from 1950 to date (see various issues of *Farm Cost Situation*). The rise in land prices could, of course, also be due to the capitalization of the effects of the government support programs. All that is claimed in the text is that the subsequent events do not contradict the implications of the production function study. An additional bit of evidence in the same direction is the continued rise in the percentage of total farm purchases made for "farm enlargement" purposes, from about 22 percent of all sales in 1950 to 45 percent of the total in 1960 (see various issues of *Farm Real Estate Situation*).

28 See Lyle P. Fettig, "Price indexes for new farm tractors in the postwar period" (unpublished PhD dissertation, University of Chicago, 1963); and my "Measuring inputs in agriculture . . . ," op. cit., and "Notes on the measurement of price and quality changes," in National Bureau of Economic Research, *Conference on Models of Income Determination* ("Studies

in Income and Wealth" (Princeton, NJ: Princeton University Press, 1963), (forthcoming).

29 A relatively minor adjustment had to be performed on the official output and input series to make them comparable to the concepts used in estimating the cross-sectional production function. There, the output measure includes sales to other farmers and the input measures include purchases from other farms. The official time series exclude, however, interfarm sales of feed, seed, and livestock from their definition of aggregate output and include only the "value added by the non-farm sector" in their input measure. The adjustment consisted of adding the difference between the two feed., seed and livestock concepts (in 1947–9 prices) to both output and input. This had very little effect on the final results.

30 For additional discussion of this problem see my "Notes on the measurement of price and quality changes," op. cit.

31 These estimates are based on the coefficients of cross-sectional regressions of tractor prices on their respective specifications. See Fettig, op. cit., for details.

32 The above adjustments affect only about half of the machinery and power index – the services of the stock of capital component. The other half – maintenance and operation of machinery – was left unadjusted. These series should probably have been also adjusted upward, but not enough data are available to do this adequately. The 1955 Farm Expenditure Survey revealed a 26 percent underestimate of this category of expenditures, but this finding has not been apparently incorporated in the subsequent revisions.

33 See my "Hedonic price indexes for automobiles: an econometric analysis of quality change" (Bureau of the Budget–NBER Price Statistics Review Committee, Staff Report no. 3), in United States Congress, Joint Economic Committee, *Government Price Statistics Hearings . . . January 24, 1961* (Washington, 1961), pp. 173–96; also reprinted as National Bureau of Economic Research, *The Price Statistics of the Federal Government* (General Series no. 73 (New York, 1961)), and "Notes on the measurement of price and quality changes," op. cit. The adjustments suggested in these papers were not applied to this case, in the belief that most of the "qualities" measured there had little relevance for automobiles as a production good (rather than as a consumption good) in agriculture.

34 Repair costs, if adequately measured, may rise with age and should be subtracted from the flow-of-services estimate. The available repair and maintenance statistics do not allow, however, for any growth due to the ageing of the stock of capital equipment.

35 For additional discussion of some of these issues see my "Capital stock investment functions: some problems of concept and measurement," in Don Patinkin (ed.), *Measurement in Economics* ("Studies in Mathematical Economics and Econometrics in Memory of Yehuda Grunfeld" (Stanford, Calif.: Stanford University Press, 1963)).

36 The available data indicate that the *average* life expectancy of farm machinery is about 18 years, and that, at least for tractors, 15-year-old machines are still working as many as 88 percent of the average hours worked by new machines. These figures are based on unpublished tabulations from the 1956

National Farm Machinery Survey. Thus, the assumption of 15 years of relatively constant service does not appear to be extreme.

37 Again, this adjustment affects only the "depreciation and interest" component (about half) of the machinery and power input index.

38 Almost all of this is accounted for by "quality-change" adjustments. The adjustment for differences in capital concept affects only the timing of the estimated "residual," shifting part of the estimated productivity increases from the post- to the pre-1947–9 period.

39 Keeping the same relative weights but using 1.2 as the sum of the coefficients (instead of the estimated 1.36) would have just about accounted for all of the measured productivity increases without leaving an embarrassment of a negative, albeit small, residual. Given the estimated variance-covariance matrix of the coefficients of the production function 1.36 is not significantly different from 1.2 at conventional significance levels.

40 Throughout this investigation, we have tried to adjust inputs for quality change, but we have made no comparable adjustments in the output series. This has been based on the assumption that, in agriculture, the problem of quality change has been much less serious for output measures than for input measures. This assumption has not been investigated, however. The whole question of measuring aggregate agricultural output probably should be also reopened. Once this is done, it is quite likely that we would find that the growth in output has also been underestimated.

41 A possible source of bias may be hidden in our assumption that the various effects are additive. To take advantage of economies of scale, one may have to use a qualitatively different set of inputs. In the cross-sectional study, we did not allow for differences in the quality of inputs used in different areas (except in the case of labor). If larger-scale farms use inputs of higher quality, some of the measured returns to scale may actually be due to quality change. If this is the case, applying the cross-sectional returns to scale estimate to quality-adjusted input time series may involve us in some double counting. Also, some of these economies of scale may be external in the sense that they reflect the lower cost of purchasing certain services in larger bundles. If, for example, larger horsepower tractors are sold at a lower price per horsepower unit and horsepower is the important dimension that enters into the production function, then some of the apparent economies of scale are really external and specific to particular inputs.

42 As in Solow, "Technical progress, capital formation, and economic growth," op. cit.

43 As in Denison, op. cit.

18
Research Expenditures, Education, and the Aggregate Agricultural Production Function*

Two previous papers [5, 6] summarized the results of estimating an aggregate agricultural production function based on 1949 data for 68 regions of the United States. The main findings of that study were the importance and significance of education as a factor affecting output and an indication of the existence of substantial economies of scale in agriculture. The present paper extends this study to cover the years 1949, 1954, and 1959, using per-farm state averages as its units of observation, and introduces explicitly the level of public expenditures on agricultural research and extension (the dissemination of research results) as a variable in the aggregate production function. The results of the current study indicate that these expenditures affect the level of agricultural output "significantly" and that their social rate of return is quite high.

The Approach and the Data

The basic approach of this paper consists of estimating an unrestricted production function of the Cobb–Douglas type, using separate variables for each of the five major input categories, and introducing, in addition, a measure of education per worker and a measure of public expenditures on research and extension (R&E) per farm into the estimating equation. The data used are derived mainly from various US Department of Agriculture publications; they are expressed as per-farm averages for 39 "states,"[1] and deflated by various national price indexes with 1949 (= 100) as a base.[2]

This study differs from the previous one in that it investigates three cross-sections of data simultaneously. This allows it to ask questions which could not be answered by the previous cross-section study. On the other hand, the

* From *American Economic Review*, vol. LIV, no. 6, December 1964.

1949 study was based on data for 68 "productivity regions," while this study has to restrict itself to observing only 39 "states" at any one point of time. The lower number and the conglomerate nature of these states reduce sharply the observed range of variation in some of the "independent" variables and, hence, also the sharpness with which their individual influence on output can be estimated.

This study differs also from the previous one in the definition and measurement of some of the variables. It treats fertilizer as a separate variable (measuring it by weighted plant nutrients rather than expenditures), while combining other current expenditures with expenditures on purchased livestock and feed. It uses also a measure of land in constant prices which is free from contemporaneous influences of trends in product prices or urbanization and a measure of machinery services based on the aggregation of inventory data on individual machines, using prices of "new" machines as weights. The latter procedure results in a machinery-services measure based on "gross-stock" concepts.[3]

A Digression on the Elasticity of Substitution

Before we examine the main results of this study, it is worth while to investigate whether there is strong a priori evidence against the assumption of a unitary elasticity of substitution between labor and all other inputs implicit in the Cobb–Douglas form of the production function. This question can be investigated by fitting a log-linear relationship between output (or value-added) per man and the wage rate. Given the assumption of a constant elasticity of substitution (CES) production function [1], competitive product and factor markets and equilibrium, correct measurement of all the variables, and an exogenous wage rate, the slope coefficient in this equation is an estimate of the elasticity of substitution between labor and all other capital inputs (if value-added is used as the output variable).

The results of fitting such a relationship to various combinations of years and different forms of the equation and data are summarized in table 18.1. The direct estimate of this equation (lines 1a, 2a, 3a) results in coefficients that are somewhat above unity. The other lines in table 18.1 reflect various attempts to deal with possible sourcs of bias in this estimate. An attempt to take into account differences in the quality of labor in different areas (the neglect of which would bias the estimated coefficient towards unity) by introducing a measure of education per man into the equation does reduce the estimated elasticity of substitution, but not consistently or significantly below unity. The possibility that the farm labor market is not in instantaneous equilibrium is investigated by fitting a distributed lag model of the Koyck–Nerlove type (line 3b). The "long run" estimate of the elasticity of substitution is again above unity (but not significantly so). The possibility that the relationship may be due to the effect of omitting some regionally stable

Table 18.1 Estimates of the elasticity of substitution in US agriculture (col. 1) based on data for 39 "states" and the years 1949, 1954, and 1959

Years and method	Wage rate	Education	Time dummies 1954	Time dummies 1959	Lagged value-added per man	R^2	Residual standard error
(1a) 1949–54–9 combined	1.212 (0.068)		0.028 (0.020)	0.046 (0.022)		0.802	0.087
(1b)	0.745 (0.133)	1.476 (0.369)	0.038 (0.019)	0.070 (0.022)		0.827	0.082
(2a) 1949–54	1.324 (0.075)	(0.019)	0.020			0.824	0.080
(2b)	1.005 (0.160)	0.995 (0.445)	0.027 (0.019)			0.835	0.078
(3a) 1954–9	1.226 (0.087)			0.017 (0.022)		0.752	0.092
(3b)	0.465[a] (0.129)			0.010 (0.020)	0.600 (0.086)	0.850	0.072
(4) First differences: 1949–54 and 1954–9 combined	1.140 (0.250)			−0.010 (0.020)		0.221	0.072
(5a) 1954	1.131 (0.240)	1.029 (0.691)				0.851	0.078
(5b) 1954 I.V.	1.041 (0.299)	1.265 (0.843)				0.820	0.086

The basic equation estimated is of the form $\log (VA/L) = a + b \log W + Zc + u$, where VA is "real" value-added equaling sales plus inventory change plus home consumption plus government payments minus all current purchases of goods and services (except labor) from both outside and inside the sector, all individually deflated (1949 = 100) before subtraction.

L is a man-days-worked measure of labor.

W is the "real" wage rate per hour: composite hourly wage rate for hired farm labor, deflated in each state separately by the implicit output deflator.

Zc is a shorthand notation for a matrix of other variables and the associated coefficient vector, u is a disturbance, usually assumed to be random and uncorrelated with the "independent" variables.

"Education" is a measure of formal schooling per male in the rural farm population, constructed by weighting the distribution of males by school years completed within each state by all US mean incomes in each of these categories. See the data appendix and [6] for more details on the derivation of this and other variables.

"Time dummies" are shift variables that take the value of one for all states in their "reference" year and zero in all the others. Their introduction allows each cross-section to have a mean level (intercept) of its own, while preserving the equality of slope coefficients. They allow for common time trends (and measurement errors) in the mean level of all the variables. Under the assumption of the CES model and neutral technical change, the coefficients of the dummies are an estimate of $(1 - \sigma)\lambda_t$, where σ is the elasticity of substitution and $\lambda_t t$ is the rate of technical advance relative to the base period (in this case 1949). If one assumes that $\lambda > 0$, the slightly positive coefficients imply a σ of somewhat less than unity.

Lagged value-added per man is the lagged value of the dependent variable from the preceding cross-section. That is, the lag is five years.

I.V. are estimates based on the method of instrumental variables, using the lagged $(t-5)$ wage rate instead of the current one as an instrumental variable.

The number of observation equals 117 for set 1, 78 each for sets 2, 3, and 4, and 39 for set 5. The numbers in parentheses are the estimated standard errors of the respective coefficients.

[a] The implied "long run" estimate of $\sigma = 0.465/(1 - 0.600) = 1.162$.

variables is investigated by fitting the same equation in the form of first differences, relating *changes* between cross-sections in output per man to changes in the wage rate (line 4). Finally, the possibility that the results may be due to simultaneous-equations bias, since the farm wage rate may in its turn be affected by the same forces that determine the level of output per man, is investigated by fitting the same equation, using the method of instrumental variables (and the lagged real wage rate as the instrumental variable) instead of the method of least squares (lines 5a and 5b). Neither of these alternative forms of the model results in estimates of the elasticity of substitution which are significantly different from unity. Thus, at this level of aggregation, there does not seem to be any strong prima facie evidence against the Cobb–Douglas form, which will therefore be used in the subsequent discussion.

The Main Results

The main results of fitting a Cobb–Douglas-type production function to data on per-farm output and input averages for 39 states are summarized in table 18.2. Preliminary investigations showed that estimates for different cross-sections do not differ significantly and, hence, most of the discussion that follows will be based on results obtained by combining the data from several cross-sections and imposing the assumption of constancy of coefficients throughout the 1949–59 period.[4]

The main findings of the previous study are confirmed by the results of this study. The education of the farm labor force is again a "significant" variable, entering the equation with a coefficient which is not significantly different from that of the labor variable.[5] Thus one can combine both the quantity and quality dimensions of labor into one variable, which simplifies some of the subsequent analysis. Again, this study finds evidence of substantial economies of scale. The sum of the estimated coefficients (excluding those of education, research and extension, and time dummies) is consistently above unity with 1.2 being a reasonable point estimate.[6] Nor is this finding contradicted when an attempt is made to guard against the possibility of simultaneous-equations bias using the method of instrumental variables to estimate the coefficients of the production function (col. 7 of table 18.2). In general, the results of using the instrumental variables approach for both the 1954 and 1959 cross-sections (the latter results are not reproduced here) differ little from those based on single-equation least-squares procedures.[7]

The main novelty of this study is the introduction of a variable to reflect the contribution of public expenditures on agricultural research and on the dissemination of its results on the level of agricultural productivity in different states. The measure used is quite crude: it is the sum of all the relevant entries in the annual budget of the respective state agricultural experiment stations and extension services. To allow for some lags in the effect of these expenditures, this variable was defined as an average of the flow of expenditures

Table 18.2 Estimates of the aggregate agricultural production function, US, 1949-54-9

Coefficients of	1949-54-9			1949-54	1954-9	1954	I.V.
"Other"	0.393	0.367	0.342	0.312	0.371	.345	0.349
	(0.024)	(0.028)	(0.027)	(0.035)	(0.034)	(0.056)	(0.082)
Land and building	0.152	0.146	0.145	0.186	0.119	0.157	0.158
	(0.022)	(0.022)	(0.021)	(0.028)	(0.026)	(0.043)	(0.058)
Fertilizer	0.107	0.100	0.095	0.101	0.120	0.140	0.144
	(0.012)	(0.013)	(0.013)	(0.016)	(0.019)	(0.030)	(0.047)
Machinery	0.200	0.158	0.164	0.188	0.185	0.241	0.250
	(0.046)	(0.037)	(0.034)	(0.045)	(0.043)	(0.072)	(0.104)
Labor	0.426	0.511					
	(0.051)	(0.060)					
Education		0.405					
		(0.161)					
Labor × education			0.448	0.415	0.398	0.364	0.311
			(0.063)	(0.084)	(0.077)	(0.123)	(0.189)
R&E			0.059	0.052	0.062	0.044	0.050
			(0.021)	(0.027)	(0.027)	(0.043)	(0.061)
$T54$	−0.011	−0.003	−0.017	−0.019			
	(0.011)	(0.011)	(0.012)	(0.014)			
$T59$	0.005	0.019	−0.006		−0.002		
	(0.014)	(0.015)	(0.017)		(0.020)		
R^2	0.980	0.981	0.983	0.981	0.980	0.977	0.959
SE	0.036	0.035	0.034	0.035	0.035	0.039	0.052
Sum of coefficients	1.278	1.282	1.197	1.202	1.193	1.247	1.212

The equation estimated is linear in the logarithms of the variables (except for the time dummies). The output and input variables are per-farm state averages. A total of 39 "states" is used in each cross section, some of these "states" being a combination of several smaller adjoining states (see note 1 for details).

Output equals value of sales, inventory change, home consumption, and government payments. The separate components are deflated in being by US price indexes with 1949 = 100.

Land and Buildings: The value of buildings is computed as the product of the percentage of value that is due to buildings and the total value of land and buildings per farm and and deflated by an index of building materials prices with 1949 = 100. "Land" is a constant price volume measure constructed by David Boyne [2], using 1940 relative prices of different categories of land in each state and the changing number of acres in these categories in the subsequent years. It is inflated to add up in 1949 (for the US total) to the total value of land and buildings. The resulting measure of land takes into account quality differences in land at 1940 relative state prices, but does not allow subsequent appreciation due to produce price trends and other changes to be reflected in the "quantity" measure. The result is a measure of the *stock* of land and buildings per farm.

Machinery: A measure of the stock of machinery is constructed by valuing a relatively detailed list of machines on farms by the average US price for such new machines in 1949, allowing for some regional differences in the average price of a machine, due to regional differences in the average size of machines, where data were available. The result is an estimate of the gross (underpreciated) stock of machines on farms in 1949 prices. This is converted into a flow measure by multiplying it by 0.15 (the fraction that flow is of gross stock under the assumption of an in the previous year and the level 6 years previously. Thus, the average of 1958 and 1953 is used in the 1959 cross-section, the average of 1953 and 1948 is used in the 1954 cross-section, and the average of 1948 and 1945 (since the data for 1943 were not strictly comparable) is used in the 1949 cross-section. Given the crudeness of this measure, it is surprising that it is as significant as it is. Also, its coefficient remains remarkably stable when cross-sections are added or subtracted, and when other variables are introduced or the measurement and definition of included variables are changed. While the estimated coefficient of R&E may appear to be rather small – a doubling in the level of public R&E expenditures per farm would lead to an increase in output of only about 5 percent – the implied absolute effect is very large. The average (geometric) public expenditure on R&E during the whole 1949–54–9 period was only about $32 per farm per year, while the average gross output per farm (in 1949 prices) was $7,205 per year during the same period. Using 0.059 for the coefficient of R&E (from column 3, table 18.2) leads to an estimate of its marginal product of $0.059 \times 7,205/32$ or approximately 13 dollars of output per year for an additional dollar of R&E expenditures per year. This finding implies the fantastically high gross rate of return of about 1,300 percent for social investment in agricultural research and extension. Even if one allows that much of it is the result of research

---

average length of life of 10 years and an 8 percent rate of interest) and added to the deflated values of gasoline, oil, and other machinery operation and repair expenditures to arrive at an estimate of the total flow of services associated with the use of farm machinery.

Fertilizer: weighted plant nutrients. The sum of nitrogen, phosphoric acid, and potash consumption weighted by 1955 relative prices for these ingredients (1.62, 0.93, and 0.45 respectively).

Other: The sum of deflated expenditures on purchased feed and livestock, interest on the livestock inventory (at 8 percent), and expenditures on seed and other current inputs not elsewhere classified.

Labor: A measure of labor-days-worked constructed by adding together total expenditures on hired workers divided by the average wage rate per day and the number of family workers times the average number of days worked on the farm (constructed by subtracting from 300 the estimated number of days worked off the farm by operators). Additional adjustments were made for operators over 65 years of age (they were entered with a weight of 0.6) and unpaid family workers (who were entered with a weight of 0.65).

Education: weighted school years per man (25 years and older in the rural farm population), using 1950 mean incomes for the respective education categories (for all US males) as weights. See [6] for additional details on the construction of this variable.

Research and Extension: undeflated sum of total expenditures on research and extension by the respective agricultural experiment stations and extension services based on all sources of funds (including USDA appropriations). Averages for the previous year and 5 years previously, except for the 1949 cross-section where these expenditures are averages for 1948 and 1945.

Time dummies take the value of unity for observations in the specified year and zero for all the other observations.

I.V.: coefficients estimated by using the method of instrumental variables, assuming that all the "current" input variables ("other," fertilizer, and labor) are endogenous, and by using the current wage rate and the lagged values of these inputs from the previous cross-section (i.e. at $t-5$) as instrumental variables.

The "sum of the coefficients" excludes the coefficients of the education, research and extension, and time-dummies variables.

expenditures by private firms (mainly in the agricultural supplies industries),[8] and that, due to our inability to solve the agricultural problem, the social value of additional agricultural output is only about half of its market value, the gross social rate of return to R&E expenditures is still about 300 percent. This last figure is of the same order of magnitude as a previous estimate (in [4]) of a 35 to 170 percent *net* social rate of return to agricultural research based on entirely different data and a different approach.

The estimated average "marginal products" of all the inputs in 1949 and 1959 are presented in table 18.3. In general, they do not appear to conflict with other information that we have on the economic conditions and economic history of this industry. For the inputs that are measured in units of dollars per year, one would expect marginal productivity coefficients on the order of 1.2, given the previous finding of economies of scale of about this order and the assumption that the marginal products, while not equal, are still proportional to marginal costs.[9] This is, in fact, approximately the order of magnitude of the numbers in cols 1, 4, and 6 of table 18.3. The only two columns that seem to imply substantial disequilibrium at both the beginning and the end of the period are the fertilizer and the research and extension columns. A ton of weighted plant nutrients cost about $250 throughout most of this period, implying a ratio of the marginal product to factor price between 3 and 5. Farmers have not remained idle in the face of such a disequilibrium. Between 1949 and 1959 fertilizer consumption grew at the tremendous rate percent per year (more than doubling fertilizer consumption during this period) and continued to grow at the rate of 5.6 percent per year between 1959 and 1962. The estimated equilibrium gap (VMP/factor price) has declined from about 5 in 1949 to 2.7 in 1959 and 2.4 in 1962 (using the estimated 1954–9 coefficient). But even if fertilizer consumption continues to grow at about the same rate, relative to output, as it did during the 1959–62 period (4 percent per year) the disequilibrium is not likely to disappear before the early 1980s unless there are some new

Table 18.3   Estimated aggregate marginal products in US agriculture (in 1949 prices) in 1949 and 1959

			Marginal product of				
Year	"Other"	Land and buildings	Fertilizer	Machinery	Labor	Education	R&E
1949	1.22	0.12	1267	1.26	5.76	0.85	16
	(0.14)	(0.02)	(200)	(0.30)	(1.17)	(0.17)	(8)
1959	1.25	0.10	697	1.18	10.43	1.30	10
	(0.11)	(0.02)	(110)	(0.27)	(2.02)	(0.25)	(4)

Computed by multiplying the appropriate geometric average output ratio by the respective input coefficient from table 18.2 (col. 4 for 1949 and col. 5 for 1959). The numbers in brackets are a crude translation of the estimated standard errors from table 18.2 into comparable units. The units are (from right to left): dollars per dollar (both per year) for cols 1, 4, 6, and 7; dollars per year per dollar or percent per year for col. 2; dollars per weighted plant nutrient ton for col. 3; and dollars per day for col. 5.

price and technological developments. At any rate, the regression estimates confirm the existence of this disequilibrium, while the observed market behavior reflects the producers' attempt to reduce it.

Public research and extension expenditures are, of course, outside the farmers' control (at least directly). Thus there is no obvious way for them to eliminate this disequilibrium except through individual attempts to acquire more of the products of this activity, which are largely free to them (except for the nonnegligible monetary and psychological cost of doing so). But even if we make the fourfold downward adjustment discussed above and also adjust for the decline in farm product prices since 1949 (by 4 percent), we are still left with a social disequilibrium gap of over 2 (or a *net* social rate of return of more than 100 percent). Thus, perhaps, we should not be so quick to cut the agricultural research appropriations in our attempt to solve the farm problem.

The education variable is measured in dollars per year that a comparable average education mix would fetch in the nonfarm sector. When detailed adjustments are made to bring the farm and nonfarm labor force measures into comparable units, as in [7], it is found that a ratio of about 0.7 of farm to urban per capita income is consistent with equal real returns for comparable labor. Thus, a marginal product of about 0.7 or so, which would be in units of farm dollars per urban dollars, would indicate an approximate equilibrium with respect to this variable. Since the estimated coefficients are higher than that, though probably not significantly so, there is some indication that the marginal return to quality of labor in agriculture may actually exceed its opportunity cost in the nonagricultural sector.

One of the findings of the previous study was that the estimated labor coefficient appeared to be lower and the coefficient of machinery higher than the coefficients that would be implied by the factor-shares approach. This finding is not confirmed by this study. When the estimated factor shares are adjusted for the difference in the concept of output used in this study and the estimated production function coefficients are divided by their sum (to add up to unity), the two sets of coefficients are quite close to each other, and the differences that remain are not statistically significant. Thus, while the previous findings would have implied a marginal product of labor below the going wage rate, when the estimates in table 18.2 are adjusted for economies of scale and for the minor decline in farm-product prices between 1949 and 1959, the resulting marginal product of labor estimates of $4.80 and $8.30 per day are not significantly different from the US average wage rate of hired farm workers of $4.40 and $6.50 per day in 1949 and 1959 respectively.[10]

If all the variables were measured correctly, the coefficients of the time dummies would be estimates of the rate of "disembodied" technical change (relative to 1949). In practice, however, they also reflect errors of measurement in the average level of variables. Hence, the fact that they are almost never significantly different from zero may be the result of underdeflating some of the input series (or, for example, not deflating the

Table 18.4 Sources of the "residual": US agriculture, 1949–59

1. *Residual*	*Percentage change*
(a) Official, adjusted	28
(b) This study	−1
2. *Sources of difference:*	*Common logarithms*
(a) Difference in output measure	0.0096
(b) Difference in weights	0.0121
(c) Difference in conventional input measures	0.0105
(d) Contribution of education	0.0146
(e) Contribution of R&E	0.0297
(f) Economies of scale in conventional inputs	0.0320
Sum	0.1085
Residual as in 1a	0.1065
Difference (−)	0.0020

1a. From [8]. Adjusted for difference in output concept, and inputs weighted geometrically.
1b. Computed using the weights from column 3 of table 18,2 and the differences in the geometric means of all the variables between 1949 and 1959.
2b. Adjusting the regression weights to add up to unity. Logarithms are used in the second half of the table, since the items are not additive otherwise.

R&E expenditures at all). On the other hand, an attempt has been made to incorporate as many adjustments for quality change as possible into the measurement of inputs and, hence, one might expect that only a very small role would be assigned to disembodied technical change. It would be beyond the scope of this paper to repeat the analysis of sources of growth presented in [6]. But if one uses the estimated coefficients from table 18.2 and data on input and output change over time, it is possible to account for *all* of the observed growth in agricultural output without invoking the unexplained concept of (residual) technical change. Table 18.4 summarizes an analysis of the "residual" which parallels closely the results presented previously in tables 4 and 7 of [6]. As against the previous study, the current results reduce somewhat the role of economies of scale, increase somewhat the role of education and other input quality change, and assign for the first time a substantial role to the previously unmeasured contribution of public investment in agricultural research and extension to the explanation of the growth in the aggregate output of agriculture. Taking items 2c through 2f as what one could have called the "residual" in this study, the total can be divided into three roughly equal parts: the contribution of input quality change (in labor and other inputs), of economies of scale, and of investments in research and extension.

Conclusions

"Productivity," "technical change," or the "residual" are usually computed residually, as part of some accounting framework. Because of the nature of the beast, it is impossible to test *accounting* frameworks statistically. One may discuss the "usefulness" of various accounting schemes, as I have done in [6], where I suggest using a framework that *minimizes* the role of the unexplained residual, but the argument eventually reduces itself to aesthetics and research strategy. It is possible, however, to test some of the components of an accounting scheme separately if data which these concepts do not have to fit tautologically are used. The empirical work on aggregate production functions reported here has been motivated by a desire to gather statistical evidence for various theoretically justifiable adjustments to the conventional accounting framework. The study has been designed to investigate whether there is evidence for introducing a measure of education and a measure of research investment into the accounting scheme for growth and whether there is evidence for relaxing the assumption of constant returns to scale. None of these questions can be answered unequivocally by looking only at one piece of data. They are too interdependent. This is why one must repeat the analysis on different sets of data, some of them being better suited for illuminating one or another aspect of the problem.

In this paper I have reproduced some of the previously reported results on a new set of data. "Education" does belong in the production function, and there do appear to be substantial economies of scale in agriculture. A new variable, public investment in research and extension, is introduced and found to be both "significant" and important as a source of aggregate output growth. Using the estimating coefficients as weights, and the expanded and adjusted list of inputs, one can again account for most of the observed growth without being left with a large unexplained "residual." None of these conclusions is very firmly established, and some may be subject to substantial bias,[11] but the only known way of either confirming or disproving them is the slow and expensive but cumulative process of conducting additional studies of this type on different bodies of data.

Appendix: Data Sources

A complete description of the data used in this study will be given in a publication presenting a more extended version of the results of this study. Only the most important adjustments are described below.

Most of the data on output and input by states are taken from USDA, *Farm Income: A Supplement to the Farm Income Situation for July 1961, State Estimates 1949–1960*, Washington, August 1961. Output is defined as the sum of cash receipts from farm marketings, government payments, value of home consumption, and net change in farm inventories. The first item was divided into 12 commodity groups (such as meat animals, dairy products, poultry and eggs, feed crops, food grains, etc.) and each was deflated separately by an appropriate US price index (1949 = 100). In a few cases where one particular commodity accounted for a large fraction of total receipts in a state, the deflation was done on an even finer commodity level. The price indexes used are components of the Prices Received by Farmers Index and were taken from various issues of *Agricultural Prices*. The resulting procedure produced an implicit farm-marketings deflator for each state. This deflator was used in turn to deflate the government payments, value of home consumption, and net inventory change categories.

The following items were also taken from the *Farm Income* publication: Current expenses on (1) feed, (2) livestock, (3) seed, (4) repairs and operation of capital items (excluding buildings), (5) miscellaneous, and (6) hired labor. Unpublished data supplied by the US Department of Agriculture were used to exclude the building component from (4). Each of these components was deflated separately by the appropriate US subindex of the Prices Paid by Farmers Index taken from USDA, *Prices Paid by Farmers . . . , 1910–1960*, Stat. Bull. no. 319; except for (6) – hired labor – which was "deflated" separately by dividing it through by the average wage rate per day paid to hired farm labor in *each* state to arrive at a "days worked" measure. These wage rates and the hourly ones used explicitly as a variable in the regressions summarized in table 18.1 were taken from various issues of *Farm Labor*.

The "other" inputs variable was constructed as the sum of items (1), (2), (3), (5) and interest (8 percent) on the constant price value of livestock and crop inventories on farms. The latter item was estimated by multiplying the physical data on the inventory of different types of livestock and different crops stored on farms by their average 1949–US prices.

The construction of the land and buildings, machinery, and fertilizer variables is described in the notes to table 18.2. The data on the inventory of machines on farms come from USDA, *Numbers of Selected Machines and Equipment on Farms*, Stat. Bull. 258, 1960, and the respective volumes of the *Census of Agriculture* for 1949, 1954, and 1959. The prices used to value the inventory were derived largely from USDA, *Farmers' Expenditures for Motor Vehicles and Machinery . . . 1955*, Stat. Bull. 243, 1959, and converted to 1949 levels by using the appropriate series from the previously cited *Prices Paid* publication. The data on plant-nutrient use by states were taken from various reports of the Soil and Water Conservation Branch of the USDA.

The labor measure, in days, equals

$$D\{N(1-0.4A)+0.65(F-N)\}+HE/W$$

where D is the estimated average days worked on farms by operators (estimated by subtracting from 300 the calculated average number of days worked off the farm by operators), N is the number of farms, assumed also to equal the number of farm operators, A is the fraction of operators above age 65, F is the total number of family workers, $N-F$ is thus the estimated number of unpaid family workers (set to zero in the few cases where $F-N$), and HE/W is the estimated number of hired worker man-days (hired worker expense divided by the wage rate per day). N and F are taken from various issues of *Farm Labor*. "*A*" and the data on off-farm work come from the *Census of Agriculture* volumes for the respective years. The implicit weighting of operators over 65 at 0.6 and of unpaid family workers at 0.65 is the same as in [6] and is based on E. G. Strand and E. O. Heady, *Productivity of Resources Used on Commercial Farms*, USDA Tech. Bull. 1128, November 1955.

The education variable was based on the distribution of the male rural farm population by school years completed taken from the 1950 and 1960 *Census of Population*. These distributions were weighted by the mean US income for males (over 25) in 1950, based on 1950 Census of Population data. See [6] for additional details. 1954 was approximated by a simple interpolation 1950 and 1960.

The data on research and extension expenditures are taken largely from Dana G. Dalrymple, *State Appropriations for Agricultural Research and Extension*, Mich. State Univ. Department of Ag. Econ., A.E. 852, Dec. 1961 and the annual reports on "Cooperative Extension Work in Agriculture . . ." and "Agricultural Experiment Stations" by the USDA Office of Agricultural Experiment Stations.

All variables, except education, are divided by N to arrive at per-farm units.

Acknowledgements

This paper was written during the author's tenure as a Ford Foundation Faculty Research Fellow at the Econometric Institute, Rotterdam. It is part of a larger study of the econometrics of technical change supported by the National Science Foundation.

Notes

1 Because of lack of data (mainly for the labor variable), some of the smaller states were aggregated into larger units. Thus all of New England was treated as one observation, and so also were Delaware and Maryland, New Mexico and Arizona, and Wyoming–Utah–Nevada.

2 See the data appendix for a more detailed discussion of definitions and sources.

3 See [6] for a more detailed discussion of problems associated with the measurement of machinery services for productivity comparisons.

4 Equations comparable to those reported in cols 3–6 of table 18.2, but without the time dummies, were computed for each of the three cross-sections separately. Using the standard normality and independence assumptions, one could not reject the null hypothesis that the separate estimates for 1949 and 1954 did not differ "significantly" (at conventional significance levels) from the joint estimates for 1949–54. The same was also true for the 1954 and 1959 with 1954–9 comparison. (The computed F statistic is 1.1 for the 1949–54 comparison and 2.0 for the 1954–9 comparison. Given the usual assumptions the "critical" value of the F statistic, with 7 and 64 degrees of freedom, is about 2.3 at the 0.05 and 3.5 at the 0.01 significance levels.) The results are somewhat weaker but similar for the overall comparison of all three separate

equations with the pooled (but without time dummies) 1949–54–9 equations (The computed F statistic is now 2.1, while the "critical" value of this statistic, with 14 and 96 degrees of freedom, is about 2.0 at the 0.05 and 2.4 at the 0.01 level.) See [3, ch. 6] for details of such tests.

5 The computed F statistic for the test of equality of the two coefficients for the 1954–9 cross-section is 2.3. The "critical" value of F with 1 and 70 degrees of freedom is at least 4.0, even at the 0.05 level. The results for other cross-sections are similar.

6 Taking the lower estimate of the sum of the coefficients in column 3 of table 18.2, we can construct a test for its difference from unity by computing the variance of this sum. In this case, it is just the sum of the variances plus twice the sum of the covariances of all the relevant coefficients. The standard deviation of the sum of the coefficients is 0.0434. Under the null hypothesis and the standard normality and independence assumptions, the ratio of the difference of the sum from unity to its standard error $0.19715/0.0434 = 4.54$ has the "t" distribution with 108 degrees of freedom. The probability of observing a ratio as high as 4.54 if the true sum were equal to unity would be less than one in ten thousand. Thus the excess over unity is "highly significant."

7 Since lagged inputs are used as instrumental variables, the implicit reduced forms have very high R^2's, and thus the substitution of a "predicted" for the actual input series (in two-stage least-squares language) makes very little difference.

8 Assuming that private expenditures are perfectly correlated with public expenditures and are of about the same order of magnitude (see [4] for some support for the last statement).

9 If there are economies of scale, the adding-up or product-exhaustion theorem doesn't hold. Factors cannot all be paid their marginal product, since this would overexhaust the available total.

10 The differences that remain may be due to the lower-than-average quality of the workers (the hired ones) for whom the market price is measured. Actually, about two-thirds of the total work is done by family members, whose average education and also motivation are higher than those of the hired farm labor force.

11 The finding of economies of scale is particularly susceptible to bias due to incomplete or incorrect accounting. If not all of the inputs are adjusted for quality differences, and they are not, and if larger size farms tend to use higher quality inputs, then some of what is measured as economies of scale are actually due to differences in the quality of inputs. Also, these economies may be the reflection of discontinuities in the range of available machine sizes with the resultant nonhomogeneity of the production funciton. In either case, the accounting is still correct, but the interpretation of a particular category is less firm.

References

[1] K. J. Arrow, M. B. Chenery, B. S. Minhas, and R. M. Solow, "Capital-labor substitution and economic efficiency," Rev. Econ. Stat., Aug. 1961, 43, 225–50.

[2] D. H. Boyne, "Changes in the Real Wealth Position of Owners of Agricultural Assets, 1940-1960." Unpublished doctoral dissertation, Univ. Chicago, 1962.

[3] F. A. Graybil, *An Introduction to Linear Statistical Models*, vol. 1. New York, 1961.

[4] Z. Griliches, "Research costs and social returns: hybrid corn and related innovations," *Jour. Pol. Econ.,* Oct. 1958, 66, 414-31.

[5] ——, "Estimates of the aggregate agricultural production function from cross-sectional data," *Jour. Farm Econ.*, May 1963, 45, 419-28.

[6] ——, "The sources of measured productivity growth: US agriculture, 1940-1960," *Jour. Pol. Econ.*, Aug. 1963, 71, 331-46.

[7] D. Gale Johnson, "Labor Mobility and Agricultural Adjustment," in Heady et al. (eds), *Agricultural Adjustment Problems in a Growing Economy*, Ames, Iowa, 1958.

[8] US Department of Agriculture, "Changes in farm production and efficiency," Stat. Bull. no. 233, Washington; rev. July 1963.

19
Postscript on Agricultural and other Production Functions

The work on agricultural production functions and the role of education and public investments in agricultural research was continued by a number of my students and also other researchers. Evenson (1968), Peterson (1966), Lattimer (1964), Evenson and Welch (1974), and Evenson and Kislev (1975) investigated the time-shape of the effects of public research in agriculture and the measurement of their regional spillovers. Much of this and related work is reviewed in Evenson (1982). The role of schooling in agricultural production, and related specification and aggregation issues, were the topics of dissertations by Kislev (1965) and Fane (1972), and later work by Huffman (1974, 1977). See also the review of later work in Jamison and Lau (1982). Related work, on an international scale, was pursued also by Hayami and Ruttan (1985) and Binswanger and Ruttan (1978). The importance of schooling levels and public research investments in agriculture for agricultural productivity has by now been widely accepted as a well-documented social science "fact."

The issue of economies of scale is much more controversial. I myself pursued this topic further, using both US and Norwegian manufacturing data (Griliches, 1967b, 1968, and Griliches and Ringstad, 1971), so also did some of my students (Krishna, 1967; Hodgins, 1968). More recently, Kislev and Peterson (1982 and 1986) have questioned my conclusions about the importance of economies of scale in agriculture, arguing that it is primarily the result of technological advances in farm equipment which have favored larger-scale enterprises. A similar debate was conducted recently in regard to the presence or absence of economies of scale in the Bell telephone system in the US (see, for example, Evans 1983). There is also work on technical change in electricity-generating equipment, which stresses the intimate connection between scale and technology (see Belinfante, 1978; Fuss, 1977; McFadden, 1978a, for examples). The empirical evidence on this range of issues is not in very good shape, however, and much more useful work could be done in this area.

There is also the issue of functional form for the estimated production functions and the associated productivity computations. I could never take this range of issues seriously. The quality of the ingredients was always much more important than the specific characteristics of the oven used to bake a cake out of them. I discuss some of these issues in Griliches (1967b) and elsewhere. For a recent review of the theory and estimation of such relations see Diewert (1986) and Jorgenson (1986).

20
The Explanation of Productivity Change*

(With D. W. Jorgenson)

But part of the job of economics is weeding out errors.
That is much harder than making them, but also more
fun (R. M. Solow).

Introduction

Measurement of total factor productivity is based on the economic
theory of production. For this purpose the theory consists of a production
function with constant returns to scale together with the necessary
conditions for producer equilibrium. Quantities of output and input
entering the production function are identified with real product and
real factor input as measured for social accounting purposes. Marginal rates
of substitution are identified with the corresponding price ratios. Employing
data on both quantities and prices, movements along the production
function may be separated from shifts in the production function.
Shifts in the production function are identified with changes in total factor
productivity.

Our point of departure is that the economic theory underlying the
measurement of real product and real factor input has not been fully
exploited. As a result a number of significant errors of measurement have
been made in compiling data on the growth of real product and the growth
of real factor input. The result of these errors is to introduce serious biases
in the measurement of total factor productivity. The allocation of changes
in real product and real factor input between movements along a given
production function and shifts of the production function must be corrected
for bias due to errors of concept and measurement.

* Reprinted with corrections from *Review of Economic Studies*, vol. XXXIV(3), no. 99, July 1967.

The purpose of this paper is to examine a hypothesis concerning the explanation of changes in total factor productivity. This hypothesis may be stated in two alternative and equivalent ways. In the terminology of the theory of production, if quantities of output and input are measured accurately, growth in total output is largely explained by growth in total input. Associated with the theory of production is a system of social accounts for real product and real factor input. The rate of growth of total factor productivity is the difference between the rate of growth of real product and the rate of growth of real factor input. Within the framework of social accounting the hypothesis is that if real product and real factor input are accurately accounted for, the observed growth in total factor productivity is negligible.

We must emphasize that our hypothesis concerning the explanation of real output is testable. By far the largest portion of the literature on total factor productivity is devoted to problems of measurement rather than to problems of explanation. In recognition of this fact changes in total factor productivity have been given such labels as The Residual or The Measure of Our Ignorance. Identification of measured growth in total factor productivity with embodied or disembodied technical change provides methods for measuring technical change, but provides no genuine explanation of the underlying changes in real output and input.[1] Simply relabeling these changes as Technical Progress or Advance of Knowledge leaves the problem of explaining growth in total output unsolved.

The plan of this paper is as follows: We first discuss the definition of changes in total factor productivity from the point of view of the economic theory of production. Second, we provide operational definitions for the measurement of prices and quantities that enter into the economic theory of production. These definitions generate a system of social accounts for real product and real factor input and for the measurement of total factor productivity. Within this system we provide an operational definition of total factor productivity. This definition is fundamental to an empirical test of the hypothesis that if real product and real factor input are accurately accounted for, the observed rate of growth of total factor productivity is negligible.

Within our system of social accounts for real product and real factor input we can assess the consequences of errors of measurement that arise from conceptual errors in the separation of the value of transactions into price and quantity. Errors in making this separation may affect real product, real factor input, or both; for example, an error in the measurement of the price of investment goods results in a bias in total output and a bias in the capital accounts that underlie the measurement of total input. Within this system of social accounts we can suggest principles for correct aggregation of inputs and outputs and indicate the consequences of incorrect aggregation. Many of the most important errors of measurement in previous compilations of data on real product and real factor input arise from incorrect aggregation.

Given a system of social accounts for the measurement of total fac-tor productivity we attempt to correct a number of common errors of measurement of real product and real factor input by introducing data that correspond more accurately to the concepts of output and input of the economic theory of production. After correcting for errors of measurement we examine the validity of our hypothesis concerning changes in total factor productivity. We conclude with an evaluation of past research and a discussion of implications of our findings for further research.

Theory

Our definition of changes in total factor productivity is the conventional one. The rate of growth of total factor productivity is defined as the difference between the rate of growth of real product and the rate of growth of real factor input. The rates of growth of real product and real factor input are defined, in turn, as weighted averages of the rates of growth of individual products and factors. The weights are relative shares of each product in the value of total output and of each factor in the value of total input. If a production function has constant returns to scale and if all marginal rates of substitution are equal to the corresponding price ratios, a change in total factor productivity may be identified with a shift in the production function. Changes in real product and real factor input not accompanied by a change in total factor productivity may be identified with movements along a production function.

Our definition of change in total factor productivity is the same as that suggested by Abramovitz [1], namely, "the effect of 'costless' advances in applied technology managerial efficiency, and industrial organization (cost – the employment of scarce resources with alternative uses – is, after all, the touchstone of an 'input') . . ."[2] Of course, changes in total factor productivity or shifts in a given production function may be accompanied by movements along a production function. For example, changes in applied technology may be associated with the construction of new types of capital equipment. The alteration in patterns of productive activity must be separated into the part which is "costless", representing a shift in the production function, and the part which represents the employment of scarce resources with alternative uses, representing movements along the production function.

On the output side the quantities that enter into the economic theory of production correspond to real product as measured for the purposes of social accounting. Similarly, on the input side these quantities correspond to real factor input, also as measured for the purposes of social accounting. The prices that enter the economic theory of production are identified with the implicit deflators that underlie conversion of the value of total output and total input into real terms. The notion of real product is a familiar one to social accountants and has been adopted by most Western countries as the appropriate measure of the level of aggregate economic activity. The notion of real factor input is somewhat less familiar, since social accounting for factor input is usually carried out only in value terms or current prices.

However, it is obvious that income streams recorded in value terms correspond to transactions in the services of productive factors. The value of these transactions may be separated into price and quantity and the resulting data may be employed to construct social accounts for factor input in constant prices. This type of social accounting is implicit in all attempts to measure total factor productivity.

The prices and quantities that enter into the economic theory of production will be given in terms of social accounts for total output and total input in current and constant prices. We observe that our measurement of total factor productivity is subject to all the well-known limitations of social accounting. Only the results of economic activities with some counterpart in market transactions are included in the accounts. No attempt is made to measure social benefits or social costs if these diverge from the corresponding private benefits or private costs. Throughout this study we adhere to the basic framework of social accounting. The measurement of both output and input is based entirely on market transactions; all prices reflect private benefits and private costs. That part of any alteration in the pattern of productive activity that is "costless" from the point of view of market transactions is attributed to change in total factor productivity. Thus the social accounting framework provides a definition of total factor productivity as the ratio of real product to real factor input.

To represent the system of social accounts that provides the basis for measuring total factor productivity, we introduce the following notation:

Y_i – quantity of the ith output,
X_j – quantity of the jth input,
q_i – price of the ith output,
p_j – price of the jth input.

Where there are m outputs and n inputs, the fundamental identity for each accounting period is that the value of output is equal to the value of input:

$$q_1 Y_1 + q_2 Y_2 + \ldots + q_m Y_m = p_1 X_1 + p_2 X_2 + \ldots + p_n X_n. \tag{1}$$

This accounting identity is important in defining an appropriate method for measuring total factor productivity; it also provides a useful check on the consistency of any proposed definitions of total output and total input.

To define total factor productivity we first differentiate (1) totally with respect to time and divide both sides by the corresponding total value. The result is an identity between a weighted average of the sum of rates of growth of output prices and quantities and a weighted average of the sum of rates of growth of input prices and quantities:

$$\Sigma w_i \left[\frac{\dot{q}_i}{q_i} + \frac{\dot{Y}_i}{Y_i} \right] = \Sigma v_j \left[\frac{\dot{p}_j}{p_j} + \frac{\dot{X}_j}{X_j} \right], \tag{2}$$

with weights $\{w_i\}$ and $\{v_j\}$ given by the relative shares of the value of the ith output in the value of total output and the value of jth input in the value of total input:

$$w_i = \frac{q_i Y_i}{\Sigma q_i Y_i}, v_j = \frac{p_j X_j}{\Sigma p_j X_j}.$$

To verify that both sides of (2) are weighted averages, we observe that:

$$w_i \geqq 0, \ i = 1 \ \ldots \ m;$$
$$v_j \geqq 0, \ j = 1 \ \ldots \ n;$$
$$\Sigma w_i = \Sigma v_j = 1.$$

A useful index of the quantity of total output may be defined in terms of the weighted average of the rates of growth of the individual outputs from (2); denoting this index of output by Y, the rate of growth of this index is

$$\frac{\dot{Y}}{Y} = \Sigma w_i \frac{\dot{Y}_i}{Y_i};$$

an analogous index of the quantity of total input, say X, has rate of growth

$$\frac{\dot{X}}{X} = \Sigma v_j \frac{\dot{X}_j}{X_j}.$$

These quantity indexes are familiar as Divisia quantity indexes; the corresponding Divisia price indexes for total output and total input, say q and p, have rates of growth:

$$\frac{\dot{q}}{q} = \Sigma w_i \frac{\dot{q}_i}{q_i},$$

$$\frac{\dot{p}}{p} = \Sigma v_j \frac{\dot{p}_j}{p_j},$$

respectively.[3]

In terms of Divisia index numbers a natural definition of total factor productivity, say P, is the ratio of the quantity of total output to the quantity of total input:

$$P = \frac{Y}{X}. \tag{3}$$

Using the definitions of Divisia quantity indexes, Y and X, the rate of growth of total factor productivity may be expressed as:

$$\frac{\dot{P}}{P} = \frac{\dot{Y}}{Y} - \frac{\dot{X}}{X} = \Sigma w_i \frac{\dot{Y}_i}{Y_i} - \Sigma v_j \frac{\dot{X}_j}{X_j}. \tag{4}$$

or, alternatively, as:

$$\frac{\dot{P}}{P} = \frac{\dot{p}}{p} - \frac{\dot{q}}{q} = \Sigma v_j \frac{\dot{p}_j}{p_j} - \Sigma w_i \frac{\dot{q}_i}{q_i}.$$

These two definitions of total factor productivity are dual to each other and are equivalent by (2). In general, any index of total factor productivity can be computed either from indexes of the quantity of total output and total input or from the corresponding price indexes.[4]

Up to this point we have defined total factor productivity as the ratio of certain index numbers of total output and total input. An economic interpretation of this definition may be obtained from the theory of production. The theory includes a production function characterized by constant returns to scale; writing this function in implicit form, we have:

$$F(Y_1, Y_2, \ldots, Y_m; X_1, X_2, \ldots, X_n) = 0.$$

Shifts in the production function may be defined in terms of appropriate weighted average rates of growth of outputs and inputs,

$$G\dot{F} = \Sigma\left(\frac{F_iY_i}{\Sigma F_iY_i} \cdot \frac{\dot{Y}_i}{Y_i}\right) - \Sigma\left(\frac{F_jX_j}{\Sigma F_jX_j} \cdot \frac{\dot{X}_j}{X_j}\right), \tag{5}$$

where

$$F_i = \frac{\partial F}{\partial Y_i}, F_j = \frac{\partial F}{\partial X_j}$$

and

$$\frac{1}{G} = \Sigma F_iY_i = -\Sigma F_jX_j.$$

Changes in total factor productivity may be identified with shifts of the production function as opposed to movements along the production function by adding the necessary conditions for producer equilibrium – all marginal rates of transformation between pairs of inputs and outputs are equal to the corresponding price ratios –

$$\frac{\partial Y_i}{\partial X_j} = -\frac{F_j}{F_i} = \frac{p_j}{q_i}; \frac{\partial Y_i}{\partial Y_k} = -\frac{F_k}{F_i} = \frac{q_i}{q_k};$$

$$\frac{\partial X_j}{\partial X_l} = -\frac{F_l}{F_j} = \frac{p_l}{p_j}; (i, k = 1\ldots m; j, l = 1\ldots n).$$

Combining these conditions with the definition (5) of shifts in the production function, we obtain the definition (4) of total factor productivity:

$$G\dot{F} = \frac{\dot{P}}{P}.$$

The rate of growth of total factor productivity is zero if and only if the shift in the production function is zero.

The complete theory of production consists of a production function with constant returns to scale together with the necessary conditions for producer equilibrium. This theory of production implies the existence of a factor price frontier relating the prices of output to the prices input. The dual to the definition (4) of total factor productivity may be identified with shifts in the factor price frontier.[5]

The economic interpretation of the index of total factor productivity is essential in measuring changes in total factor productivity by means of Divisia index numbers. As is well known,[6] the Divisia index of total factor productivity is a line integral so that its value normally depends on the path of integration; even if the path returns to its initial value the index of total factor productivity may increase or decrease. However, if price ratios are identified with marginal rates of transformation of a production function with constant returns to scale, the index will remain constant if the shift in the production function is zero.[7]

From either of the two definitions of the index of total factor productivity we have given it is obvious that the rate of growth of this index is not zero by definition. Even for a production function characterized by constant returns to scale with all factors paid the value of their marginal products, the rate of growth of real product may exceed or fall short of the rate of growth of real factor input; similarly, the rate of growth of the price of real factor input may exceed or fall short of the rate of growth of the price of real product.[8]

The economic theory of production on which our interpretation of changes in total factor productivity rests is not the only possible theory of production. From the definition of shifts in the production function (5) it is clear that the production function may be considered in isolation from the necessary conditions for producer equilibrium, provided that alternative operational definitions of the marginal rates of transformation are introduced. Such a production function may incorporate the effects of increasing returns to scale, externalities, and disequilibrium. Changes in total factor productivity in our sense could then be interpreted as movements along the production function in this more general sense.

To provide a basis for assessing the role of errors of measurement in explaining observed changes in total factor productivity, we first set out principles for measuring total output and total input. The measurement of flows of output and labor services is, at least conceptually, straightforward. Beginning with data on the value of transactions in each type of output and each type of labor service, this value is separated into a price and a quantity. A quantity index of total output is constructed from the quantities of each output, using the relative shares of the value of each output in the value of total output as weights. Similarly, a quantity index of total labor input is constructed from the quantities of each labor service, using the relative shares of the value of each labor service in the value of all labor services as weights.

If capital services were bought and sold by distinct economic units in the same way as labor services, there would be no conceptual or empirical difference between the construction of a quantity index of total capital input and the construction of the corresponding index of total labor input. Beginning with data on the value of transactions in each type of capital service, this value could be separated into a price of capital service or rental and a quantity of capital service in, say, machine-hours. These data would correspond to the value of transactions in each type of labor service which could be separated into a price of labor service or wage and a quantity of labor service in, say, man-hours. A quantity index of total capital input would be constructed from the quantities of each type of capital service, using the relative shares of the rental value of each capital service in the rental value of all capital services as weights.

The measurement of capital services is less straightforward than the measurement of labor services because the consumer of a capital service is usually also the supplier of the service; the whole transaction is recorded only in the internal accounts of individual economic units. The obstacles to extracting this information for purposes of social accounting are almost insuperable; the information must be obtained by a relatively lengthy chain of indirect inference. The data with which the calculation begins are the values of transactions in new investment goods. These values must be separated into a price and quantity of investment goods. Second, the quantity of new investment goods reduced by the quantity of old investment goods replaced must be added to accumulated stocks. Third, the quantity of capital services corresponding to each stock must be calculated.[9]
Paralleling the calculation of quantities of capital services beginning with the quantities of new investment goods, the prices of capital services must be calculated beginning with the prices of new investment goods. Finally, a quantity index of total capital input must be constructed from the quantities of each type of capital service, using the relative shares of the implicit rental value of each capital service in the implicit rental value of all capital services as weights. The implicit rental value of each capital service is obtained by simply multiplying the quantity of that service by the corresponding price. At this final stage the construction of a quantity index of total capital input is formally identical to the construction of a quantity index of total labor input or total output. The chief difference between the construction of price and quantity indexes of total capital input and any other aggregation problem is in the circuitous route by which the necessary data are obtained.

The details of the calculation of a price and quantity of capital services from data on the values of transactions in new investment goods depend on empirical hypotheses about the rate of replacement of old investment goods and the quantity of capital services corresponding to a given stock of capital. In studies of total factor productivity it is conventional to assume that capital services are proportional to capital stock. Where independent data on rates of utilization of capital are available, this assumption can be dispensed with. A number of hypotheses about the rate of replacement of old investment goods have been used in the literature: (1) Accounting depreciation measured by the straight-line method is set equal to replacement, possibly with a correction for changes in prices. (2) Gross investment in some earlier period is set equal to replacement. (3) A weighted average of past investment with weights derived from studies of the "survival curves" of individual pieces of equipment[10] is set equal to replacement. From a formal point of view, the last of these hypotheses includes the first two as special cases.

We assume that the proportion of an investment replaced in a given interval of time declines exponentially over time. A theoretical justification for this assumption is that replacement of investment goods is a recurrent event. An initial investment generates a series of replacement investments over time; each replacement generates a new series of replacements, and so on; this process repeats itself indefinitely. The appropriate model for replacement of investment goods is not the distribution over time of replacements for a given investment, but rather the distribution over time of the infinite stream of replacements generated by a given investment. The distribution of replacements for such an infinite stream approaches a constant fraction of the accumulated stock of investment goods for any "survival curve" of individual pieces of equipment and for any initial age distribution of the accumulated stock, whether the stock is constant or growing. But this is precisely the relationship between replacement and accumulated stock if an exponentially declining proportion of any given investment is replaced in a given interval of time.

The quantity of capital services corresponding to each stock could be measured directly, at least in principle. The stock of equipment would be measured in numbers of machines while the service flow would be measured in machine hours, just as the stock of labor is measured in numbers of men while the flow of labor services is measured in man-hours. While the stock of equipment may be calculated by cumulating the net flow of investment goods, the relative utilization of this equipment must be estimated in order to convert stocks into flows of equipment services. For the purposes of this study we assume that the relative utilization of all capital goods is the same; we estimate the relative utilization of capital from the relative utilization of power sources. An adjustment for the relative utilization of equipment is essential in order to preserve comparability among our measurements of output, labor input, and capital input.

To represent the capital accounts which provide the basis for measuring total capital input, we introduce the following notation:

I_k – quantity of output of the kth investment good,

K_k – quantity of input of the kth capital service.

As before, we use the notation:

q_k – price of the kth investment good,

p_k – price of the kth capital service.

Under the assumption that the proportion of an investment replaced in a given interval of time declines exponentially, the cumulated stock of past investments in the kth capital good, net of replacements, satisfies the well-known relationship:

$$I_k = \dot{K}_k + \delta_k K_k, \tag{6}$$

where δ_k is the instantaneous rate of replacement of the kth investment good. Similarly, in the absence of direct taxation the price of the kth capital service satisfies the relationship:

$$p_k = q_k \left[r + \delta_k - \frac{\dot{q}_k}{q_k} \right], \tag{7}$$

where r is the rate of return on all capital, δ_k is the rate of replacement of the kth investment good, and $\dot{q}_k / q_k$ is the rate of capital gain on that good.

Given these relationships between the price and quantity of investment goods and the price and quantity of the corresponding capital services, the only data beyond values of transactions in new investment goods required for the construction of price and quantity indexes of total capital input are rates of replacement for each distinct investment good and the rate of return on all capital. We turn now to the problem of measuring the rate of return.

First, to measure the values of output and input it is customary to exclude the value of capital gains from the value of input rather than to include the value of such gains in the value of output. This convention has the virtue that the value of output may be calculated directly from the values of transactions. Second, to measure total factor productivity, depreciation is frequently excluded from both input and output; this convention is adopted, for example, by Kendrick [37]. Exclusion of depreciation on capital introduces an entirely arbitrary distinction between labor input and capital input, since the corresponding exclusion of depreciation of the stock of labor services is not carried out.[11] To calculate the rate of return on all capital, our procedure is to subtract from the value of output plus capital gains the value of labor input and of replacement. This results in the rate of return multiplied by the value of accumulated stocks. The rate of return is calculated by dividing this quantity by the value of the stock.[12] The implicit rental value of the kth capital good is:

$$p_k K_k = q_k \left[r + \delta_k - \frac{\dot{q}_k}{q_k} \right] K_k.$$

To calculate price and quantity indexes for total capital input, the prices and quantities of each type of capital service are aggregated, using the relative shares of the implicit rental value of each capital service in the implicit rental value of all capital services as weights.

An almost universal conceptual error in the measurement of capital input is to confuse the aggregation of capital stock with the aggregation of capital service. This error may be exemplified by the following passage from a recent paper by Kendrick [38] devoted to theoretical aspects of capital measurement:

the prices of the underlying capital goods, as established in markets or imputed by owners, can be appropriately combined (with variable weights) to provide a deflator to convert capital values into physical volumes of the various types of underlying capital goods at base-period prices. Or the result can be achieved directly by weighting quantities by constant prices.

As I view it, this is the most meaningful way to measure "real capital stock," since the weighted aggregate measures the physical complex of capital goods in terms of its estimated ability to contribute to production as of the base period.[13]

The "ability to contribute to production" is, of course, measured by the price of capital services, not the price of investment goods.[14]

We have already noted that direct observations are usually available only for values of transactions; the separation of these values into prices and quantities is based on much less complete information and usually involves indirect inferences; the presence of systematic errors in this separation is widely recognized. For output of consumption goods or input of labor services an error in separating the value of transactions into price and quantity results in an error in measurement of the price and quantity of total output or total labor input and in the measurement of total factor productivity. For example, suppose that the rate of growth of the price of a particular type of labor service is measured with an error; since all relative value shares remain the same, the resulting error in the price of total labor input has a rate of growth equal to the rate of growth of the error multiplied by the relative share of the labor service. The quantity of total labor input is measured with an error which is equal in magnitude but opposite in sign. The error in measurement of the rate of growth of total factor productivity is equal to the negative of the rate of growth of the error in the quantity of total labor input multiplied by the relative share of labor. The effects of an error in the rate of growth of the price of a particular type of consumption good are entirely analogous; of course, an upward bias in the rate of growth of output increases the measured rate of growth of total factor productivity, while an upward bias in the rate of growth of input decreases the measured rate of growth.

An error in the separation of the value of transactions in new investment goods into the price and quantity of investment goods will result in errors in measurement of the price and quantity of investment goods, of the price and quantity of capital services and of total factor productivity. To measure the bias in the rate of growth of the quantity of investment goods, we let Q^* be the relative error in the measurement of the price of investment goods, I^* the "quantity" of investment goods output, calculated using the erroneous "price" of investment goods, and I the actual quantity of investment goods output. The bias in the rate of growth of investment goods output is then:

$$\frac{\dot{I}^*}{I^*} - \frac{\dot{I}}{I} = -\frac{\dot{Q}^*}{Q^*}. \tag{8}$$

The rate of growth of this bias is negative if the rate of growth of the error in measurement of the price of investment goods is positive, and vice-versa. If we let K^* be the "quantity" of capital calculated using the erroneous "price" of investment goods and K the actual quantity of capital:

$$K^* = \int_{-\infty}^{t} e^{-\delta(t-s)} I^*(s)ds = \int_{-\infty}^{t} e^{-\delta(t-s)} \frac{I(s)}{Q^*(s)} ds.$$

The bias in the rate of growth of the quantity of capital services is then:

$$\frac{\dot{K}^*}{K^*} - \frac{\dot{K}}{K} = \frac{I}{Q^*K^*} - \frac{I}{K} = \frac{I}{\int_{-\infty}^{t} e^{-\delta(t-s)} \frac{Q^*(t)}{Q^*(s)} I(s)ds} - \frac{I}{\int_{-\infty}^{t} e^{-\delta(t-s)} I(s)ds}, \tag{9}$$

which is negative if the rate of growth of the error in measurement of the price of investment goods is positive, and vice-versa.

To calculate the error of measurement in total factor productivity, we let C represent the quantity of consumption goods and L the quantity of labor input; second, we let w_I represent the relative share of the value of investment goods in the value of total output and w_C the relative share of consumption goods; finally, we let v_K represent the relative share of the value of capital input in the value of total input and v_L the relative share of labor. The rate of growth of total factor productivity may be represented as:

$$\frac{\dot{P}}{P} = w_I \frac{\dot{I}}{I} + w_C \frac{\dot{C}}{C} - v_K \frac{\dot{K}}{K} - v_L \frac{\dot{L}}{L}.$$

If we let P^* represent the measured index of total factor productivity using the erroneous "price" of investment goods:

$$\frac{\dot{P}^*}{P^*} = w_I \frac{\dot{I}^*}{I^*} + w_C \frac{\dot{C}}{C} - v_K \frac{\dot{K}^*}{K^*} - v_L \frac{\dot{L}}{L}.$$

Subtracting the first of these expressions from the second we obtain the bias in the rate of growth of total factor productivity:

$$\frac{\dot{P}^*}{P^*} - \frac{\dot{P}}{P} = w_I \left[\frac{\dot{I}^*}{I^*} - \frac{\dot{I}}{I} \right] - v_K \left[\frac{\dot{K}^*}{K^*} - \frac{\dot{K}}{K} \right].$$

Substituting expressions (9) and (8) for the biases in the measured rates of growth of capital input and the output of investment goods, we have:

$$\frac{\dot{P}^*}{P^*} - \frac{\dot{P}}{P} = - w_I \frac{\dot{Q}^*}{Q^*} - v_K \left(\frac{I}{\int_{-\infty}^{t} e^{-\delta(t-s)} \frac{Q^*(t)}{Q^*(s)} I(s)ds} - \frac{I}{\int_{-\infty}^{t} e^{-\delta(t-s)} I(s)ds} \right). \tag{10}$$

If investment and the error in measurement are growing at constant rates, the biases in the rates of growth of the quantity of investment goods produced and the quantity of capital services are equal, so that the net effect is equal to the rate of growth in the error in measurement of the price of investment goods multiplied by the difference between the capital share in total input and the investment share in total output.[15]

A second source of errors in measurement arises from limitations on the number of separate inputs that may be distinguished empirically. The choice of commodity groups to serve as distinct "inputs" and "outputs" involves aggregation within each group by simply adding together the quantities of all commodities within the group and aggregation among groups by computation of the usual Divisia quantity index. The resulting price and quantity indexes are Divisia price and quantity indexes of the individual commodities only if the rates of growth either of prices or of quantities within each group are identical.

Errors of aggregation in studies of total factor productivity have not gone unnoticed; however, these errors are frequently mislabeled as "quality change." Quality change in this sense occurs whenever the rates of growth of quantities within each separate group are not identical. For example, if high quality items grow faster than items of low quality, the rate of growth of the group is biased downward relative to an index treating high and low quality items as separate commodities. To eliminate this bias it is necessary to construct the index of input or output for the group as a Divisia index of the individual items within the group. Elimination of "quality change" in the sense of aggregation bias is essential to accurate social accounting and to measurement of changes in total factor productivity. Separate accounts should be maintained for as many product and factor input categories as possible. An attempt should be made to exploit available detail in any empirical measurement of real product, real factor input, and total factor productivity.

In some contexts the choice of an appropriate unit for the measurement of quantities of real product or real factor input is not obvious. For example, fuel may be measured in tons or in BTU equivalents, tractor services may be measured in tractor hours or in horsepower hours, and so on. Measures of real product and real factor input may be adjusted for "quality change" by converting one unit of measurement to another. This procedure conforms to the principles of social accounting we have outlined and their interpretation in terms of the economic theory of production if the adjustment for quality change corrects errors of aggregation. In the examples we have given, if the marginal products of different types of fuel always move in proportion when fuel is measured in BTU equivalents but fail to do so when fuel is measured in tons, the appropriate unit for the measurement of fuel is the BTU. Similarly, if the marginal products of tractor services measured in horsepower hours always move in proportion, but when measured in tractor hours fail to do so, tractor services should be measured in horsepower hours.

The appropriateness of any proposed adjustment for quality change may be confronted with empirical evidence on the marginal products of individual items within a commodity group. Under the assumption that these products are equal to the corresponding price ratios this evidence takes the form of data on relative price movements for the individual items. Under a more general set of assumptions the marginal products might be calculated from an econometric production function. The latter treatment would be especially useful for "linking in" new factors and products since the relevant prices cannot be observed until the new factors and products appear in the market. Any change in measured total factor productivity resulting from adjustments for quality change is explained by evidence on the movement of marginal products and is not the result of an arbitrary choice of definitions. The choice of appropriate units for measurement of real product and real factor input may go beyond selection among alternative scalar measures such as BTU equivalents or tons; a commodity may be regarded as multi-dimensional and an appropriate unit of measurement may be defined implicitly by taking prices as given by so-called "hedonic" price indexes. The critical property of such price indexes is that when prices are given by a "hedonic" price index for the commodities within a group, all such commodities have marginal rates of transformation *vis-à-vis* commodities outside the group that move in proportion to each other. Insofar as this property is substantiated by empirical evidence, adjustment of the commodity group for "quality change" by means of such a price index is entirely legitimate and amounts to correcting an error of aggregation.[16] This is not to say that any proposed adjustment for quality change is legitimate. The appropriateness of each adjustment must be judged on the basis of the evidence. If no fresh evidence is employed, the choice of appropriate units is entirely arbitrary and any change in measured total factor productivity resulting from adjustment for "quality change" is simply definitional.

"Quality change" is sometimes used to describe a special type of aggregation error, namely, the error that arises in aggregating investment goods of different vintages by simply adding together quantities of investment goods of each vintage. If the quality of investment goods, as measured by the marginal productivity of capital, is not constant over all vintages, this procedure results in aggregation errors. An appropriate index of capital services may be constructed by treating each vintage of investment goods as a separate commodity. To construct such an index empirically, data on the marginal productivity of capital of each vintage at each point of time are required. If independent data on relative prices of capital services of different vintages are used in the construction of such a capital services index, any resulting reduction in measured productivity growth is not tautological. Only where the change in quality is measured indirectly from the resulting increase in total factor productivity, as suggested by Solow [60], does such a procedure result in the elimination of productivity change by definition.[17]

Measurement

Initial Estimates

We can now investigate the extent to which measured changes in total factor productivity are due to errors of measurement. We begin by constructing indexes of total output and total input for the United States for the 20-year period following World War II, 1945–65, without correcting for errors of measurement. As an initial index of total output we take US private domestic product in constant prices as measured in the US national product accounts [48]. As an index of total input we take the sum of labor and capital services in constant prices. Labor and capital services are assumed to be proportional to stocks of labor and capital, respectively. The stock of labor is taken to be the number of persons engaged in the private domestic sector of the US economy. The stock of capital is the sum of land, plant, equipment, and inventories employed in this sector.[18] The rate of growth of total factor productivity is equal to the difference in the rates of growth of total output and total input.

Indexes of total output, total input, and total factor productivity are given in table 20.1. The average annual rate of growth of total output over the period 1945–65 is 3.49 percent. The average rate of growth of total input is 1.83 percent. The average rate of growth of total factor productivity is 1.60 percent. The rate of growth of total input explains 52.4 percent of the growth in output; the remainder is explained by changes in total factor productivity.

Errors of Aggregation

The first error of measurement to be eliminated is an error of aggregation. This error results from aggregating labor and capital services by summing quantities in constant prices. To eliminate the error, we replace our initial index of total input by a Divisia index of labor and capital input, as suggested by Solow [61]. A similar error results from aggregating consumption and investment goods output by adding together quantities in constant prices. This error may be eliminated by replacing our initial index of total output by a Divisia index of consumption and investment goods output. Indexes of

Table 20.1 Total output, input, and factor productivity, US private domestic economy, 1945–65, initial estimates

	1	2	3
1945	0.699	0.786	0.891
1946	0.680	0.817	0.836
1947	0.695	0.854	0.818
1948	0.729	0.876	0.836
1949	0.726	0.867	0.841
1950	0.801	0.891	0.901
1951	0.852	0.928	0.919
1952	0.873	0.947	0.924
1953	0.917	0.966	0.951
1954	0.904	0.954	0.949
1955	0.981	0.976	1.005
1956	0.999	1.001	0.998
1957	1.013	1.012	1.000
1958	1.000	1.000	1.000
1959	1.069	1.019	1.048
1960	1.096	1.036	1.057
1961	1.115	1.039	1.072
1962	1.189	1.057	1.123
1963	1.240	1.074	1.152
1964	1.307	1.097	1.188
1965	1.387	1.129	1.224

1. Output 2. Input. 3. Productivity

total output, total input, and total factor productivity with these errors of aggregation eliminated are presented in table 20.2.

The average annual rate of growth of total output over the period 1945–65 with the error in aggregation of consumption and investment goods output eliminated is 3.39 percent. The average rate of growth of total input with the error in aggregation of labor and capital services eliminated is 1.84 percent. The resulting rate of growth of total factor productivity is 1.49 percent. We conclude that these errors in aggregation result in an overstatement of the initial rate of growth of total factor productivity. With these errors eliminated total input explains 54.3 percent of the growth in total output. This result may be compared with the 52.4 percent of the growth in total output explained initially.

Investment Goods Prices

We have demonstrated that an error in the measurement of investment goods prices results in errors in the measurement of total output, total input, and total factor productivity. Roughly speaking, a positive bias in the rate of growth of the investment goods price index results in a positive bias in the rate of growth of total factor productivity, provided that the share of capital

Table 20.2 Total output, input, and factor productivity, US private domestic economy, 1945–65, errors of aggregation eliminated

	1	2	3
1945	0.713	0.783	0.912
1946	0.679	0.810	0.841
1947	0.694	0.847	0.824
1948	0.727	0.870	0.840
1949	0.727	0.864	0.845
1950	0.800	0.888	0.903
1951	0.851	0.925	0.921
1952	0.873	0.945	0.926
1953	0.918	0.964	0.953
1954	0.905	0.954	0.950
1955	0.981	0.976	1.005
1956	0.999	1.001	0.998
1957	1.013	1.012	1.000
1958	1.000	1.000	1.000
1959	1.070	1.019	1.049
1960	1.096	1.036	1.057
1961	1.115	1.038	1.073
1962	1.189	1.057	1.124
1963	1.240	1.073	1.153
1964	1.307	1.096	1.189
1965	1.387	1.128	1.225

1. Output. 2. Input. 3. Productivity.

in the value of input exceeds the share of investment in the value of output. This condition is fulfilled for the US private domestic sector throughout the period, 1945–65. Hence, we must examine the indexes of investment goods prices that underlie our measurement for possible sources of bias.

Except for the price index for road construction the price indexes for structures that underlie the US national accounts are indexes of the cost of input rather than the price of output. In the absence of changes in total factor productivity properly constructed price indexes for construction input would parallel the movements of price indexes for output. This is assured by the dual to the usual definition of total factor productivity (3). Dacy [12] has shown that the rate of growth of the price of inputs in highway construction is considerably greater than that of the price of construction output. Dacy's output price index grows from 0.805 to 0.982 from 1947 through 1959, while the input price index grows from 0.615 to 1.024 in the same period, both on a base 1.000 in 1958.[19] This empirical finding is simply another way of looking at the positive residual between rates of growth of total output and total input where total factor productivity is measured with error. Input price indexes are subject to the same errors of aggregation as the corresponding quantity indexes. Since input quantity indexes grow too slowly, input price indexes grow too rapidly.

The use of input prices in place of output prices for structures results in an important error of measurement. To eliminate this error it is necessary to use an output price index in measuring prices of both investment goods output and capital services input. An index of this type has been constructed for the OBE 1966 Capital Stock Study [49]. Components of this index include the Bureau of Public Roads price index for highway structures, the Bell System price index for telephone buildings, and the Bureau of Reclamation price indexes for pumping plants and power plants. The resulting composite index may be compared with the implicit deflator for new construction from the US national accounts [48]. The implicit deflator grows from 0.686 to 1.029 during the period 1947 through 1959 while the OBE Capital Goods Study price index for new construction output grows from 0.762 to 0.958 during the same period. Thus the relative bias in the input price index for all new construction as a measure of the price of construction output is roughly comparable to the relative bias in Dacy's input price index for highway construction as a measure of the price of highway construction output. The input price index, labeled Structures I, and the output price index, labeled Structures II, are given in table 20.3.

The price indexes for equipment that underlie the US national accounts are based primarily on data from the wholesale price index of the Bureau of Labor Statistics [6]. Since expenditures on the wholesale price index are less than those on the consumers' price index [4], adjustments for quality change are less frequent and less detailed. A direct comparison of the durables components of the wholesale and consumers' price indexes gives some notion of the relative bias. The wholesale price index increases from 0.646 to 1.023 and the consumers' price index increases from 0.858 to 1.022 over the period

Table 20.3 Alternative investment deflators

	1	2	3	4	5	6
1945	0.544	0.510	0.759	0.517	0.633	0.357
1946	0.594	0.570	0.768	0.575	0.705	0.638
1947	0.721	0.686	0.827	0.646	0.786	2.310
1948	0.749	0.770	0.863	0.703	0.827	1.023
1949	0.743	0.755	0.868	0.736	0.818	0.788
1950	0.763	0.791	0.878	0.752	0.823	0.818
1951	0.836	0.847	0.942	0.809	0.879	0.945
1952	0.881	0.876	0.954	0.822	0.896	0.949
1953	0.895	0.889	0.943	0.835	0.903	0.497
1954	0.897	0.886	0.929	0.840	0.914	0.772
1955	0.902	0.910	0.919	0.859	0.921	0.931
1956	0.959	0.956	0.949	0.918	0.945	0.978
1957	1.001	0.992	0.984	0.975	0.978	1.113
1958	1.000	1.000	1.000	1.000	1.000	0.994
1959	1.006	1.029	1.014	1.020	1.012	0.991
1960	1.005	1.042	1.009	1.022	1.026	1.020
1961	1.008	1.053	1.006	1.021	1.037	1.011
1962	1.024	1.069	1.008	1.023	1.048	1.001
1963	1.038	1.089	1.004	1.023	1.059	1.011
1964	1.059	1.119	1.004	1.031	1.071	1.014
1965	1.089	1.149	0.995	1.038	1.089	1.032

1. Structures II. 2. Structures I. 3. Equipment II.
4. Equipment I. 5. Inventories II. 6. Inventories I.

1947 to 1959, both on a base of 1.000 in 1958. A direct comparison of components common to both indexes reveals essentially the same relationship. To correct for bias in the implicit deflator for producers' durables, we substitute for this deflator the implicit deflator for consumers' durables. The deflator for producers' durables increased from 0.646 in 1947 to 1.020 in 1959. Over this same period the deflator for consumers' durables increased from 0.827 to 1.014, both on a base of 1.000 in 1958. Thus the relative bias in the producers' durables price index as revealed by a comparison with components common to the wholesale and consumers' price indexes may be corrected by simply substituting the implicit deflator for consumers' durables for the producers' durables deflator. Both indexes are given in 20.3; the producers' durables index is labeled Equipment I while the consumers' durables index is labeled Equipment II.

The durables component for the consumers' price index was itself subject to considerable upward bias in recent years. The consumers' price index for new automobiles increased 62 percent from 1947 to 1959. It has been estimated that correcting this index for quality change would reduce this increase to only 31 percent in the same period.[20] In view of the upward bias in the consumers' price index our adjustment for bias in the producers' durables price index is conservative. In order to reduce the error of measurement further, detailed research like that already carried out for automobiles is required for each class of producers' durable equipment.

The price indexes for change in business inventories from the US national accounts contain year-to-year fluctuations that result from changes in the composition of investment in inventories; these changes are much more substantial than the corresponding changes in the composition of inventory stocks. The implicit deflator for change in inventories is not published; however, it may be computed from data on change in inventories in current and constant dollars. Changes that amount to nearly doubling or halving the index occur from 1946 to 1947, 1947 to 1948, and 1951 to 1952. The value of the index is 0.357 in 1945, 0.638 in 1946 and 2.310 in 1947, all on a base of 1.000 (or, to be exact, 0.994) in 1958. The index drops to 1.023 in 1948 and 0.788 in 1949. A less extreme but equally substantial movement in the index occurs from 1952 through 1957. Changes in the implicit deflator of this magnitude cannot represent movements in the price of all stocks of inventories considered as investment goods. To represent these movements more accurately, we replace the implicit deflator for change in inventories by the deflator for private domestic consumption expenditures. The level of this index generally coincides with that of the implicit deflator for change in business inventories; however, the fluctuations are much less. Both indexes are given in table 20.3; the implicit deflator for change in business inventories is labeled Inventories I while the implicit deflator for private domestic consumption expenditures is labeled Inventories II.

Indexes of total input, total output, and total factor productivity with errors in the measurement of prices of investment goods eliminated are presented in table 20.4. The average rate of growth of total output over the period 1945–65 with these errors of measurement removed is 3.59 percent. This rate of growth may be compared with the original rate of growth of total output of 3.49 percent or with the rate of growth of 3.39 percent for total output with errors of aggregation removed. The average rate of growth of total input over this period is 2.19 percent. The original rate of growth of total input is 1.83 percent; with errors of aggregation removed the rate of growth of total input is 1.84 percent. The rate of growth of total factor productivity is 1.41 percent. With errors in measurement of the prices of investment goods eliminated the rate of growth of total input explains 61.0 percent of the rate of growth of total output.

Measurement of Services

Up to this point we have assumed that labor and capital services are proportional to stocks of labor and capital. This assumption is obviously incorrect. In principle flows of capital and labor services could be measured directly. In fact it is necessary to infer the relative utilization of stocks of capital and labor from somewhat fragmentary data. Okun [50] has attempted to circumvent the problem of direct observation of labor and capital services by assuming that the relative utilization of both labor and capital is a function of the unemployment rate for labor so that the gap between actual and "potential" output, that is, output at full utilization of both factors, may be expressed in terms of the unemployment rate. A similar notion has been used by Solow [62] to adjust stocks of labor and capital for relative utilization. Most of the available capacity utilization measures are based on the relationship of actual output to output at full utilization of both labor and capital, so that these measures also attempt to adjust both labor and capital simultaneously.

Our approach to the problem of relative utilization is somewhat more direct in that we attempt to adjust capital and labor for relative utilization separately. Of course, this adjustment gives rise to a new concept of "potential" or capacity output, but we do not pursue this notion further in this paper. Our first assumption is that the relative utilization of capital is the same for all capital goods; while this is a very strong assumption it is weaker than the assumption underlying the Okun–Solow approach in which the relative utilization of capital and labor depends on that of labor. We estimate the relative utilization of capital from the relative utilization of power sources.[21] Data on the relative utilization of electric motors provide an indicator of the relative utilization of capital in manufacturing, since electric

Table 20.4 Total output, input, and factor productivity, US private domestic economy, 1945–65, errors in investment goods prices eliminated

	1	2	3
1945	0.692	0.759	0.913
1946	0.662	0.786	0.846
1947	0.679	0.822	0.829
1948	0.718	0.845	0.853
1949	0.717	0.842	0.854
1950	0.798	0.867	0.922
1951	0.839	0.908	0.925
1952	0.858	0.930	0.925
1953	0.905	0.950	0.954
1954	0.900	0.942	0.957
1955	0.982	0.966	1.016
1956	0.995	0.996	0.999
1957	1.009	1.010	1.000
1958	1.000	1.000	1.000
1959	1.076	1.022	1.052
1960	1.107	1.042	1.061
1961	1.127	1.049	1.073
1962	1.199	1.071	1.117
1963	1.249	1.091	1.142
1964	1.319	1.117	1.177
1965	1.400	1.153	1.209

1. Output   2. Input.   3. Productivity.

motors are the predominant source of power there. We assume that relative utilization of capital goods in the manufacturing and non-manufacturing sectors is the same. When more complete data become available this assumption can be replaced by less restrictive assumptions. Unfortunately, this adjustment allows only for the trend in the relative utilization of capital; it does not adjust for short-term cyclical variations in capacity utilization. Thus we are unable to attain the objective of complete comparability between measures of labor and capital input.

The assumption that labor services are proportional to the stock of labor is obviously incorrect. On the other hand, the assumption that labor services can be measured directly from data on man-hours is equally incorrect, as Denison [14] has pointed out. The intensity of effort varies with the number of hours worked per week, so that labor input can be measured accurately only if data on man-hours are corrected for the effects of variations in the number of hours per man on labor intensity. Denison [15] suggests that the stock of labor provides an upper bound for labor services while the number of man-hours provides a lower bound. He estimates labor input by correcting man-hours for variations in labor intensity. We employ Denison's correction for intensity, but we apply this correction to actual hours per man rather than potential hours per man. Thus, our measure of labor input reflects short-run variations in labor intensity.

Table 20.5 Total input and factor productivity, US private domestic economy, 1945–65, errors in relative utilization eliminated

	1	2
1945	0.716	0.968
1946	0.742	0.895
1947	0.777	0.877
1948	0.801	0.899
1949	0.802	0.897
1950	0.830	0.963
1951	0.873	0.963
1952	0.899	0.956
1953	0.924	0.980
1954	0.923	0.976
1955	0.959	1.023
1956	0.994	1.001
1957	1.009	1.000
1958	1.000	1.000
1959	1.035	1.038
1960	1.057	1.046
1961	1.067	1.054
1962	1.089	1.098
1963	1.114	1.118
1964	1.146	1.147
1965	1.189	1.172

1. Input.   2. Productivity.

The assumption that labor and capital services are proportional to stocks of labor and capital results in an error in separating a given value of transactions into a price and a quantity. To correct this error we multiply the number of persons engaged by hours per man. The resulting index of man-hours is then corrected for variations in labor intensity. The corresponding error for capital is corrected by multiplying the stock of capital by the relative utilization of capital. Indexes of total input and total factor productivity after these errors have been eliminated are presented for the period 1945–65 in table 20.5. The average annual rate of growth of total output is the same as before these corrections, 3.59 percent per year. The average rate of growth of total input is 2.57 percent. The resulting average rate of growth of total factor productivity is 0.96 percent. Total input now explains 71.6 percent of the rate of growth in total output.

Capital Services

In converting estimates of capital stock into estimates of capital services we have disregarded an important conceptual error in the aggregation of capital services. While investment goods output must be aggregated by means of investment goods or asset prices, capital services must be aggregated by means of service prices.

The prices of capital services are related to the prices of the corresponding investment goods; in fact, the asset price is simply the discounted value of all future capital services. Asset prices for different investment goods are not proportional to service prices because of differences in rates of replacement and rates of capital gain or loss among capital goods. Implicitly, we have assumed that these prices are proportional; to eliminate the resulting error in measurement, it is necessary to compute service prices and to use these prices in aggregating capital services.

We have already outlined a method for computing the price of capital services in the absence of direct taxation of business income. In the presence of direct taxes we may distinguish between the price of capital services before and after taxes. The expression (7) given above for the price of capital services is the price after taxes. The price of capital services before taxes is:

$$p_k = q_k \left[\frac{1-uv}{1-u} r + \frac{1-uw}{1-u} \delta_k - \frac{1-ux}{1-u} \frac{\dot{q}_h}{q_k} \right] \qquad (11)$$

where u is the rate of direct taxation, v the proportion of return to capital allowable as a charge against income for tax purposes, w the proportion of replacement allowable for tax purposes, and x the proportion of capital gains included in income for tax purposes.

We estimate the variables describing the tax structure as follows: The rate of direct taxation is the ratio of profits tax liability to profits before taxes. The proportion of the return to capital allowable for tax purposes is the ratio of net interest to the total return to capital. Total return to capital is the after tax rate of return, r, multiplied by the current value of capital stock. The proportion of replacement allowable for tax purposes is the ratio of capital consumption allowances to the current value of replacement. The proportion of capital gains included in income is zero by the conventions of the US national accounts. Given the value of direct taxes we estimate the after tax rate of return by subtracting from the value of output plus capital gains the value of labor input, replacement, and direct taxes. This results in the total return to capital. The rate of return is calculated by dividing this quantity by the current value of the stock of capital. Given data on the rate of return and the variables describing the tax structure, we calculate the price of capital services before taxes for each investment good.[22] These prices of capital services are used in the calculation of indexes of capital input, total input, and total factor productivity.

For the US private domestic economy it is possible to distinguish five classes of investment goods – land, residential and non-residential structures, equipment, and inventories. Although it is also possible to distinguish a number of sub-classes within each of these groupings, we will employ only the five major groups in calculating an index of total capital input. For each group we first compute a before tax service price analogous to (11). We then compute an index of capital input as a Divisia index of the services of land,

Table 20.6 Total input and factor productivity, US private domestic economy, 1945–65, errors in aggregation of capital input eliminated; implicit rate of return after taxes

	1	2	3	4
1945	0.692	0.671	1.030	0.158
1946	0.661	0.698	0.950	0.198
1947	0.678	0.735	0.926	0.237
1948	0.717	0.765	0.940	0.223
1949	0.716	0.773	0.930	0.126
1950	0.797	0.804	0.992	0.095
1951	0.837	0.850	0.986	0.242
1952	0.857	0.880	0.976	0.143
1953	0.905	0.908	0.997	0.091
1954	0.900	0.911	0.988	0.078
1955	0.982	0.951	1.032	0.113
1956	0.995	0.987	1.008	0.175
1957	1.009	1.005	1.004	0.138
1958	1.000	1.000	1.000	0.107
1959	1.077	1.039	1.035	0.097
1960	1.107	1.063	1.040	0.105
1961	1.127	1.076	1.046	0.118
1962	1.199	1.099	1.089	0.138
1963	1.250	1.126	1.107	0.131
1964	1.320	1.160	1.134	0.127
1965	1.401	1.206	1.157	0.141

1. Output. 2. Input. 3. Productivity. 4. Rate of return.

structures, equipment and inventories. In constructing this index we eliminate the conceptual error that arises from the implicit assumption that service prices are proportional to asset prices for different investment goods. In eliminating this conceptual error we also eliminate the error of aggregation that results from adding together capital services in constant prices to obtain an index of total capital input. To eliminate the corresponding error in our index of investment goods output we replace our initial index by a Divisia index of investment in structures, equipment, and inventories. Indexes of total output, total input and total factor productivity resulting from the elimination of these errors are presented in table 20.6. The after tax rate of return implicit in the new index of capital input is also given in table 20.6.

The average rate of growth of total output over the period 1945–65 with the error in aggregation of investment goods eliminated is 3.59. This rate of growth is essentially the same as for total output with errors in the aggregation of consumption and investment goods and errors in the measurement of investment goods prices eliminated. The average rate of growth of total input with errors in aggregation of capital services eliminated is 2.97 percent. This rate of growth may be compared with the initial rate of growth of 1.83 percent.

The resulting rate of growth of total factor productivity is 0.58 percent. The index of total factor productivity with these errors eliminated is presented in table 20.6. With these errors eliminated total input explains 82.7 percent of the growth in total output. The original index of total input explains 52.4 percent of this growth.

Labor Services

We have eliminated errors of aggregation that arise in combining capital services into an index of total capital input. Similar errors arise in combining different categories of labor services into an index of total labor input. Implicitly, we have assumed that the price per man-hour for each category of labor services is the same; to eliminate the resulting error of measurement it is necessary to use prices per man-hour for each category in computing an index of total labor input. Second, to eliminate the error of aggregation that results from adding together labor services in constant prices, we replace our initial index of labor input by a Divisia index of the individual categories of labor services.

The Divisia index of total labor input is based on a weighted average of the rates of growth of different categories of labor, using the relative shares in total labor compensations as weights. To represent our index of total labor input we let L_i represent the quantity of input of the lth labor service, measured in man-hours. The rate of growth of the index of total labor input, say L, is:

$$\frac{\dot{L}}{L} = \Sigma v_l \frac{\dot{L_l}}{L_l}$$

where v_l is the relative share of the lth category of labor in the total value of labor input. The number of man-hours for each labor service is the product of the number of men, say n_l, and hours per man, say h_l; using this notation the index of total labor input may be rewritten:

$$\frac{\dot{L}}{L} = \Sigma v_l \frac{\dot{n}_l}{n_l} + \Sigma v_l \frac{\dot{h}_l}{h_l} .$$

For comparison with our total indexes of labor input we separate the rate of growth of the index of labor input into three components – change in the total number of men, change in hours per man, and change in labor input per man-hour. We have assumed that the number of hours per man is the same for all categories of labor services, say H. Letting N represent the total number of men and e_l the proportion of the workers in the lth category of labor services, we may write the index of total labor input in the form:

$$\frac{\dot{L}}{L} = \frac{\dot{H}}{H} + \frac{\dot{N}}{N} + \Sigma v_l \frac{\dot{e}_l}{e_l} . \tag{12}$$

Our initial index of labor input was simply N, the number of persons engaged; we corrected this index by taking into account the number of hours per man, H. To eliminate the remaining errors of aggregation we must correct the rate of growth of man-hours by adding to it an index of labor input per man-hour. The third item in the expression (12) for total labor input given above provides such an index. We will let E represent this index, so that:

$$\frac{\dot{E}}{E} = \Sigma v_l \frac{\dot{e}_l}{e_l} . \tag{13}$$

For computational purposes it is convenient to note that the index may be rewritten in the form:

$$\frac{\dot{E}}{E} = \Sigma \frac{p_l}{\Sigma p_l e_l} \dot{e}_l = \Sigma p_l' \dot{e}_l ,$$

where p_l is the price of the lth category of labor services and p_l' is the relative price. The relative price is the ratio of the price of the lth category of labor services to the average price of labor services, $\Sigma p_l e_l$.

In principle it would be desirable to distinguish among categories of labor services classified by age, sex, occupation, number of years schooling completed, industry of employment, and so on. An index of labor input per man-hour based on such a breakdown requires detailed research far beyond the scope of this study. We will compute such an index only for males and only for categories of labor broken down by the number of school years completed. The basic computation is presented in table 20.7. Data on relative prices for labor services are available for the years 1939, 1949, 1956, 1958, 1959, and 1963.[23] Combining these prices with changes in the distribution of the labor force provides a measure of the change in labor input per man-hour.[24]

Table 20.7 Relative prices,[a] changes in distribution of the labor force, and indexes of labor-input per man-hour, US males, the civilian labor force, 1940–64

School year completed	p'_i 1939	Δe_i 1940–48	p'_i 1949	Δe_i 1948–52	p'_i 1956	Δe_i 1952–7	p'_i 1958	Δe_i 1957–9	p'_i 1959	Δe_i 1959–62	p'_i 1963	Δe_i 1962–5
Elementary 0–4	0.497	−2.3	0.521	−0.3	0.452	−1.3	0.409	−0.8	0.498	−0.8	0.407	−0.8
5–6 or 5–7	0.672	−3.1	0.685	−0.5	0.624	−0.2	0.565	−1.0	0.688	−0.9	0.562	−1.5
7–8 or 8	0.887	−6.8	0.813	−1.8	0.796	−3.3	0.753	−1.2	0.801	−1.9	0.731	−1.2
High school 1–3	1.030	2.4	0.974	−1.3	0.955	0.7	0.923	0.6	0.912	−0.6	0.886	−0.3
4	1.241	7.0	1.143	1.0	1.159	2.6	1.113	0.9	1.039	1.6	1.087	3.2
College 1–3	1.442	1.4	1.336	1.2	1.356	0.2	1.392	0.7	1.255	1.3	1.269	0.0
4+ or 4	1.947	1.3	1.866	1.6	1.810	1.3	1.840	0.9	1.569	1.0	1.571	0.2
5+	—	—	—	—	—	—	—	—	1.888	0.3	1.730	0.4
Percentage change in labor input per man-hour		6.45		2.50		2.97		2.39		2.36		2.13
Annual percentage change		0.78		0.62		0.59		1.20		0.79		0.72

Source: derived from tables 20.11 and 20.12, statistical appendix.
[a] The relative prices are computed using the appropriate beginning period distribution of the labor force as weights.

Table 20.8 Total input and factor productivity, US private domestic economy 1945–65, errors in aggregation of labor input eliminated

	1	2
1945	0.634	1.090
1946	0.661	1.001
1947	0.700	0.971
1948	0.732	0.981
1949	0.743	0.966
1950	0.776	1.026
1951	0.823	1.017
1952	0.857	1.002
1953	0.887	1.020
1954	0.894	1.007
1955	0.936	1.048
1956	0.976	1.019
1957	0.997	1.012
1958	1.000	1.000
1959	1.047	1.027
1960	1.077	1.027
1961	1.096	1.027
1962	1.125	1.064
1963	1.158	1.076
1964	1.200	1.096
1965	1.255	1.112

1. Input. 2. Productivity.

Indexes of total input and total factor productivity with errors in the aggregation of labor services eliminated are presented in table 20.8. The average rate of growth of total input over the period 1945–65 with the error in aggregation of labor services eliminated is 3.47. This rate of growth may be compared with the initial rate of growth of total input or 1.83 percent. The resulting rate of growth of total factor productivity is 0.10 percent. With these errors eliminated total input explains 96.7 percent of the growth in total output.

Summary

The purpose of this paper has been to examine the hypothesis that if quantities of output and input are measured accurately, growth in total output may be largely explained by growth in total input. The results are given in table 20.9 and figure 20.1. We first present our initial estimates of rates of growth of output, input, and total factor productivity. These estimates include many of the errors made in attempts to measure total factor productivity without fully exploiting the economic theory underlying the social accounting concepts of real product and real factor input. We begin by eliminating errors of

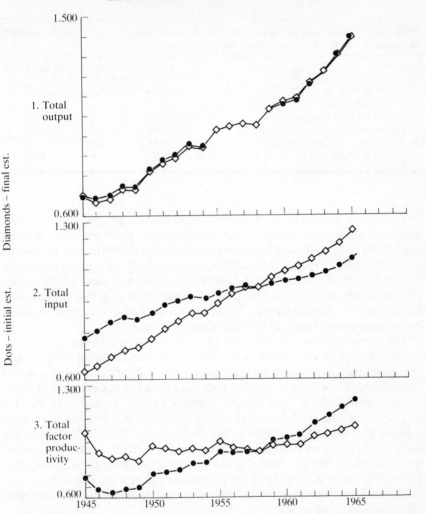

Figure 20.1 Indexes of total output, total input, and total factor productivity (1958 = 1.0), US private domestic economy, 1945–65.

aggregation in combining investment and consumption goods and labor and capital services. We then eliminate errors of measurement in the prices of investment goods arising from the use of prices for inputs into the investment goods sector rather than outputs from this sector. We remove errors arising from the assumption that the flow of services is proportional to stocks of labor and capital by introducing direct observations on the rates of utilization of labor and capital stock. We present rates of growth that result from correct aggregation of investment goods and capital services. Finally, we give rates of growth that result from correcting the aggregation of labor services.

Table 20.9 Total output, input, and factor productivity, US private domestic economy, 1945–65, average annual rates of growth

	Output	Input	Productivity
1. Initial estimates	3.49	1.83	1.60
Estimates after correction for:			
2. Errors of aggregation	3.39	1.84	1.49
3. Errors in investment goods prices	3.59	2.12	1.41
4. Errors in relative utilization	3.59	2.57	0.96
5. Errors in aggregation of capital services	3.59	2.97	0.58
6. Errors in aggregation of labor services	3.59	3.47	0.10

The rate of growth of input initially explains 52.4 percent of the rate of growth of output. After elimination of aggregation errors and correction for changes in rates of utilization of labor and capital stock the rate of growth of input explains 96.7 percent of the rate of growth of output; change in total factor productivity explains the rest. In the terminology of the theory of production, movements along a given production function explain 96.7 percent of the observed changes in the pattern of productivity activity; shifts in the production function explain what remains.

This computation is based on the 1945–65 period, measuring total factor productivity peak to peak. If one were to choose a different set of years, the numerical results would be slightly different, but their main thrust would be the same. For example, starting with the post-Korean peak year of 1953, the rate of growth of input initially explains only 37.3 percent of the rate of growth of output. After all the corrections the rate of growth of input explains 79.2 percent of the growth in output between 1953 and 1965, reducing the estimated rate of change in total factor productivity from 2.12 percent per year to 0.72. We conclude that our hypothesis is consistent with the facts. If the economic theory underlying the measurement of real product and real factor input is properly exploited, the role to be assigned to growth in total factor productivity is small.

Evaluation of Past Research

Our conclusion that most of the growth in total output may be explained by growth in total input is just the reverse of the conclusion drawn from the great body of past research on total factor productivity, the research of Schmookler [55], Mills [46], Fabricant [23], Abramovitz [2], Solow [61], and Kendrick [37]. These conclusions, stated by Abramovitz, are "that to explain a very large part of the growth of total output and the great bulk of output *per capita*, we must explain the increase in output per unit of conventionally measured inputs . . .".[25] This conclusion results from inadequacies in the basic economic theory underlying the social accounts employed in productivity measurements. The increase in output per unit of conventionally measured inputs is characterized by very substantial errors of measurement, equal in magnitude to the alleged increase in productivity. We have given a concrete and detailed list of errors of this type.

Our results differ from those of Denison [15] in that we correct changes in total factor productivity for errors in the measurement of output, capital services and labor services, while Denison corrects only for errors in the measurement of labor services. To get some idea of the relative importance of errors in the measurement of labor and errors in the measurement of output and capital, we may observe that the rate of growth of total factor productivity is reduced from 1.60 percent per year to 0.10 percent per year. Of the total reduction of 1.50 percent per year errors in the measurement of output and capital account for 1.17 percent per year while errors in the measurement of labor account for 0.33 percent per year. We conclude that errors of measurement of the type left uncorrected by Denison are far more important than the type of errors he corrects.[26]

Our results suggest that the residual change in total factor productivity, which Denison attributes to Advance in Knowledge, is small. Our conclusion is not that advances in knowledge are negligible, but that the accumulation of knowledge is governed by the same economic laws as any other process of capital accumulation. Costs must be incurred if benefits are to be achieved. Although we have made no attempt to isolate the effects of expenditures on research and development from expenditures on other types of current inputs or investment goods, our results suggest that social rates of return to this type of investment are comparable to rates of return on other types of investment. Of course, our inference is indirect and a better test of this proposition could be provided by direct observation of private and social rates to investment in scientific research and development activities. Unfortunately, many of the direct observations on these rates of return available in the literature attribute all or part of the measured increase in total factor productivity to investment in research and development;[27] since these measured increases are subject to all the errors of measurement we have enumerated, satisfactory direct tests of the hypothesis that private and social rates of return to research and development investment are equal to private rates of return to other types of investment are not yet available.

Another implication of our results is that discrepancies between private and social returns to investment in physical capital may play a relatively minor role in explaining economic growth. Under the operational definitions of total factor productivity we have adopted, a positive discrepancy between social and private rates of return would appear as a downward bias in the rate of growth of input, hence an upward bias in the rate of growth of total factor productivity. The effects of such discrepancies are lumped together with the effects of other sources of growth in total factor productivity we have measured. The fact that the growth of the resulting index is small indicates that the contribution of investment to economic growth is largely compensated by the private returns to investment. This implication of our findings is inconsistent with explanations of economic growth such as Arrow's model of learning by doing [3], which are based on a higher social than private rate of return to physical capital.[28]

Of course, ours is not the first explanation of productivity change that does not rely primarily on discrepancies between private and social rates of return. An explanation of this type has been proposed by Solow [60], namely, embodied technical change. As Solow [59] points out, explanation of measured changes in total factor productivity as embodied technical change does not require discrepancies between private and social rates of return: "the fact of expectable obsolescence reduces the private rate of return on saving below the marginal product of capital as one might ordinarily calculate it. But this discrepancy is fully reflected in a parallel difference between the marginal product of capital and the social rate of return on saving. So . . . the private and social rates of return coincide."[29] In referring to "capital as one might ordinarily calculate it," Solow explicitly does not identify quality-corrected or "surrogate" capital with capital input and "surrogate" investment with investment goods output. In Solow's framework the marginal product of "surrogate" capital is precisely equal to the private and social rate of return on saving. The difference between Solow's point of view and ours is that the private and social rates of return are equal by definition in his framework, where the equality between private and social rates of return is a testable hypothesis within our framework.[30]

Implications for Future Research

The problem of measuring total factor productivity is, at bottom, the same as the estimation of national product and national factor input in constant prices. The implication of our findings is that the predominant part of economic growth may be explained within a conventional social accounting framework. Of course, precise measurement of productivity change requires attention to reliability as well as accuracy. Our catalogue of errors of measurement could serve as an agenda for correction of errors in the measurement of output and for incorporation of the measurement of input into a unified social accounting framework. Given time and resources we could attempt to raise all of our measurements to the high standards of the US National Product Accounts in current prices. This could be done with some difficulty for rates of relative utilization of labor and capital stock and the prices of investment goods, which require the introduction of new data into the social accounts. The elimination of aggregation errors in measuring capital services and investment goods requires a conceptual change to bring these concepts into closer correspondence with the economic theory of production. The measurement of appropriate indexes of labor input, corrected for errors of aggregation, necessitates fuller exploitation of existing data on wage differentials by education, occupation, sex, and so on.

The most serious weakness of the present study is in the use of long-term trends in the relative utilization of capital and labor to adjust capital input and labor input to concepts appropriate to the underlying theory of production. As a result of discrepancies between these trends and year-to-year variations in relative utilization of capital and labor, substantial errors of measurement have remained in the resulting index of total factor productivity. Examination of any of the alternative indexes we have presented reveals substantial unexplained cyclical variation in total factor productivity. An item of highest priority in future research is to incorporate more accurate data on annual variations in relative utilization. Hopefully, elimination of these remaining errors will make it possible to explain cyclical changes in total factor productivity along the same lines as our present explanation of secular changes. Cyclical changes are very substantial, so that even our secular measurements could be improved with better data. For example, the use of the period 1945–58, a peak in total factor productivity to a trough, reveals a drop in total factor productivity of 9 percent; the use of the period 1949–65, a trough to a peak, yields an increase in total factor productivity of 11½ percent.

In compiling data on labor input we have relied upon observed prices of different types of labor services. Given a broader accounting framework it would be possible to treat human capital in a manner that is symmetric with our measurement of physical capital. Investment in human capital could be cumulated into stocks along the lines suggested by Schultz [56]. The flow of investment could be treated as part of total output. The rate of return to this investment could then be measured and compared with the rate of return to physical capital. Similarly, investment in scientific research and development could be separated from expenditures on current account and cumulated into stocks. The rate of return to research activity could then be computed. In both of these calculations it would be important not to rely on erroneously measured residual growth in total output for measurement of the social return to investment.

It is obvious that further disaggregation of our measurements would be valuable in order to provide a more stringent test of the basic hypothesis that growth in output may be explained by growth in input. The most important disaggregation of this type is to estimate levels of output and input by individual industries. The statistical raw material for dis-aggregation by industry is already available for stocks of labor and capital and levels of output. However, data for relative utilization of labor and capital, and for disaggregation of different types of labor and capital within industry groups, would have to be developed. Once these data are available it will be possible to estimate rates of return to capital for individual industries and to study the effects of the distribution of productive factors among industries along the lines suggested by Massell [43]. The fact that past observations do not reveal significant changes in productivity does not imply that the existing allocation of productive resources is efficient relative to allocations that could be brought about by policy changes. In such a study it might be useful to extend the scope of productivity measurements to include the government sector. This would be particularly desirable if educational investment, which is largely in that sector, is to be incorporated into total output.

Finally, our results suggest a new point of departure for econometric studies of production function at every level of aggregation. While some existing studies [29, 30] employ data on output, labor, and capital corrected for errors of measurement along the lines we have suggested, most estimates of production functions are based on substantial errors of measurement. Econometric production functions are not an alternative to our methods for measuring total factor productivity, but rather supplement these methods in a number of important respects. Such production functions provide one means of testing the assumptions of constant returns to scale and equality between price ratios and marginal rates of transformation that underlie our measurement. A complete test of the hypothesis that growth in total output may be explained by growth in total input requires the measurement of input within a unified social accounting framework, the measurement of rates of return to both human and physical capital, further disaggregation, and new econometric studies of production functions. A start has been made on this task, but much interesting and potentially fruitful research remains to be done.

Statistical Appendix

1. As our initial estimate of output we employ gross private domestic product which is defined as gross national product less gross product, general government, and gross product, rest of the world, all in constant prices of 1958. These data are obtained from the US national accounts. Our second estimate of output requires data on gross private domestic investment and gross private domestic consumption, defined as gross private domestic product less gross private domestic investment, in both current and constant prices of 1958. These data are also obtained from the US national accounts.

As our initial estimate of labor input we employ private domestic persons engaged, defined as persons engaged for the national economy less persons engaged, general government, and persons engaged, rest of the world. These data are obtained from the US national accounts [48]. Our initial estimate of capital input is obtained by the perpetual inventory method based on double declining balance estimates of replacement. For structures and equipment the lifetimes of individual assets are based on the "Bulletin F lives" employed by Jaszi, Wasson, and Grose [33]. Data for gross private domestic investment prior to 1929 are unpublished estimates that underlie the capital stock estimates by Jaszi, Wasson, and Grose [33]. For inventories and land, the initial values of capital stock in constant prices of 1958 are derived from Goldsmith [25]. The stock of land in constant prices is assumed to be unchanged throughout the period we consider. Estimates of the value of land in current prices are obtained from Goldsmith [25].

The estimates of gross private domestic investment are subsequently revised by introducing alternative deflators to those employed in the US national accounts. These deflators are given in table 20.3. Gross private domestic consumption is left unchanged in this calculation. We compute stocks of land, structures, residential and non-residential, equipment, and inventories separately for each set of deflators. The basic formula is:

$$K_{t+1} = I_t + (1 - \delta)K_t,$$ (14)

where I_t is the value of gross private domestic investment for each category in constant prices. The initial (1929) value of capital stock in constant prices of 1958 and the depreciation rates are as shown in table 20.9.

2. In dropping the assumption that services are proportional to stock for both labor and capital, we require data on hours/man and hours/machine. The data on hours/man are derived from Kendrick's data on man-hours in the US private domestic economy, extended through 1965.

To estimate hours/machine we first estimate the relative utilization of electric motors in manufacturing. Estimates have been given by Foss [24] for 1929, 1939, and 1954. We have updated these estimates to 1962. The basic computation is given in table 20.11. The 1954 data and the basic method of computation are taken from Foss [24, table II, p. 11]. The 1954 data differ from the figures given by Foss due to a revision of the 1954 horsepower data by the Bureau of the Census and omission of the "fractional horsepower motors" adjustment. The latter, applied to both 1954 and 1962, would not have affected the estimated change in relative utilization. The horsepower data for 1962 and 1954 are from the 1963 *Census of Manufactures* [7], "Power equipment in manufacturing industries," MC63(1)-6. Consumption of electric energy is taken from the 1962 *Survey of Manufactures* [11], chapter 6. The 1962 total (388.2) is reduced by the consumption of electric power for nuclear energy (51.5) as shown in Series S81-93 of Bureau of the Census, *Continuation to 1962 of Historical Statistics of the U.S.* [9].

3. To estimate service prices for capital from the formula (11) given in the text we require data on the tax structure and on the rate of return. The

Table 20.10

	National accounts deflators		Alternative deflators	
	K_{1929}	δ	K_{1929}	δ
Land	254,700	0	254,700	0
Structures				
Residential	183,234	0.0386	162,708	0.0384
Non-residential	163,205	0.0513	142,670	0.0509
Equipment	74,851	0.1325	51,701	0.1226
Inventories	48,504	0	48,504	0

variable u, the rate of direct taxation, is the ratio of corporate profits tax liability to total net private property income. These data are from the US national accounts. The variable v, the proportion of return to capital allowable as a charge against income for tax purposes, is the ratio of private domestic net interest to the after tax rate of return, r, multiplied by the current value of capital stock. Private domestic net interest is net interest less net interest for the rest of the world sector. These data are taken from the US national accounts. We discuss estimation of the after tax rate of return below. The current value of capital stock is the sum of stock in land, structures, equipment, and inventories. Each of the four components is the product of the corresponding stock in constant prices of 1958, multiplied by the investment deflator for the component. Finally the variable w, the proportion of replacement allowable for tax purposes, is the ratio of capital consumption allowances to the current value of replacement. Capital consumption allowances are taken from the US national accounts. The current value of replacement is the sum of replacement in current prices for structures and equipment. Replacement in current prices is the product of replacement in constant prices of 1958 and the investment deflator for the corresponding component. Replacement

Table 20.11 Relative utilization of electric motors, manufacturing, 1954 and 1962

	Unit	1954	1962
1. Horsepower of electric motors, total	Thousand horsepower	91,505	126,783
2. Available kilowatt-hours of motors (line 1 × 7261)	Billions of kilowatt-hours	664.4	920.6
3. Electric power actually consumed, all purposes	Billions of kilowatt-hours	222.1	336.7
4. Percent power used for electric motors	—	64.6	65.6
5. Power consumed by motors (line 3 × line 4)	Billions of kilowatt-hours	143.5	220.9
6. Percent utilization (line 5/ line2 × 100)	—	21.6	24.0
7. Number of equivalent 40 hour weeks (line 6 × 4.2/100)	0.907	1.008	
8. Index	1954 = 100	100.0	111.1

Line 2: The adjustment is derived as follows: It is assumed "that each electric motor could work continuously throughout the year . . . , 8760. . . . Horsepower hours are converted to kilowatt-hours; . . . 1 horsepower-hour = 0.746 kilowatt-hours. The result [is] . . . adjusted upward by dividing through 0.9, since modern electric motors have an efficiency of approximately 90 percent. . . ." Foss [23, p. 11]. 8760 × 0.746/0.9 = 7261.

Line 4: Percentage power used for electric motors in 1962 computed using the industry distribution in 1945 given by Foss [24] in his table I, and the 1962 consumption of total electric power by industries from the 1962 *Survey of Manufactures* [11, chapter 6].

Line 7: There are 4.2 forty-hour shifts in a full week of 168 hours.

in constant prices is a by-product of the calculation of capital stock by formula (14) given above. Replacement is simply δK_t, where K_t is capital stock in constant prices.

To estimate the rate of return we define the value of capital services for land, structures, equipment, and inventories as the product of the service price (11) and the corresponding stock in constant prices. Setting this equal to total income from property, we solve for the rate of return. Total income from property is gross private domestic product in current prices less private domestic labor income. Private domestic labor income is private domestic compensation of employees from the US national accounts multiplied by the ratio of private domestic persons engaged in production to private domestic full-time equivalent employees, both *The National Income and Product Accounts of the United States, 1929–1965* [49]. This amounts to assuming that self-employed individuals have the same average labor income as employees.

The final formula for the rate of return is then the ratio of total income from property less profits tax liability less the current value of replacement plus the current value of capital gain to the current value of capital stock. The current value of capital gain is the sum of capital gains for all assets; the capital gain for each asset is the product of the rate of growth of the corresponding investment deflator and the value of the asset in constant prices of 1958.

Table 20.12 Civilian labor force, males 18 to 64 years old, by educational attainment, percentage distribution by years of school completed

School year completed	1940	1948	1952	1957	1959	1959†	1962†	1965[b]
Elementary 0–4	10.2	7.9	7.6	6.3	5.5	5.9	5.1	4.3
5–6 or 5–7[a]	10.2	7.1	6.6	11.6	11.4	10.4	10.7	9.8	8.3
7–8 or 8[a]	33.7	26.9	25.1	20.1	16.8	15.6	15.8	13.9	12.7
High school 1–3	18.3	20.7	19.4	20.1	20.7	19.8	19.2	18.9
4	16.6	23.6	24.6	27.2	28.1	27.5	29.1	32.3
College 1–3	5.7	7.1	8.3	8.5	9.2	9.4	10.6	10.6
4+ or 4	5.4	6.7	8.3	9.6	10.5	6.3	7.3	7.5
5+	—	—	—	—	—	4.7	5.0	5.4

Source: The basic data for columns 1, 2, 3, 4, 5, and 6 are taken from US Department of Labor, *Special Labor Force Report* [5], no. 1 "Educational attainment of workers, 1959." The 5–8 years class is broken down into the 5–7 and 8 (5–6 and 7–8 for 1940, 1948, and 1952) on the basis of data provided in *Current Population Report* [10], Series P-50, nos 14, 49, and 78. The 1940 data were broken down using the 1940 *Census of Population* [8], vol. III, part 1, table 13. The 1952 breakdown for translating the 5–7 class into 5–6 and 7–8 was done using the information on the educational attainment of all males by single years of school completed from the 1950 *Census of Population* [8], Detailed Characteristics, US Summary. The 1962 data are from *Special Labor Force Report* [5], no. 30, and the 1965 figures are from *Special Labor Force Report* [11], no. 65, "Education attainment of workers, March 1965."

[a] 5–6 and 7–8 for 1940, 1948, and the first part of 1952, 5–7 and 8 thereafter.

[b] Employed, 18 years and over.

Table 20.13 Mean annual earnings of males, 25 years and over by school years completed, selected years

School year completed	1939	1949	1956	1958	1959	1963
Elementary 0–4	665	1724	2127	2046	2935	2465
5–6 or 5–7	900	2268	2927	2829	4058	3409
7–8 or 8	1188	2693 2829	3732	3769	4725	4432
High school 1–3	1379	3226	4480	4618	5379	5370
4	1661	3784	5439	5567	6132	6588
College 1–3	1931	4423	6363	6966	7401	7693
4 + or 4	2607	6179	8490	9206	9255	9523
5 +	—	—	—	—	11,136	10,487

Source: Columns 1, 2, 3, 4, H. P. Miller [45, table 1, p. 966]. Column 5 from 1960 Census of Population [8], PC(2)-7B "Occupation by Earnings and Education." Column 6 computed from *Current Population Reports* [10], Series P-60, no. 43, table 22, using midpoints of class intervals and $44,000 for the over $25,000 class. The total elementary figure in 1940 broken down on the basis of data from the 1940 *Census of Population* [8]. The "less than 8 years" figure in 1949 split on the basis of data given in H. S. Houthakker [32]. In 1956, 1958, 1959, and 1963, split on the basis of data on earnings of males 25–64 from the 1959 1-in-a-1000 Census sample. We are indebted to G. Hanoch for providing us with this tabulation.

Earnings in 1939 and 1959; total income in 1949, 1958, and 1963.

4. The basic sources of data underlying table 20.8 are summarized in tables 20.12 and 20.13. Table 20.12 presents estimates of the distribution of the male labor force by school years completed for 1940, 1948, 1952, 1957, 1959, 1962, and 1964. These data are taken from various issues of the *Special Labor Force Reports* [5] and *Current Population Reports* [10], with some additional data from the 1940, 1950, and 1960 *Census of Population* [8] used to break down several classes into sub-classes. We could have used data from the 1950 and 1960 Censuses on educational attainment. The increase in the number of links did not seem to offset the decrease in comparability that would be introduced by the use of different sources of data. Table 20.2 presents estimates of the mean incomes of males (25 years and over) for these classes. These data are largely taken from Miller [45], supplemented by Census and *Current Population Reports* [10] data. Table 20.7 presents the relative incomes, the first differences of the educational distribution, and the computation of an appropriate index of the change in the average education per man.

Acknowledgements

The author's work was supported by grants from the National Science and Ford Foundations.

Notes

1 See Jorgenson [35] for details.
2 Abramovitz [1, p. 764].

3 Divisia [17, 19]. Application of these indexes to the measurement of total factor productivity is suggested by Divisia in a later publication [18, pp. 53–4]. The economic interpretation of Divisia indexes of total factor productivity has been discussed by Solow [61] and Richter [52].

4 The basic duality relationship for indexes of total factor productivity has been discussed by Siegel, [57, 58].

5 The notion of a factor price frontier has been discussed by Samuelson [54]; the factor price frontier is employed in defining changes in total factor productivity by Diamond [16] and by Phelps and Phelps]51[.

6 See, for example, Wold [64].

7 See Richter [52]. We are indebted to W. M. Gorman for bringing this fact to our attention.

8 It is essential to distinguish our basic hypothesis from a misinterpretation of it recently advanced by Denison:

Since advances in knowledge cannot increase national product without raising the marginal product of one or more factors of production, they of course disappear as a source of growth if an increase in a factor's marginal factor resulting from the advance of knowledge is counted as an increase in the quantity of factor input [14, p. 76].

In terms of our social accounting framework Denison suggests that we measure factor input as the sum of the increase in both prices and quantities; denoting the index of input implied by Denison's interpretation by X^D, gives:

$$\frac{\dot{X}^D}{X^D} = \Sigma v_j \frac{\dot{p}_j}{p_j} + \Sigma v_j \frac{\dot{X}_j}{X_j} \; ;$$

the corresponding index of output, say Y^D, would then be defined as:

$$\frac{\dot{Y}^D}{Y^D} = \Sigma w_i \frac{\dot{q}_i}{\dot{q}_i} + \Sigma w_i \frac{\dot{Y}i}{Y_i};$$

The resulting index of total factor productivity, say P^D, is constant by definition:

$$\frac{\dot{P}^D}{P^D} = \frac{\dot{Y}^D}{Y^D} - \frac{\dot{X}^D}{X^D} = 0.$$

By comparing this definition with our definition (4), the error in Denison's interpretation of our hypothesis is easily seen.

9 Here we assume that the "quantity" of a particular type of capital as an asset is proportional to its "quantity" as a service, whatever the age of the capital. If this condition is not satisfied, capital of each distinct age must be treated as a distinct asset and service. Output at each point of time consists of the usual output plus "aged" capital stock.

10 Studies in which these three methods have been employed are (1) Jaszi, Wasson, and Grose [33], Goldsmith [25], and Kuznets [39]; (2) Meyer and Kuh [44] and Denison [15]; (3) Terborgh [63].

11 This point is made by Domar [21].

12 Domar's procedure [21, p. 717, fn. 3] fails to correct for capital gains. Implicitly, Domar is assuming either no capital gains or that all capital gains are included in the value of output, whether realized or not.

13 Kendrick [38, p. 106]; see the comments by Griliches [27, p. 129]. Kendrick takes a similar position in a more recent paper [36]; see the comments by Jorgenson [35]. The treatment of capital input outlined above is based on our earlier paper [31]. The data have been revised to reflect recent revisions in the US national accounts.

14 The answer to Mrs Robinson's [53] rhetorical question, "what units is capital measured in?" is dual to the measurement of the price of capital services. Given either an appropriate measure of the flow of capital services or a measure of its price, the other measure may be obtained from the value of income from capital. Since this procedure is valid only if the necessary conditions for producer equilibrium are satisfied, the resulting quantity of capital may not be employed to *test* the marginal productivity theory of distribution, as Mrs Robinson and others have pointed out.

15 Domar [22, p. 587, formula (5)] considers a special case of this problem in which capital "is imported from the outside." This specialization is unnecessary, as suggested in the text. A more detailed discussion of this issue is presented by Jorgenson [35].

For constant rates of growth of the relative error in the investment goods price index and the level of investment, formula (10) may be expressed in closed form:

$$\frac{\dot{P}^*}{P^*} - \frac{\dot{P}}{P} = -w_I \frac{\dot{Q}^*}{Q^*} + v_K \frac{\dot{Q}^*}{Q^*},$$

$$= (v_K - w_I) \frac{\dot{Q}^*}{Q^*}.$$

16 See Griliches [28] and the references given there.

17 Jorgenson [35].

18 To make stocks of labor and capital precisely analogous, it would be necessary to go even further. Unemployed workers should be included in the stock of labor since unemployed machines are included in the stock of capital. Workers should be aggregated by means of discounted lifetime incomes since capital goods are aggregated by means of asset prices.

19 The growth of the output price index may be compared with that for personal consumption expenditures, which grows from 76.5 to 108.6 from 1947 through 1959. The close parallel between the output price index for construction and the price of consumption goods suggests an explanation for the difference in rates of growth of prices of consumption and investment goods described by Gordon [26]. This difference results from the error of measurement in using an input price index in place of an output price index for investment goods. If this error is corrected the difference vanishes.

20 Griliches [28, table 8, last column, p. 397].

21 Foss [24]. See the statistical appendix for further details.

22 Further details are given in the statistical appendix.

23 Additional details on relative prices for labor services are presented in the statistical appendix, table 20.12.
24 Additional details on the distribution of the labor force are presented in the statistical appendix, table 20.11.
25 Abramovitz [1, p. 776].
26 Errors in the aggregation of labor services account for 0.48 percent per year, but this is offset by errors of measurement in the relative utilization of labor of −0.15 percent per year so that the net correction for errors of measurement of labor is 0.33 percent per year.

An alternative interpretation of our results may be provided by analogy with the conceptual framework for technical change discussed by Diamond [16]. Errors of measurement in the growth of labor services may be denoted labor-diminishing errors of measurement; capital-diminishing errors of measurement may be separated into embodied and disembodied errors. Errors in capital due to errors in the measurement of prices of investment goods are analogous to embodied technical change. Finally, some of the errors in measurement affect levels of output; these errors may be denoted output-diminishing errors of measurement.

A decomposition of total errors of measurement into labor-diminishing, capital-diminishing, embodied and disembodied, and output-diminishing is as follows: Labor-diminishing errors of measurement contribute 0.33 percent per year to the initial measured rate of growth of total factor productivity. Embodied capital-diminishing errors contribute 0.28 percent per year and disembodied capital-diminishing errors contribute 0.99 percent per year. Finally, output-diminishing errors of measurement of 0.10 percent per year must be set off against the input-diminishing errors totaling 1.60 percent per year.

27 See, for example, the studies of Minasian [47] and Mansfield [42].
28 See Levhari [40, 41] for an elaboration of this point.
29 Solow [59, p. 58–9].
30 For further discussion of this point see Jorgenson [35].

References

[1] Abramovitz, Moses, "Economic growth in the United States," *American Economic Review*, 52(4), (September 1962), 762–82.
[2] Abramovitz, Moses, *Resource and Output Trends in the United States since 1870.* Occasional Paper 63, New York, National Bureau of Economic Research, 1950.
[3] Arrow, K. J., "The economic implications of learning by doing," *Review of Economic Studies*, 29(3), no. 80 (June 1962), 155–73.
[4] Bureau of Labor Statistics, *Consumers' Price Index.* Washington, DC: US Department of Labor, various monthly issues.
[5] ——, *Special Labor Force Reports.* Washington, DC: US Government Printing Office.
[6] ——, *Wholesale Prices and Price Indexes.* Washington, DC: US Department of Labor, various monthly issues.

[7] Bureau of the Census, *Census of Manufactures*. Washington, DC: US Government Printing Office.

[8] ——, *Census of Population*. Washington, DC: US Government Printing Office.

[9] ——, *Continuation to* 1962 *of Historical Statistics of the U.S.* Washington, DC: US Government Printing Office.

[10] ——, *Current Population Reports*. Washington, DC: US Government Printing Office.

[11] ——, *Survey of Manufactures*. Washington, DC: US Government Printing Office.

[12] Dacy, D., "A price and productivity index for a nonhomogeneous product," *Journal of the American Statistical Association*, 59(306), (June 1964), 469–80.

[13] Denison, E. F., "Discussion," *American Economic Review*, 66(2), (May 1966), 76–8.

[14] ——, "Measurement of labor input: some questions of definition and the adequacy of data," in Conference on Research in Income and Wealth, *Output, Input, and Productivity Measurement*, Studies in Income and Wealth, vol. 25. Princeton: Princeton University Press, 1961, pp. 347–72.

[15] ——, *The Source of Economic Growth in the United States and the Alternatives Before Us*, Supplementary Paper no. 13. New York: Committee for Economic Development, 1962.

[16] Diamond, P. A., "Technical change and the measurement of capital and output," *Review of Economic Studies*, 32(4), no. 92 (October 1965), 289–98.

[17] Divisia, F., *Économique Rationnelle*. Paris: Gaston Doin et Cie, 1928.

[18] ——, *Exposés d'économique*, vol. I. Paris: Dunod, 1952.

[19] ——, "L'indice monétaire et la théorie de la monnaie," *Revue d'Économie Politique*, 39e Année, no. 4, 5, 6; Juillet–Août, Septembre–Octobre, Novembre–Décembre, 1925, pp. 842–61, 980–1008, 1121–51.

[20] Ibid., 40e Année, no. 1, Janvier–Février, pp. 49–81.

[21] Domar, E. D., "On the measurement of technological change," *Economic Journal*, 71(284), (December 1961), 709–29.

[22] ——, "Total productivity and the quality of capital," *Journal of Political Economy*, 71(6), (December 1963), 586–8.

[23] Fabricant, S., *Basic Facts on Productivity Change*. Occasional Paper 63, New York, National Bureau of Economic Research, 1959.

[24] Foss, M., "The utilization of capital equipment," *Survey of Current Business*, 43(6), (June 1963), 8–16.

[25] Goldsmith, R., *A Study of Saving in the United States*. Princeton: Princeton University Press, 1955.

[26] Gordon, R. A., "Price changes: consumers' and capital goods," *American Economic Review*, 51(5), (December 1961), 937–57.

[27] Griliches, Z., "Comment," *American Review*, 51(2), (May 1961), 127–30.

[28] ——, "Notes on the measurement of price and quality changes," in Conference on Research in Income and Wealth, *Models of Income Determination*. Princeton: Princeton University Press, 1964, pp. 381–404.

[29] ——, "Production functions in manufacturing: some preliminary results," in M. Brown (ed.), *The Theory and Empirical Analysis of Production*

[31] NBER, Studies in Income and Wealth, vol. 31. New York: Columbia University Press, 1967, pp. 275-340.

[30] ——, "The sources of measured productivity growth: United States agricultural, 1940-60," *Journal of Political Economy*, 71(4), (August 1963), 331-46.

[31] ——, and Jorgenson, D., "Sources of measured productivity change: capital input," *American Economic Review*, 56(2), (May 1966), 50-61.

[32] Houthakker, H. S., "Education and income," *Review of Economics and Statistics*, 41(1), (February 1959), 24-8.

[33] Jaszi, G., Wasson, R., and Grose, L., "Expansion of business fixed capital in the United States," *Survey of Current Business*, 42, (November 1962), 9-18.

[34] Jorgenson, D., "Comment" on J. W. Kendrick "Industry Changes in Nonlabor Costs," in J. W. Kendrick (ed.), *The Industrial Composition of Income and Product*, NBER, Studies in Income and Wealth, vol. 32. New York: Columbia University Press, 1968, pp. 176-84.

[35] ——, "The embodiment hypothesis," *Journal of Political Economy*, 74(1), (February 1966), 1-17.

[36] Kendrick, J. W., "Industry Changes in Nonlabor Costs," in J. W. Kendrick (ed.), *The Industrial Composition of Income and Product*, NBER, Studies in Income and Wealth, vol. 32, New York: Columbia University Press, 1968, pp. 151-76.

[37] ——, *Productivity Trends in the United States*. Princeton: Princeton University Press, 1961.

[38] ——, "Some theoretical aspects of capital measurement," *American Economic Review*, 51(2), (May 1961), 102-11.

[39] Kuznets, S., *Capital in the American Economy*. Princeton: Princeton University Press, 1962.

[40] Levhari, D., "Extensions of Arrow's 'learning by doing'," *Review of Economic Studies*, 33(2), 94 (April 1966), 117-32.

[41] —— "Further implications of learning by doing," *Review of Economic Studies*, 33(1), 93 (January 1966), 31-8.

[42] Mansfield, E., "Rates of return from industrial research and development," *American Economic Review*, 55(2) (May 1965), 310-22.

[43] Massell, B. F., "A disaggregated view of technical change," *Journal of Political Economy*, 69(6) (December 1961), 547-57.

[44] Meyer, J. and Kuh, E., *The Investment Decision*. Cambridge: Harvard University Press, 1957.

[45] Miller, H. P., "Annual and lifetime income in relation to education," *American Economic Review*, 50(5) (December 1960), 962-86.

[46] Mills, F. C., *Productivity and Economic Progress*, Occasional Paper 38. New York: National Bureau of Economic Research, 1952.

[47] Minasian, J., "The economics of research and development," in Universities-National Bureau Committee for Economic Research, *The Rate and Direction of Inventive Activity*. Princeton: Princeton University Press, 1962, pp. 93-142.

350   EXPLANATIONS OF PRODUCTIVITY GROWTH

[48] Office of Business Economics, *1966 Capital Stock Study*. Washington, DC: US Department of Commerce, no date.
[49] ——, *The National Income and Product Accounts of the United States, 1929–1965. A Supplement to the Survey of Current Business*. Washington, DC: US Department of Commerce, 1966.
[50] Okun, A. M., "Potential GNP: its measurement and significance," *Proceedings of the Business and Economic Statistics Section of the American Statistical Association*, 1962, pp. 98–104.
[51] Phelps, E. S., and Phelps, C., "Factor-price-frontier estimation of a 'vintage' production model of the postwar US nonfarm business sector," *Review of Economics and Statistics*, 48(3) (August 1966), 251–65.
[52] Richter, M. K., "Invariance axioms and economic indexes," *Econometrica*, forthcoming.
[53] Robinson, J., "The production function and the theory of capital," *Review of Economic Studies*, 21(2), 55 (1953–4), 81–106.
[54] Samuelson, P. A., "Parable and realism in capital theory: the surrogate production function," *Review of Economic Studies*, 29(3), 80 (June 1962), 193–206.
[55] Schmookler, J., "The changing efficiency of the American economy, 1869–1938," *Review of Economics and Statistics*, 34(3) (August 1952), 214–31.
[56] Schultz, T. W., "Education and economic growth." In N. B. Henry (ed.), *Social Forces Influencing American Education*. Chicago: University of Chicago Press, 1961.
[57] Siegel, I. H., *Concepts and Measurement of Production and Productivity*. US Bureau of Labor Statistics, March 1952.
[58] ——, "On the design of consistent output and input indexes for productivity measurement," in Conference on Research in Income and Wealth, *Output, Input and Productivity Measurement*, Studies in Income and Wealth, vol. 25. Princeton: Princeton University Press, 1961, pp. 23–41.
[59] Solow, R. M., *Capital Theory and the Rate of Return*. Chicago: Rand-McNally, 1964.
[60] ——, "Investment and technical progress." In K. J. Arrow, S. Karlin and P. Suppes (eds), *Mathematical Methods in the Social Sciences*, 1959, Stanford, Stanford University Press, 1960, pp. 89–104.
[61] ——, "Technical change and the aggregate production function," *Review of Economics and Statistics*, 39(3) (August 1957), 312–20.
[62] ——, "Technical progress, capital formation, and economic growth," *American Economic Review*, 52(3) (May 1962), 76–86.
[63] Terborgh, G., *Sixty Years of Business Capital Formation*. Washington, DC: Machinery and Allied Products Institute, 1960.
[64] Wold, H., *Demand Analysis*. New York: Wiley and Sons, 1953.

21
Excerpts from: "Issues in Growth Accounting: A Reply to Edward F. Denison"*
(With D. W. Jorgenson)

3.3 Price of Investment Goods

The price indexes used by Christensen and Jorgenson in constructing the capital stock series differ from our original ones in using the national income implicit deflator for producers' durable equipment and the WPI as the deflator of the *stock* of inventories. There is enough evidence that the various official capital deflator series are biased upward during this period for us to be unwilling to concede that our original attempt to substitute something else (the CPI durables index) for the official equipment investment deflator was an error. While this is not the place to go into great detail, there is ample evidence that components of the WPI, which in turn are a major source of deflators for the producers' durables investment, are (or at least have been) rather poor measures of price change. The WPI is based almost entirely on company and trade papers and association reports. Moreover, for a variety of reasons, it has had much less resources devoted to it relative to the CPI. All this has combined to produce what we believe to be a significant upward drift in components of this index during the post-World War II period.[1]

Our example of consumer durables was not intended to claim that the particular items were representative of most of the producers' durables but rather that such a comparison allowed one to detect the magnitude of the drift in the WPI which was due to the particular way in which its data were collected. The difference between the movement of prices for these identical items in the two index sources was interpreted not as property of the particular items, but as an estimate of the bias introduced by the basic procedure used in collecting the wholesale price data. The latter, we assumed, was generalizable to most of the other WPI items.

*From *Survey of Current Business*, May 1972, pp. 65–94.

Actually, there is quite a bit more evidence on this point than was alluded to in our original paper and some of it is presented in table 21.1. The first line recapitulates the CPI–WPI identical durables comparison. The other comparisons can be divided into three groups: (1) transaction price data (circuit breakers and power transformers from the Dean-DePodwin study[10][1] and tubes and batteries prices from Flueck's staff report [14]); (2) more detailed attention to quality change and/or more analysis of the changing specifications of the priced items, sometimes via regression techniques (Dean-DePodwin and Census on steam generators, Barzel on electric equipment, the Association of American Railroads on railroad equipment prices, and Fettig on tractor prices); and (3) wider coverage and transaction pricing (Census unit values data).

The last, Census-based, set of data (summarized in table 21.2) is particularly interesting since one might have expected that unit values would themselves be upward biased due to the secular shift to more elaborate, higher "quality" models. In fact, they and all the other additional comparisons point strongly to the existence of an upward bias in the comparable WPI components, at least in the recent past. Our implied estimate of this upward drift of 1.4 percent per year between 1950 and 1962 is quite consistent with the new evidence presented in this table. While it is not used in the productivity computations we borrow from Christensen and Jorgenson we are willing to stand by this part of our original estimates.[2]

Our substitution of the new OBE "constant cost 2" construction deflator for the comparable implicit GNP deflator component is not ideal and could be improved on. The "constant cost 2" deflator is an average, implicitly, of the Bureau of Public Roads highway structures, the Bureau of Reclamation pumping and power plant indexes, and the AT&T and Turner construction cost indexes. The latter two are basically input price rather than output price indexes with some feeble adjustment for productivity changes.[3] The Bureau of Reclamation indexes are hard to interpret and seem to be based, to a large extent, on list prices of raw materials. A recent study by Gordon [40] indicates that the constant cost 2 index may also be biased upward to an unknown degree.[4] It is likely, therefore, that if a more accurate construction price index were used it would imply a higher rate of growth in the structures component of capital input than was estimated in our original paper and is also used in this one. In short, more remains to be done in this area but we believe that our original procedures were on the right track. The estimates we borrow from Christensen and Jorgenson are conservative in their choice of investment deflators.

4.1 Introduction

It has been common to assume that one may be able to approximate the unemployment of capital by the unemployment of labor. Solow [26] assumed that there is a proportionality relationship between these concepts (and his

Table 21.1 Evidence on drift in components of WPI

Item	Reference	Period	Approximate drift in percent per year[a]
Identical consumer durables[b] (10 items)	CPI	1947-9-58	1.9
Circuit breakers	Dean-DePodwin[c]	1954-9	4.0
Power transformers	Dean-DePodwin	1954-9	0.7
Power transformers	Census[d]	1954-63	1.2
Steam generators	Dean-DePodwin	1954-9	1.9
Steam generators	Census[a]	1954-63	6.4
Electric equipment	Dean-DePodwin	1954-9	1.2
Electric equipment	Census[d]	1954-63	1.9
Electric equipment	Barzel[f]	1949-59	4.4
Railroad equipment	Association of American Railroads[g]	1961-7	0.8
Tractors	Fettig[h]	1950-62	0.6
Tubes, automobile	Flueck[i]	1955-9	1.4
Batteries, vehicle	Flueck[i]	1949-60	6.3
Storage batteries	Census[d]	1954-63	2.9
Plumbing and heating	Census[d]	1954-63	1.2
Oil burners	Census[d]	1954-63	2.8
Warm air furnaces	Census[d]	1954-63	1.1
Metal doors	Census[d]	1954-63	0.7
Bolts and nuts	Census[d]	1954-63	2.3
Internal combustion engines	Census[d]	1954-63	1.8
Elevators and escalators	Census[d]	1954-63	1.1
Pumps and compressors	Census[d]	1954-63	2.0
Integrating instruments	Census[d]	1954-63	3.1
Electric welding	Census[d]	1954-63	− 1.1
Electric lamps	Census[d]	1954-63	1.1
Trucks	Census[d]	1954-63	0.3

[a] Last column is the average change, over the specified period, in the particular WPI component relative to the estimated price change over the same period in the alternative source.
[b] The following items were compared for this period: automobiles, tires, radios, refrigerators, sewing machines, ranges, washing machines, vacuum cleaners, toasters, and furniture.
[c] Dean and DePodwin [10] and an unpublished appendix to the original General Electric version.
[d] 1963 *Census of Manufactures* [4], vol. IV, *Indexes of Production*, appendix A.
[e] Census unit values, adjusted for capacity and horsepower differences, 1963 Census of Manufactures [5], vol. IV, *Indexes of Production*, appendix A.
[f] Barzel [2]. Indexes in table 3 holding size constant are essentially flat throughout this period. A similar story is also told by the indexes in table 6, where size is taken into account.
[g] Joint Equipment Committee Report [27] shows no significant increase in the "cost" of locomotives and freight and passenger cars during this period.
[h] Fettig [13], table 6, p. 609.
[i] J. Flueck [14].

Table 21.2 A comparison of OBE producers' durables investment deflators with census unit value indexes, 1962 (1954 = 100)

Category	Percentage direct[a] coverage by data from Census	Census[a] (cross weights)	OBE[b]	Drift in percentage per year
Furniture and fixtures	42	110.9	119.1	0.8
Fabricated metal products	34	117.3	121.7	0.4
Engines and turbines	54	93.3	134.7	4.2
Construction machinery	20	126.2	132.0	0.5
Metalworking machinery	42	122.9	137.2	1.2
Special industry machinery	20	119.3	138.7	1.7
General industry machinery[c]	15	116.9	131.4	1.3
Service industry machinery	27	82.3	100.9	2.3
Electric machinery	27	98.7	112.0	1.4
Trucks and buses[d]	91	118.0	122.5	0.4
Ships and boats	27	100.1	116.6	1.7
Railroad equipment	46	132.1	128.3	−0.3

[a] 1963 *Census of Manufactures* [15], vol. IV, *Indexes of Production*, appendix A.
[b] NIP [24], table 8.8. For tractors, agricultural machinery, mining and oil field machinery, office equipment, passenger cars, aircraft, and instruments Census unit values are based on less than 15 percent coverage from Census sources. For a comparison of tractor price indexes see table 21.1.
[c] OBE definition includes also materials handling machinery.
[d] Four separate Census categories aggregated using 1963 shipments as weights.

capital measure included land and buildings, too!) while Okun [25] suggested a nonlinear relationship between the two. It appeared to us that the unemployment of capital can be better approximated by the "unemployment" of one kind of capital (power-driven equipment), implicitly assuming a proportionality relationship between this type of capital and other capital, than by the assumption of proportionality between the employment of all labor and of all capital.

It is our assumption, for which we have no explicit evidence, that our measure of utilization measures not only the utilization of power-driven equipment but also the fraction of calendar time that establishments or plants are in actual operation. That is, machine-hours per week are interpreted as a proxy for total hours per week operated by an establishment or industry. This, of course, is not an unambiguous concept, but it does explain why we were and still are willing to apply this estimated utilization rate not only to equipment but also to buildings. We are also willing, for lack of any better evidence, to extrapolate this to all industrial and agricultural equipment and structures and also to structures and equipment in the service industries. There is some scattered evidence that the hours operated per week by various retail establishments have increased in recent years.

4.2   Measurement of Relative Utilization

In measuring the change in utilization between 1945 and 1954 by the average estimated change in utilization (per annum) between 1939 and 1954, we overestimated the former. The estimates used in this paper (also taken from Christensen and Jorgenson) solve this problem by adding a cyclical adjustment to the previously computed secular one. The benchmark years are now used only to derive the ratio of installed horsepower to potential capital. This ratio is assumed to change slowly and is interpolated linearly between benchmarks. Installed horsepower is then estimated as the product of this ratio and our index of potential flow of (business) capital services. The ratio of electric power consumed by motors to this estimate of installed horsepower is our new measure of relative utilization. The resulting series grows at a significantly lower rate, 0.54 percent per year, during the 1950–62 period than the utilization index used in our original study (which rose at 1.06 percent per year).

Denison suggests that the weighting of utilization estimates for industry groups should be done by something other than the total horsepower of electric motors. Since we use it as a proxy for the utilization of all capital, the appropriate weights would be estimates of the value of capital services at the two-digit level. The closest we can come to it is to use weights based on the distribution of total fixed assets in 1962. Recomputing our estimates separately for each two-digit industry and then weighting them with these weights doesn't really change the numbers significantly (see table 21.3). If anything, it makes them slightly higher. The same is also true for mining during the 1954 to 1963 period (see table 21.4). The resulting weighted utilization index is still quite high and of the same order of magnitude as the manufacturing one (if allowance is made for the cyclical difference between 1963 and 1962). We conclude, therefore, that the unweighted figures we used are rather close to what the weighted figures would have been had we computed them.

Thus, except for the over-estimate of the rate of change of utilization from 1945 to 1954, our estimates appear to be reasonably good estimates of the rate of utilization of electric motors in manufacturing. Similar estimates were presented for mining in table 21.4. An entirely different set of estimates, based on actual machine-hours worked for three textile subindustries, is presented in table 21.5. They, too, indicate an upward trend in utilization in the post-World War II period of about the same order of magnitude. Thus, there is something in these data. They are measuring something, at least as far as the utilization of electric motors in manufacturing and mining is concerned.

Given our data, it was an error on our part (and on the part of those who preceded us on this path) to adjust the residential housing, land, and inventories components by this measure of capacity utilization. Until better evidence comes along, however, we are willing to hazard the very strong assumption that the capacity utilization of all *business* equipment and

Table 21.3 Relative utilization of electric motors, US manufacturing, 1962

| Industry[a] | *Indexes, 1954 = 1.000* | | | |
	Horsepower of electric motors[b] *(1)*	*Total electricity consumption*[c] *(2)*	*Utilization*[d] *(3)*	*Total fixed assets weight*[e] *(4)*
20	1.420	1.539	1.084	0.103
21	1.446	1.794	1.241	0.004
22	1.155	1.229	1.064	0.036
24	1.543	1.289	0.835	0.023
25	1.247	1.438	1.153	0.008
26	1.616	1.624	1.005	0.070
27	1.833	2.385	1.301	0.034
28	1.552	1.769	1.140	0.122
29	1.537	1.765	1.148	0.069
30	1.554	1.579	1.016	0.024
31	1.158	1.335	1.153	0.004
32	1.529	1.447	0.944	0.055
33	1.289	1.394	1.081	0.165
34	1.289	1.488	1.154	0.049
35 and 36	1.344	1.713	1.275	0.119
37	1.173	1.505	1.283	0.076
38	1.234	2.187	1.773	0.012
39 and 19	1.082	1.336	1.235	0.014
Total[f]	1.386	1.567	1.131	—
Total weighted[g]	—	—	1.135	—

[a] "Two-digit" manufacturing industries. Industry 23 apparel, excluded because no horsepower figures were asked for in 1954.
[b] Horsepower of electric motors from 1963 *Census of Manufactures* [7], "Power equipment in manufacturing industries as of December 31, 1962", MC 63 (1)-6, table 2.
[c] Electricity, total purchased and generated minus sold, from 1963 *Census of Manufactures* [7], "Fuels and electric energy consumed in manufacturing industries: 1962", MC 63 (1)-7, table 3.
[d] Utilization: column 2/column 1.
[e] 1962 fixed assets weights computed from 1964 *Annual Survey of Manufacturers* [6], M 65 (AS)-6.
[f] Numbers differ from table X in Jorgenson and Griliches [21], because no allowance could be made at the two-digit level for electricity consumption in nuclear energy installations. The comparable utilization index for total manufacturing allowing for this is 1.111.
[g] Σ (column 3 × column 4)/0.987, where $0.987 = \Sigma$ column 4.

structures may be approximated by our estimate of capacity utilization of power-driven equipment in manufacturing (and mining). Business equipment and structures account for about 46 percent of our total capital input. Applying this to the reduced rate of growth in utilization leads to a utilization adjustment on the order of 16 percent of our previous adjustment.

Table 21.4 Equipment utilization indexes, mining industries, 1963 (1954 = 100)

Industry	Horsepower of electric motors[a] (1)	Electricity consumption[b] (2)	Utilization index[c] (3)	Depreciable assets weights[d] (4)
Metal mining	111.3	175.0	157.2	0.246
Anthracite	42.4	51.7	122.0	0.014[e]
Bituminous coal	99.4	134.5	135.3	0.134[e]
Oil and gas	224.0	229.6	102.5	0.432
Nonmetallic minerals	152.2	156.9	103.1	0.174
Total mining	126.6	149.3	117.9	—
Adjusted	—	—	117.6	—
Weighted	—	—	120.7	—

[a] 1963 Census of Mining [5] chapter 7, table 1.
[b] 1963 Census of Mining [5] chapter 6, table 1; purchased and used.
[c] Column 2/column 1.
[d] From US Internal Revenue Service, *1963 Statistics of Income* [55], *Corporation Income Tax Returns*, table 37, col. 3, p. 264.
[e] Total "coal mining" weight allocated on the basis of 1954 data for total capital given in Creamer [9], table B–11, p. 318.
[f] Adjusted for a small implied change in percentage of electric power used by electric motors (from 93.5 to 93.3) using the 1945 percentages given by Foss [15] and the 1954 and 1963 total electricity consumption as weights.
[g] Σ (column 3 × column 4).

7.9   Conclusions and Suggestions for Further Research

We have summarized the differences among our estimates of the rate of growth of total factor productivity for the period 1950–62, based on the results of Christensen and Jorgenson [8], our original estimates [21], and Denison's estimates [12]. At this point it is useful to compare these alternative estimates and to attempt a reconciliation among them; a partial reconciliation is given in table 21.6. From this comparison it is apparent that our new estimates represent a compromise between our original position and Denison's position. Referring to table 21.6, we may now summarize our conclusions. From an empirical point of view the greatest differences among our original estimates, our revised estimates, and Denison's estimates are in the adjustment for utilization of resources. Denison estimates that the utilization of resources declines between 1950 and 1962. We estimate that utilization increased, but by considerably less than we originally suggested. The revision in our adjustment for relative utilization accounts for 0.47 percent per year of the total discrepancy of 0.73 percent per year between our original estimate of the rate of growth of total factor productivity and our revised estimate.

From a conceptual point of view the greatest difference among alternative procedures is in the allocation of income from property among its

Table 21.5   Selected utilization measures

Year	Cotton broad woven goods: Average loom hours per loom in place[a]	Cotton-system spindle hours per spindle in place[b]	Manmade fiber broadwoven goods: average loom hours per loom in place[a]
1947	5,042	5,074	5,220
1948	5,161	5,305	5,408
1949	4,689	4,433	4,991
1950	5,547	5,048	5,532
1951	5,276	5,823	5,045
1952	5,046	4,919	4,970
1953	5,579	5,513	5,240
1954	5,431	5,141	4,802
1955	5,658	5,501	5,326
1956	5,837	5,783	5,036
1957	5,425	5,512	5,463
1958	5,499	5,311	5,397
1959	6,114	5,853	5,718
1960	6,145	6,216	5,844
1961	6,020	5,830	5,717
1962	6,061	6,283	6,042
1963	6,124	6,074	6,105
1964	6,450	6,243	6,412
1965	6,741	6,489	6,513
Rates of growth, percentage per year:			
1950–62	0.8	1.8	0.7
1947–65	1.6	1.4	1.7

[a] Computed from various issues of *Current Industrial Reports* [7], series M22T.1 and M22T.2 1947–53: Looms in place are averages of quarterly data as of the end of the quarter; 1954–64: Looms in place are averages of beginning and of year figures; 1965 for cotton broadwoven goods extrapolated on the basis of averages of monthly data on average hours per loom per week from the American Textile Manufacturers Institute [1], for manmade fibers based on looms in place at the end of 1964.
[b] Bureau of the Census, *Cotton Production and Distribution* [6] page 37. This is a more variable series, since the denominator is available only once during each year.

components. Except for our assumption that replacement requirements should be estimated by the double declining balance formula, our estimates of capital stock for each class of assets are very similar to Denison's estimates. Our estimates of capital input differ very substantially from his due to differences in treatment of the tax structure for property income, the use of real rates of return rather than nominal rates for each class of assets, and the use of declining balance depreciation and replacement. Part of the unexplained residual between our version of Denison's estimate of total factor productivity and his own is accounted for by his separation of assets among those held by housing, agricultural, and all other sectors of the economy. This separation

Table 21.6 Reconciliation of alternative estimates of growth in total factor productivity, 1950–62 (percentage per year)

Denison, adjusted for utilization, his data		1.41
Denison's utilization adjustment	−0.04	
Denison, unadjusted, his data		1.37
Unexplained difference	0.07	
Denison, unadjusted, our data		1.44
Capital input:		
Quality change	0.30	
Our utilization adjustment	0.11	
Jorgenson–Griliches, adjusted, revised		1.03
Revision in utilization adjustment	0.47	
Other revisions	0.26	
Jorgenson–Griliches, adjusted, original		0.30

goes part of the way toward a satisfactory treatment of the tax structure, but should be replaced, in our view, by a breakdown by legal form of organization.

In revising our original computations we have made a number of conservative assumptions and did not correct for some obvious errors in the data where the data base for such adjustments appeared to be too scanty. This is particularly true of the deflators of capital expenditures that we used and of our measure of land input. More research is needed on these and on the magnitude and sources of changes in utilization rates, on capital deterioration and replacement rates, and on the changing characteristics of the labor force.

While better data may decrease further the role of total factor productivity in accounting for the observed growth in output, they are unlikely to eliminate it entirely. It is probably impossible to achieve our original program of accounting for all the sources of growth within the current conventions of national income accounting. But this is no reason to accept the current estimates of total factor productivity as final. Their residual nature makes them intrinsically unsatisfactory for the understanding of actual growth processes and useless for policy purposes.

To make further progress in explaining productivity change will require the extension of such accounts in at least three different directions: (1) allowing rates of return to differ not only by legal form of organization but also by industry and type of asset; (2) incorporating the educational sector into a total economy-wide accounting framework; and (3) constructing measures of research (and other intangible) capital and incorporating them into such productivity accounts.

To allow rates of return to differ among industries and assets would require a much more detailed data base than is currently available and would introduce the notion of disequilibrium (at least in the short and intermediate runs) into such accounts. Such a framework would be consistent with a more general view of sources of growth[5] and would introduce explicitly the changing industrial composition of output as one such source.

In measuring labor input, OBE data on persons engaged should include estimates of the number of unpaid family workers, such as those of Kendrick [22, 23]. Estimates of man-hours for different components of the labor force should be compiled on a basis consistent with data on persons engaged as Kendrick has done. Although Denison [12] has given additional evidence in support of his adjustment of labor input for intensity of effort, a satisfactory treatment of this adjustment requires data on income by hours of work, holding other characteristics of the labor force constant. Until such data become available it may be best to exclude this adjustment from the measure of real labor input incorporated into the national accounts. Quality adjustments for labor input based on such characteristics of the labor force as age, race, sex, occupation, and education should be incorporated into the labor input measure.

The basic accounting framework should also be expanded to incorporate investment in human capital along with investment in physical capital. Investment in human capital is primarily a product of the educational sector, which is not included in the private domestic sector of the economy. In addition to data on education already incorporated into the national accounts, data on physical investment and capital stock in the educational sector would be required for incorporation of investment in human capital into growth accounting.

Another issue for long-term research is the incorporation of research and development into growth accounting. At present research and development expenditures are treated as a current expenditure. Labor and capital employed in research and development activities are comingled with labor and capital used to produce marketable output. The first step in accounting for research and development is to develop data on factors of production devoted to research. The second step is to develop measures of investment in research and development.[6] The final step is to develop data on the stock of accumulated research. A similar accounting problem arises for advertising expenditures, also currently treated as a current expenditure.

Both education and investment in research and development are heavily subsidized in the United States, so that private costs and returns are not equal to social costs and returns. The effects of these subsidies would have to be taken into account in measuring the effects of human capital and accumulated research on productivity in the private sector. If the output of research activities is associated with external benefits in use, these externalities would not be reflected in the private cost of investment in research. Some way must be found to measure these externalities. Once such measures are developed and the growth accounts expanded accordingly, this would result in a significant departure from the conventions of national accounting, more far-reaching than the departures contemplated in our original paper. A new accounting system is required to comprehend the whole range of possible sources of economic growth.

Notes

1 Detailed evidence on the quality of the price quotations underlying the WPI is presented by Flueck [14].
2 See Gordon [16] for additional evidence supporting this position.
3 The AT&T structures index uses American Appraisal Company indexes with essentially negligible productivity adjustments since 1955.
4 Gordon's "final price of structures" index rises by 11 percent less between 1950 and 1965 than the constant cost 2 deflator. See Gordon [17] table A-1, pp. 427-8. Gordon errs, in a paper published a year later than ours, in failing to notice that the final version of our paper did not incorporate the Bureau of Public Roads index as a deflator but used the more representative but still imperfect OBE constant cost 2 index.
5 See Johnson [20] for an outline of a similar position.
6 See Griliches [18] for further discussion of this topic, and for some order-of-magnitude estimates.

References

[1] American Textile Manufacturers Institute, *Textile Highlights*, various monthly issues.
[2] Barzel, Y., 1964, "The production function and technical change in the steam-power industry," *Journal of Political Economy*, 72, 133-50.
[3] Bureau of the Census, *Annual Survey of Manufactures*. Washington, DC: US Government Printing Office.
[4] ——, *Census of Manufactures, 1963*. Washington, DC: US Government Printing Office.
[5] ——, *Census of Mining*, 1963. Washington, DC: US Government Printing Office.
[6] ——, 1966, *Cotton Production and Distribution*, Bulletin 202. Washington, DC: US Government Printing Office.
[7] ——, *Current Industrial Reports*. Washington, DC: US Government Printing Office.
[8] Christensen, L. R. and Jorgenson, D. W., 1970. "U.S. real product and real factor input, 1929-1967, *Review of Income and Wealth*", Series 16, pp. 19-50.
[9] Creamer, D., Dobrovolsky, S. P., and Berenstein, I., 1960, *Capital in Manufacturing and Mining: Its Formation and Financing*. Princeton: Princeton University Press.
[10] Dean, C. R., and DePodwin, H. J., 1961, "Product variation and price indexes: a case study of electrical apparatus," *Proceedings of the Business and Economic Statistics Section of the American Statistical Association*, 271-9.

[11] Denison, E. F., 1969, "Some major issues in productivity analysis: an examination of estimates by Jorgenson and Griliches," *Survey of Current Business*, 49, part II, 1–27.

[12] ——, 1967, *Why Growth Rates Differ: Postwar Experience in Nine Western Countries*. Washington, DC: The Brookings Institution.

[13] Fettig, L., 1963, "Adjusting farm tractor prices for quality changes, 1950–1962, "*Journal of Farm Economics*, 45, 599–611.

[14] Flueck, J., 1961, "A study in validity: BLS wholesale price quotations," in G. J. Stigler (ed.), *Price Statistics of the Federal Government*. Washington, DC: US Government Printing Office, pp. 419–58.

[15] Foss, M., 1963, "The utilization of capital equipment, "*Survey of Current Business*, 43, 8–16.

[16] Gordon, R. J., 1971, "Measurement bias in price indexes for capital goods," *The Review of Income and Wealth*, 17, 121–74.

[17] ——, 1968, "A new view of real investment in structures, 1917–66," *Review of Economics and Statistics*, 50, 417–28.

[18] Griliches, Z., 1973, "Research expenditures and growth accounting," in B. R. Williams (ed.) *Science and Technology in Economic Growth*, London, MacMillan Press, pp. 59–95.

[19] Internal Revenue Service, *Statistics of Income, 1963, Corporation Income Tax Returns*, Washington, DC: US Government Printing Office.

[20] Johnson, H. G., 1964, "Comment." In OECD, *The Residual Factor and Economic Growth*. Paris, pp. 219–27.

[21] Jorgenson, D. W., and Griliches, Z., 1967, "The explanation of productivity change," *Review of Economic Studies*, 34, 249–83.

[22] Kendrick, J. W., 1961, *Productivity Trends in the United States*. Princeton: Princeton University Press.

[23] ——, 1973, *Postwar Productivity Trends in the United States, 1948–1969*. New York: National Bureau of Economic Research, Columbia University Press.

[24] Office of Business Economics, 1966, *The National Income and Product Accounts of the United States, 1929–65, A Supplement to the Survey of Current Business*. Washington, DC: US Department of Commerce.

[25] Okun, A., 1962, "Potential GNP: its measurement and significance," *Proceedings of the Business and Economic Statistics Section of the American Statistical Association*, pp. 98–104.

[26] Solow, R. M., 1962, "Technical progress, capital formation, and economic growth," *American Economic Review*, 52, 76–86.

[27] Joint Equipment Committee Reprint, 1986, "Costs of railroad equipment and machinery." Washington, DC: Association of American Railroads, 1 October.

22
Postscript on Productivity Studies

The work on productivity measurement has entered the stream of "normal" science and is too wide now to be reviewed or summarized effectively in this context. The Jorgenson and Griliches paper was extended and transmuted in a number of subsequent papers by Jorgenson and his co-workers. The most important landmarks here are Christensen and Jorgenson (1973), Gollop and Jorgenson (1980), Fraumeni and Jorgenson (1980), Jorgenson and Nishimizu (1978) and Jorgenson et al. (1987). Denison's work is summarized in Denison (1985), and Kendrick's in Kendrick and Grossman (1980) and Kendrick (1983). The BLS has recently also started producing "official" multi-factor productivity measures (see BLS, 1983). For overall reviews see Nadiri (1970) and Norsworthy (1984) and also Griliches (1987). Despite all of this work there is still no general agreement on what the computed productivity measures actually measure, how they are to be interpreted, and what are the major sources of their fluctuation and growth. For a recent example of the state of confusion in this area, see Romer (1987).

Additional References

Adelman, I. and Z. Griliches 1961, "On an index of quality change," *Journal of the American Statistical Association*, 56, 535–48.

Arrow, K., 1973, "Higher education as a filter," *Journal of Public Economics* II, 193–216.

Atkinson, A. B. and J. E. Stiglitz, 1969, "A new view of technological change," *Economic Journal*, LXXIX, 573–87.

Bartik, T. J. and V. Kerry Smith, 1987, "Urban amenities and public policy." In E. S. Mills (ed.), *Handbook of Regional and Urban Economics*, vol. II, chapter 31. Amsterdam: Elsevier, pp. 505–52.

Behrman, J. R., Z. Hrubec, P. Taubman, and T. J. Wales, 1980, *Socioeconomic Success*. Contributions to Economic Analysis 128. Amsterdam: North Holland.

——, P. Taubman, and T. Wales, 1977, "Controlling for and measuring the effects of genetics and family environment in equations for schooling and labor market success." In P. Taubman, (ed.), *Kinometrics*. Amsterdam: North Holland, pp. 35–96.

Belinfante, A., 1978, "The identification of technical change in the electricity generating industry." In M. Fuss and D. McFadden (eds), *Production Economics: A Dual Approach to Theory and Applications*, vol. 2. Amsterdam: North Holland, pp. 149–186.

Berndt, E. R., 1983, "Quality adjustment, hedonics, and modern empirical demand analysis." In W. E. Diewert and C. Montemarquette (eds), *Price Level Measurement*, Ottawa: Statistics Canada, pp. 817–63.

——, and M. Fuss, 1986, "Productivity measurement with adjustments for variations in capacity utilization and other forms of temporary equilibrium," *Journal of Econometrics*, 33(1/2), 7–29.

Bernstein, J. and M. I. Nadiri, 1986, "Rates of return on physical and R&D capital and the structure of the production process: cross section and time series evidence." In M. I. Nadiri and E. R. Berndt (eds), *Temporary Equilibrium and Cost of Adjustment in the Theory of the Firm*, forthcoming.

Bhattacharya, S., K. Chatterjee, and Larry Samuelson, 1986, "Sequential research and the adoption of innovations," *Oxford Economic Papers*, New Series 38, 219–43.

Bielby, W. T., R. M. Hauser, and D. C. Featherman, 1976, "Response errors of nonblack males in models of the stratification process." In D. J. Aigner and A. S. Goldberger (eds), *Latent Variables in Socio-economic Models*. Amsterdam: North Holland, pp. 227–252.

Binswanger, H. and V. W. Ruttan, 1978, *Induced Innovation*. Baltimore: Johns Hopkins University Press.

Bishop, J. H., 1974, "Biases in measurement of the productivity benefits of human capital investments." Institute for Research on Poverty, Discussion Paper no. 323–74, Madison, Wisconsin.

Blau, P. M. and O. D. Duncan, 1967, *The American Occupational Structure*, New York: Wiley.

Bound, J., Z. Griliches, and B. H. Hall, 1986, "Wages, schooling and IQ of brothers and sisters: do the family factors differ?" *International Economic Review*, 27(1), 77–105.

Brandner, L. and M. A. Strauss, 1959, "Congruence versus profitability in the diffusion of hybrid sorghum," *Rural Sociology*, 24, 381–3.

Brown, J. M., 1983, "Structural estimation in implicit markets." In J. E. Triplett (ed.), *The Measurement of Labor Cost. Studies in Income and Wealth*, vol. 48. Chicago: University of Chicago Press, pp. 123–152.

Bureau of Labor Statistics, 1983, *Trends in Multifactor Productivity, 1948–81*. Washington, DC: US Department of Labor, Bulletin 2178.

Cardell, N. S., 1977, "Methodology for predicting the demand for new electricity using goods." Charles River Associates Report 244, Cambridge, MA (unpublished).

Chamberlain, G., 1977, "Education, income, and ability revisited," *Journal of Econometrics*, 5, 241–57.

——, and Z. Griliches, 1975, "Unobservables with a variance-components structure: ability, schooling, and the economic success of brothers," *International Economic Review*, XVI(2), 422–9.

——, ——, 1977, "More on brothers." In P. Taubman (ed.), *Kinometrics: The Determinants of Socio-economic Success Within and Between Families*. Amsterdam: North Holland, pp. 97–124.

Chinloy, P. T., 1980, "Sources of quality change in labor input," *American Economic Review*, 70(1), 108–19.

Christensen, L. R. and D. W. Jorgenson, 1973, "Measuring the performance of the private sector of the U.S. economy, 1929–69." In M. Moss (ed.), *Measuring Economic and Social Performance*. New York: NBER, pp. 233–338.

Cole, R., Y. C. Chen, J. A. Barquin-Stolleman, E. Dulberger, N. Helvacian, and J. H. Hodge, 1986, "Quality adjusted price indexes for computer processors and selected peripheral equipment," *Survey of Current Business*, 66(1), 41–50.

Coleman, J. S., E. Katz, and H. Menzel, 1966, *Medical Innovation: A Diffusion Study*. Indianapolis: Bobbs-Merrill.

David, P. A. 1969, "A contribution to the theory of diffusion." Stanford Center for Research in Economic Growth, Memorandum no. 71, June.

Davies, S., 1979, *The Diffusion of Process Innovations*. London: Cambridge University Press.

Denison, E. F., 1964, "Measuring the contribution of education." In OECD, *The Residual Factor and Economic Growth*. Paris, pp. 13–55.

——, 1969, "Some major issues in productivity analysis: an examination of estimates by Jorgenson and Griliches," *Survey of Current Business*, 49 (5, part II), 1–27.

——, 1974, *Accounting for United States Economic Growth 1929–1969*. Washington, DC: The Brookings Institution.

——, 1979, *Accounting for Slower Economic Growth*. Washington, DC: The Brookings Institution.

——, 1984, "Accounting for slower economic growth: an update." In J. W. Kendrick (ed.), *International Comparisons of Productivity and Causes of the Slowdown*. Cambridge, MA: Ballinger, pp. 1–46.

——, 1985, *Trends in American Economic Growth, 1929–82*. Washington, DC: The Brookings Institution.

Diewert, W. E., 1976, "Exact and superlative index numbers," *Journal of Econometrics*, 4, 115–45.

——, 1986, *The Measurement of the Economic Benefits of Infrastructure Services*. New York: Springer-Verlag.

Dixon, R., 1980, "Hybrid corn revisited," *Econometrica*, 48, 1451–61.

Dougherty, C. R. S., 1972, "Estimates of labour aggregation functions," *Journal of Political Economy*, 80(6), 1101–19.

Eisner, R., 1985, "Total incomes system of accounts," *Survey of Current Business*, 65, 24–48.

Epple, D., 1987, "Hedonic prices and implicit markets: estimating demand and supply functions for differentiated products, *Journal of Political Economy*, 95(1), 59–80.

Evans, D. S., 1983 (ed.), *Breaking Up Bell: Essays on Industrial Organization Regulation*. Amsterdam: North Holland.

Evenson, R. E., 1968, "The contribution of agricultural research and extension to agricultural productivity." PhD dissertation, University of Chicago.

——, 1982, "Agriculture." In R. R. Nelson (ed.), *Government and Technical Progress*. New York: Pergamon Press, pp. 233–82.

——, and Y. Kislev, 1975, *Agricultural Research and Productivity*. New Haven: Yale University Press.

——, and F. Welch, 1974, "U.S. agricultural productivity: studies in technical change and allocative efficiency." Economic Growth Center, Yale University (unpublished manuscript).

——, P. E. Waggoner, and V. W. Ruttan, 1979, "Economic benefits from research: an example from agriculture," *Science*, 205, 1101–7.

Fallon, P. and L. Layard, 1975, "Capital–skill complementarity, income distribution and output accounting," *Journal of Political Economy*, 83(2), 279–301.

Fane, C. G., 1972, "The productive value of education in agriculture in the U.S. corn belt, 1964." Unpublished PhD thesis, Harvard University.

Feder, G., R. Just, and O. Zilberman, 1985, "Adoption of agricultural innovations in developing countries: a survey," *Economic Development and Cultural Change*, 33(2), 255–98.

Fraumeni, B. M. and D. W. Jorgenson, 1980, "The role of capital in U.S. economic growth, 1948–76." In G. von Furstenberg (ed.), *Capital Efficiency and Growth*. Cambridge, MA: Ballinger, pp. 9–250.

Freeman, R. B., 1973, "Changes in the labor market for black Americans, 1948–72," *Brookings Papers on Economic Activity*, no. 1, pp. 67–132.

——, 1976, *The Overeducated American*. New York: Academic ·Press.

Fuss, M. A., 1977, "The structure of technology over time: a model for testing the 'putty–clay' hypothesis," *Econometrica*, 45(8), 1797–1821.

——, D. McFadden, and Y. Mundlak, 1978, "A survey of functional forms in the economic analysis of production." In M. Fuss and D. McFadden (eds), *Production Economics: A Dual Approach to Theory and Application*, vol. 1. Amsterdam: North Holland, pp. 219–68.

Gollop, F. M. and D. W. Jorgenson, 1980, "U.S. productivity growth by industry, 1947–73." In J. W. Kendrick and B. Vaccara (eds), *New Developments in Productivity Measurement. Studies in Income and Wealth*, vol. 44. Chicago: University of Chicago Press, pp. 17–136.

Gordon, R. J., 1982, "Energy efficiency, user-cost charge, and the measurement of durable goods prices." In M. F. Foss (ed.), *The U.S. National Income and Product Accounts, Studies in Income and Wealth*, vol. 47. NBER, Chicago: The University of Chicago Press, pp. 205–68.

——, 1986, "The measurement of durable goods prices." Unpublished manuscript.

——, 1987, "The postwar evolution of computer prices." NBER Working Paper no. 2227, April.

Gordon, R. H., M. Schankerman, and R. Spady, 1986, "Estimating the effects of R&D on Bell system productivity." In M. Peston and R. Quandt (eds), *Festschrift in Honor of William Baumol*. Oxford: Philip Allan, pp. 164–88.

Grabowski, H. G. and D. C. Mueller, 1978, "Industrial research and development, intangible capital stocks, and firm profit rates," *Bell Journal of Economics*, 9(2), 328–43.

——, and J. M. Vernon, 1982, "The pharmaceutical industry." In R. R. Nelson (ed.), *Government and Technical Progress*. New York: Pergamon Press, pp. 283–360.

Griliches, Zvi, 1958, "The demand for fertilizer: an econometric interpretation of a technical change," *Journal of Farm Economics*, XL(3), 591–606.

——, 1959, "Distributed lags, disaggregation, and regional demand functions for fertilizer," *Journal of Farm Economics*, LXI(1), 90–102.

——, 1960, "The demand for a durable input: U.S. farm tractors, 1929–57." In A. C. Harberger (ed.), *The Demand for Durable Goods*. Chicago: University of Chicago Press, pp. 181–207.

——, 1961, "A note on serial correlation bias in estimates of distributed lags," *Econometrica*, 29(1), 65–73.

——, 1962, "Profitability versus interaction: another false dichotomy," *Rural Sociology*, 27(3), 327–30.

——, 1964. "Notes on the measurement of price and quality changes." In *Models of Income, Determination. Studies in Income and Wealth*, NBER, Princeton: Princeton University Press, vol. 28, pp. 381–418.

——, 1967a, "Distributed lags: a survey," *Econometrica*, 35(1), 16–49.

——, 1967b, "Production functions in manufacturing: some preliminary results." In M. Brown (ed.), *The Theory and Empirical Analysis of Production. Studies in Income and Wealth*, vol. 31. NBER. New York: Columbia University Press, pp. 275–340.

——, 1968, "Production functions in manufacturing: some additional results," *Southern Economic Journal*, XXXV(2), 151–6.

—— (ed.), 1971, *Price Indexes and Quality Change*. Cambridge, MA: Harvard University Press.

——, 1974, "Errors in variables and other unobservables," *Econometrica*, 42(6), 971–98.

——, 1976a, "The changing economics of education." In M. Allingham and M. L. Burnstein (eds), *Resource Allocation and Economic Policy*. London: Macmillan Press, pp. 35–43.

——, 1976b, "Wages of very young men," *Journal of Political Economy*, 84 (4, part 2), S69–S85.

——, 1977, "Estimating the returns to schooling: some econometric problems," *Econometrica*, 45(1), 1–22.

——, 1978, "Earnings of very young men." In Z. Griliches, W. Krelle, H-J. Krupp, and O. Kyn (eds), *Income Distribution and Economic Inequality*. Frankfurt: Campus Verlag, pp. 209–19.

——, 1979a, "Sibling models and data in economics: beginnings of a survey," *Journal of Political Economy*, 87(5, part 2), S37–S64.

——, 1979b, "Issues in assessing the contribution of research and development to productivity growth," *Bell Journal of Economics*, 10(1), 92–116.

——, 1980a, "Schooling interruption, work while in school and the returns from schooling," *Scandinavian Journal of Economics*, 82, 291–303.

——, 1980b, "Expectations, realizations, and the aging of young men." In R. G. Ehrenberg (ed.), *Research in Labor Economics*, vol. 3. Greenwich, Connecticut. JAI Press, pp. 1–21.

——, 1980c, "Hybrid corn revisited: a reply," *Econometrica*, 48(6), 1463–5.

——, 1980d, "Returns to research and development expenditures in the private sector." In J. W. Kendrick and B. Vaccara (eds), *New Developments in Productivity Measurement. Studies in Income and Wealth*, vol. 44. NBER. Chicago: University of Chicago Press, pp. 419–54.

——, 1980e, "R&D and the productivity slowdown," *American Economic Review*, 70(2), 343–8.

——, 1984, "Introduction." In Z. Griliches (ed.), *R&D, Patents, and Productivity*. Chicago: University of Chicago Press, pp. 1–19.

——, 1986, "Productivity, R&D, and basic research at the firm level in the 1970s," *American Economic Review*, 76(1), 141–54.

——, 1987, "Productivity: measurement problems." In J. Eatwell, M. Milgate, and P. Newman (eds), *The New Palgrave: A Dictionary of Economics*. London: Macmillan. Vol. 7, 1010–13.

——, and F. Lichtenberg, 1984, "R&D and productivity growth at the industry level: is there still a relationship?" In Z. Griliches (ed.), *R&D, Patents, and Productivity*. Chicago: University of Chicago Press, pp. 465–96.

——, and J. Mairesse, 1983, "Comparing productivity growth: an exploration of French and U.S. industrial firm data," *European Economic Review*, 21, 89–119.

——, ——, 1984, "Productivity and R&D at the firm level," in Z. Griliches (ed.), *R&D, Patents, and Productivity*. Chicago: University of Chicago Press, pp. 339–94.

——, and V. Ringstad, 1971, *Economies of Scale and the Form of the Production Function*. Amsterdam: North Holland.

——, A. Pakes and B. H. Hall, 1987, "The value of patents as indicators of inventive activity." In: P. Dasgupta and P. Stoneman (eds) *Economic Policy and Technological Performance*, 97–124. Cambridge: Cambridge University Press, 1987.

Hall, R. E., 1986, "Market structure and macroeconomic fluctuations," *Brookings Papers on Economic Activity*, 2, 285–322.

Hauser, R. M. and T. N. Daymont, 1977, "Schooling, ability, and earnings: cross-sectional findings 8 to 14 years after high-school graduation," *Sociology of Education*, 50(3), 182–206.

——, and D. Featherman, 1977, *The Process of Stratification*. New York: Academic Press.

——, and W. H. Sewell, 1986, "Family effects in simple models of education, occupational status, and earnings: findings from the Wisconsin and Kalamazoo studies," *Journal of Labor Economics*, 4(3, part 2), S83–S115.

Havens, A. E. and E. M. Rogers, 1961, "Adoption of hybrid corn: profitability and interaction effects," *Rural Sociology*, 26, 409–14.

Hayami, Y. and V. Ruttan, 1985, *Agricultural Development: An International Perspective*, rev. edn. Baltimore: Johns Hopkins University Press.

Hodgins, C., 1968, "On estimating the economies of large-scale production: Some tests on data for the Canadian manufacturing sector." PhD dissertation, University of Chicago.

Huffman, W., 1974, "Decision making: the role of education," *American Journal of Agricultural Economics*, 56(1), 85–97.

——, 1977, "Allocative efficiency: the role of human capital," *Quarterly Journal of Economics*, XCI(1), 59–79.

Jaffe, A., 1986, "Technological opportunity and spillovers of R&D," *American Economic Review*, 76(5), 984–1001.

Jamison, D. T. and L. J. Lau, 1982, *Farm Education and Farm Efficiency*. Baltimore: Johns Hopkins University Press.

Jencks, C., 1972, *Inequality*. New York: Basic Books.

——, and M. Brown, 1977, "Genes and social stratification." In P. Taubman (ed.), *Kinometrics*, Amsterdam: North Holland, pp. 169–234.

——, et al., 1979, *Who Gets Ahead?* New York: Basic Books.

Jensen, R., 1982, "Adoption and diffusion of an innovation of uncertain profitability," *Journal of Economic Theory*, 27, 182–93.

Jorgenson, D. W., 1984, "The contribution of education to U.S. economic growth, 1948–73." In E. Dean (ed.), *Education and Economic Productivity*. Cambridge, MA: Ballinger, pp. 95–162.

——, 1986, "Econometric methods for modelling producer behavior." In Z. Griliches and M. Intriligator (eds), *Handbook of Econometrics*, vol. III. Amsterdam: North Holland, pp. 1841–1915.

——, and M. Nishimizu, 1978, "U.S. and Japanese economic growth, 1952–74: an international comparison," *Economic Journal*, 88(352), 707–26.

——, and A. Pachon, 1983, "The accumulation of human and non-human capital." In R. Hemming and F. Modigliani (eds), *The Determinants of National Saving and Wealth*. London: Macmillan, pp. 302–50.

Jorgenson, D. W., F. Gollop, and B. Fraumeni, 1987, *Productivity and U.S. Economic Growth, 1987.* Cambridge, MA: Harvard University Press.

Kahn, J. A., 1986, "Gasoline prices and the used automobile market: a rational expectations asset price approach," *Quarterly Journal of Economics,* 101(2), 323–40.

Kendrick, J. W., 1976, *The Formation and Stocks of Total Capital.* NBER, General Series, No. 100. New York: Columbia University Press.

——, 1983, "International comparisons of recent productivity trends." In S. H. Shurr (ed.), *Energy, Productivity, and Economic Growth.* Cambridge, MA: Delgeschlager & Co., pp. 71–120.

——, and E. S. Grossman, 1980, *Productivity in the U.S.: Trends and Cycles.* Baltimore: Johns Hopkins University Press.

King, A. T., 1976, "The demand for housing: integrating the roles of journey-to-work, neighborhood quality, and prices." In N. Terleckyj (ed.), *Household Production and Consumption. Studies in Income and Wealth,* vol. 40, New York: Columbia University Press for NBER, pp. 451–34.

Kislev, Y., 1965, "Estimating a production function from 1959 U.S. census of agriculture data." Unpublished PhD dissertation, University of Chicago.

——, and W. Peterson, 1982, "Prices, technology and farm size," *Journal of Political Economy,* 90, 578–95.

——, ——, 1986, "Economies of scale in agriculture," The World Bank (unpublished paper).

Koyck, L. M., 1954, *Distributed Lags and Investment Analysis.* Amsterdam: North Holland.

Krishna, K. L., 1967, "Production relations in manufacturing plants: an exploratory study." Unpublished PhD dissertation, University of Chicago.

Kroch, E. A., and K. Sjoblom, 1986, "Education as a signal: some evidence." Center for the Study of Organizational Innovation, University of Pennsylvania, Discussion Paper no. 197.

Lancaster, K., 1971, *Consumer Demand: A New Approach.* New York: Columbia University Press.

Lattimer, R. G., 1964, "Some economic aspects of agricultural research and education in the U.S." Unpublished PhD dissertation, Purdue University.

Link, A. N., 1981, *Research and Development Activity in U.S. Manufacturing.* New York: Praeger.

Lucas, R. E. B., 1975, "Hedonic price functions," *Economic Inquiry,* 13(2), 157–78.

Mahajan, V. and R. A. Peterson, 1985, *Models for Innovation Diffusion.* Sage University Papers, Series: Quantitative Applications in the Social Sciences, Beverly Hills: Sage Publications.

Mansfield, E., 1961, "Technical change and the rate of imitation," *Econometrica,* 29, 741–66.

——, 1968, *Industrial Research and Technological Innovation.* New York: Norton.

——, 1984, "R&D and innovation: some empirical findings." In Z. Griliches (ed.), *R&D, Patents, and Productivity.* Chicago: University of Chicago Press, pp. 127–54.

——, J. Rapoport, A. Romeo, and G. Beardsley, 1977a, "Social and private rates of return from industrial innovations," *Quarterly Journal of Economics,* 91(2), 221–40.

——, ——, ——, E. Villani, S. Wagner, and F. Husic, 1977b, *The Production and Application of New Industrial Technology.* New York: Norton.

McFadden, D., 1974. "Conditional Logit Analysis of Qualitative Choice Behavior," in P. Zarembka (ed.), *Frontiers in Econometrics,* New York: Academic Press, 105–42.

——, 1978a, "Estimation techniques for elasticity of substitution and other production parameters." In M. Fuss and D. McFadden (eds), *Production Economics: A Dual Approach to Theory and Applications,* vol. 2. Amsterdam: North Holland, pp. 73–124.

——, 1978b, "Modelling the choice of residential location." In A. Karlquist et al. (eds), *Spatial Interaction Theory and Residential Location.* Amsterdam: North Holland, pp. 75–96.

Miller, E. M., 1983, "Capital aggregation in the presence of obsolescence-inducing technical change," *Review of Income and Wealth,* 29(23), 283–96.

Mohnen, P., M. I. Nadiri, and I. R. Prucha, "R&D, production structure and rates of return in the U.S., Japanese and German manufacturing sectors: a non-separable dynamic factor demand model," *European Economic Review,* 30(4), 749–71.

Morrison, C. J., 1985, "On the economic interpretation and measurement of optimal capacity utilization with anticipatory expectations," *Review of Economic Studies,* 52, 295–310.

Morrison, C. and E. Berndt, 1981, "Short-run labor productivity in a dynamic model," *Journal of Econometrics,* 16(3), 339–66.

Muellbauer, J., 1974, "Household production theory, quality and 'hedonic technique'," *American Economic Review,* 64(6), 977–94.

Nabseth, L. and G. Ray, 1974, *The Diffusion of New Industrial Processes.* London: Cambridge University Press.

Nadiri, M. I., 1970, "Some approaches to the theory and measurement of total factor productivity: a survey," *Journal of Economic Literature,* 8(4), 1137–78.

Nelson, R. R. and E. S. Phelps, 1966, "Investment in humans, technological diffusion, and economic growth," *American Economic Review,* LVI(2), 69–75.

——, M. I. Peck, and D. Kalachek, 1967, *Technology, Economic Growth and Public Policy.* Washington, DC: The Brookings Institution.

Nerlove, M., 1963, "Returns to scale in electricity supply." In C. Christ et al., *Measurement in Economics.* Stanford: Stanford University Press, pp. 167–200.

Norsworthy, J. R., 1984, "Growth accounting and productivity measurements," *Review of Income and Wealth,* 30(3), 309–29.

Ohta, M. 1975, "Production technologies in the U.S. boiler and turbo generator industries and hedonic indexes for their products: a cost function approach," *Journal of Political Economy,* 83(1), 1–26.

——, and Z. Griliches, 1976, "Automobile prices revisited: extensions of the hedonic hypothesis." In N. Terleckyj (ed.), *Household Production and Consumption, Studies in Income and Wealth,* vol. 40. New York: Columbia University Press for NBER, pp. 325–90.

——, ——, 1986, "Automobile prices and quality change: did the gasoline price increases change consumer tastes in the U.S.?", *Journal of Business and Economic Statistics,* 4(2), 187–98.

Olneck, M., 1977, "On the use of sibling data to estimate the effects of family background, cognitive skills, and schooling." In P. Taubman (ed.), *Kinometrics*, Amsterdam: North Holland, pp. 125–62.

Pakes, A., 1976, "A simple aggregate diffusion model: estimation techniques and a case study in a secondary country." Falk Institute Discussion Paper no. 7613, Jerusalem, Israel.

——, and Z. Griliches, 1984, "Estimating distributed lags in short panels with an application to the specification of depreciation patterns and capital stock constructs," *Review of Economic Studies*, LI(2), 243–62.

Peterson, W. L., "Returns to poultry research in the U.S." PhD dissertation, University of Chicago, 1966.

Plant, M. and F. Welch, 1984, "Measuring the impact of education on productivity." In E. Dean (ed.), *Education and Economic Productivity*, Cambridge, MA: Ballinger, pp. 163–94.

Psacharopoulos, G., 1981, "Returns to education: an updated international comparison," *Comparative Education*, 17(3), 321–41.

Reiss, P. C., 1985, "Economic measures of the returns to federal research and development." Paper commissioned for the National Academy of Sciences Workshop on "The Federal Role in Research and Development."

Riley, J., 1979, "Testing the educational screening hypothesis," *Journal of Political Economy*, 87(5, part 2), S227–S252.

Rogers, E. M., 1983, *Diffusion of Innovations*, 3rd edn. New York: Free Press.

Romer, P. M., 1987, "Crazy explanations for the productivity slowdown." In S. Fischer (ed.), *NBER Macroeconomics Annual*, Vol. 2, 163–202. Cambridge, MA: MIT Press.

Rosen, S., 1974, "Hedonic price and implicit markets: product differentiation in pure competition," *Journal of Political Economy*, 82(1), 34–55.

Schankerman, M., 1981, "The effects of double-counting and expensing on the measured returns to R&D," *Review of Economics and Statistics*, 63(2), 275–82.

Scherer, F. M., 1982, "Interindustry technology flows and productivity growth," *Review of Economics and Statistics*, 64, 627–34.

——, 1984a, "Using linked patent and R&D data to measure interindustry technology flows." In Z. Griliches (ed.), *R&D, Patents, and Productivity*. Chicago: University of Chicago Press, pp. 417–464.

——, 1984b, *Innovation and Growth*. Cambridge, MA: MIT Press.

Schmookler, J. 1952, "The Changing Efficiency of the American Economy, 1869–1938," *Review of Economics and Statistics*, 34:214–31.

Sewell, W. and R. Hauser, 1975, *Education, Occupation, and Earnings*. New York: Academic Press.

Smith, J. and F. Welch, 1986, "Closing the gap: forty years of economic progress for blacks," Rand Corporation, R-3330-POL.

Solow, R. M., 1957, "Technical change and the aggregate production function," *Review of Economics and Statistics*, 39(3), 312–20.

——, 1960, "Investment and technical progress." In K. Arrow et al. (eds), *Mathematical Methods in the Social Sciences*, Stanford: Stanford University Press, pp. 89–104.

Spence, M., 1974, *Market Signalling*. Cambridge, MA: Harvard University Press.

Stoneman, P. L., 1976, *Technological Diffusion and the Computer Revolution:*

The U.K. Experience. Department of Applied Economics Monographs, no. 25. London: Cambridge University Press.

Terleckyj, N. E., 1974, *Effects of R&D on the Productivity Growth of Industries*. Washington, DC: National Planning Association Report no. 140.

Thirtle, C. G. and V. W. Ruttan, 1986, *The Role of Demand and Supply in the Generation and Diffusion of Technical Change*. University of Minnesota Economic Development Center, Bulletin 86, St Paul.

Trajtenberg, M., 1979, "Quantity is all very well but two fiddles don't make a Stradivarius: aspects of consumer demand for characteristics." Maurice Falk Institute Discussion Paper no. 7910, Jerusalem, Israel.

——, 1983, "Dynamics and welfare analysis of product innovations." PhD thesis, Harvard University.

——, 1985, "The welfare analysis of product innovations with an application to CT scanners." NBER Working Paper no. 1724. October.

Triplett, J. E., 1975, "The measurement of inflation: a survey of research on the accuracy of price indexes." In Paul H. Earl (ed.), *Analysis of Inflation*. Lexington, MA: D. C. Heath & Co., pp. 19–82.

——, 1983a, "Concept of quality in input and output price measures: a resolution of the user-value resources-cost debate." In M. F. Foss, (ed.), *The U.S. National Income and Product Accounts, Studies in Income and Wealth*, vol. 47, NBER, Chicago: University of Chicago Press.

——, 1983b, "Introduction: an essay on labor cost." In J. E. Triplett (ed.), *The Measurement of Labor Cost. Studies in Income and Wealth*, vol. 48, NBER, Chicago: University of Chicago Press.

——, 1986, "The economic interpretation of hedonic methods," *Survey of Current Busines*, 86(1), 36–40.

——, 1987, "Hedonic functions and hedonic indexes," *New Palgrave Dictionary of Economic Theory*.

——, 1988, "Price and technology change in a capital good: a survey of research computers." In D. W. Jorgenson and R. Landau (eds), *Technology and Capital Formation*, forthcoming.

Waugh, F. V., 1929, *Quality as a Determinant of Vegetable Prices*. New York: Columbia University Press.

Weaver, R. D., 1985, "Federal research and development and U.S. agriculture." Paper commissioned for the National Academy of Sciences Workshop on "The Federal Role in Research and Development."

Weiss, R. 1974, "Sources of change in the occupational structures of American manufacturing industries 1950 to 1960: an application of production function analysis." PhD thesis, Harvard University.

Welch, F., 1970, "Education in production," *Journal of Political Economy*, 78(1), 35–59.

——, 1975, "Human capital theory: education, discrimination, and life-cycles," *American Economic Review*, Proceedings Issue, 65(2), 63–73.

——, K. Murphy, and M. Plant, 1985, "Cohort size and earnings" (unpublished).

Willis, R. J., 1986, "Wage determinants: a survey and reinterpretation of human capital earnings functions." In O. Aschenfelter and R. Layard (eds), *Handbook of Labor Economics*. Amsterdam: North Holland, pp. 525–602.

Wolpin, K., 1977, "Education and screening," *American Economic Review*, 67(5), 949–58.

Index